Knowledge Management in the Construction Industry:

A Socio-Technical Perspective

Abdul Samad Kazi
VTT - Technical Research Centre of Finland, Finland

IDEA GROUP PUBLISHING
Hershey • London • Melbourne • Singapore

Acquisitions Editor:	Mehdi Khosrow-Pour
Senior Managing Editor:	Jan Travers
Managing Editor:	Amanda Appicello
Development Editor:	Michele Rossi
Copy Editor:	Maria Boyer
Typesetter:	Sara Reed
Cover Design:	Lisa Tosheff
Printed at:	Yurchak Printing Inc.

Published in the United States of America by
Idea Group Publishing (an imprint of Idea Group Inc.)
701 E. Chocolate Avenue, Suite 200
Hershey PA 17033
Tel: 717-533-8845
Fax: 717-533-8661
E-mail: cust@idea-group.com
Web site: http://www.idea-group.com

and in the United Kingdom by
Idea Group Publishing (an imprint of Idea Group Inc.)
3 Henrietta Street
Covent Garden
London WC2E 8LU
Tel: 44 20 7240 0856
Fax: 44 20 7379 3313
Web site: http://www.eurospan.co.uk

Library of Congress Cataloging-in-Publication Data

Knowledge management in the construction industry : a socio-technical perspective / Abdul Samad Kazi, Editor.
p. cm.
Includes bibliographical references and index.
ISBN 1-59140-360-X (h/c) -- ISBN 1-59140-361-8 (s/c) -- ISBN 1-59140-362-6 (eisbn)
1. Construction industry--Management. 2. Knowledge management. 3. Organizational learning. 4. Construction industry--Information resources management. I. Kazi, Abdul Samad.
HD9715.A2K55 2005
624'.068'4--dc22

2004017842

British Cataloguing in Publication Data
A Cataloguing in Publication record for this book is available from the British Library.

All work contributed to this book is new, previously-unpublished material. The views expressed in this book are those of the authors, but not necessarily of the publisher.

Knowledge Management in the Construction Industry: A Socio-Technical Perspective

Table of Contents

Preface vi

Section I: Knowledge Management in Construction Practice

Chapter I
Strategies for Managing Project Generated Knowledge: A New Zealand Case Study 1
Steve de Kretser, Sinclair Knight Merz Ltd, New Zealand
Suzanne Wilkinson, The University of Auckland, New Zealand

Chapter II
HRM Strategies for Promoting Knowledge Sharing within Construction Project Organisations: A Case Study 18
Andrew R.J. Dainty, Loughborough University, UK
Jidong Qin, Loughborough University, UK
Patricia M. Carrillo, Loughborough University, UK

Chapter III
Managing the External Provision of "Knowledge Management" Services for Projects 34
Hedley Smyth, University College London, UK

Chapter IV
The Intellectual Capital Report of Danya Cebus: Smart Construction 53
Edna Pasher, Edna Pasher Ph.D. & Associates, Israel
Gil Horsky, Edna Pasher Ph.D. & Associates, Israel

Chapter V
Capturing and Reusing Building Maintenance Knowledge: A Socio-Technical Perspective 67
Patrick S.W. Fong, Hong Kong Polytechnic University, Hong Kong
Ka-chi Wong, Raymond Chan Surveyors Ltd, Hong Kong

Chapter VI
Diffusion of Information and Communication Technology: A Community of Practice Perspective **90**
Vachara Peansupap, RMIT University, Australia
Derek H.T. Walker, RMIT University, Australia

Chapter VII
Community of Practice Software Management Tools: A UK Construction Company Case Study **112**
Mark Jewell, Addept Solutions, Australia
Derek H.T. Walker, RMIT University, Australia

Section II: Learning from Experience

Chapter VIII
Knowledge Management: Facilitating Organisational Learning within the Construction Industry **130**
Graham Orange, Leeds Metropolitan University, UK
Patrick Onions, Knowledge Studio, South Africa
Alan Burke, Leeds Metropolitan University, UK
Barbara Colledge, Leeds Metropolitan University, UK

Chapter IX
Knowledge Management in Higher Education and Professional Development in the Construction Industry **150**
Fung Fai Ng, The University of Hong Kong, Hong Kong

Chapter X
Empirical Investigation of Organisational Learning Ability as a Performance Driver in Construction **166**
Irem Dikmen, Middle East Technical University, Turkey
M. Talat Birgonul, Middle East Technical University, Turkey
Tunca Ataoglu, GAMA Al Moushegah Arabia Ltd., Kingdom of Saudi Arabia

Chapter XI
Evaluating an Organisation's Learning Culture Using Learning Histories **185**
Marcus C. Jefferies, University of Newcastle, Australia
Chen Swee Eng, University of Newcastle, Australia
Ralf Zenke, Department of Health and Human Services TAS, Australia

Chapter XII
Uncertainty and Information in Construction: From the Socio-Technical Perspective 1962-1966 to Knowledge Management – What Have We Learned? **203**
Alan Wild, Consultant, Birmingham, UK

Section III: Emerging Models and Solutions

Chapter XIII
The Culture of the Knowledge Advantage (K-Adv): A Holistic Strategic Approach to the Management of Knowledge 225
Derek H.T. Walker, RMIT University, Australia
Tayyab Maqsood, RMIT University, Australia
Andrew Finegan, Charles Darwin University, Australia

Chapter XIV
Knowledge Emergence and Adaptive Management: An Exploration on the Co-Production of Project Needs and Requirements by Client-Specialist Groups 251
Michael Whelton, University of California, USA
Ari Pennanen, University of Tampere, Finland
Glenn Ballard, University of California, Berkeley, USA

Chapter XV
Decision Support Systems and their Application in Construction 276
Alaa Abdou, United Arab Emirates University, United Arab Emirates
Moh'd Radaideh, United Arab Emirates University, United Arab Emirates
John Lewis, University of Liverpool, UK

Chapter XVI
A Knowledge Management Portal System for Construction Projects Using Knowledge Map 299
H. Ping (Jack) Tserng, National Taiwan University, Taiwan
Yu-Cheng (Vincent) Lin, National Taiwan University, Taiwan

Chapter XVII
An Integrative Knowledge Management System for Environmental-Conscious Construction 323
Zhen Chen, Massey University, New Zealand
Stephen C.W. Kong, Loughborough University, UK
Heng Li, Hong Kong Polytechnic University, Hong Kong
Qian Xu, Hong Kong Polytechnic University, Hong Kong

Chapter XVIII
Configurable Environments: A Vision for Future Project Information Technologies 343
William J. O'Brien, University of Texas at Austin, USA
Raja R.A. Issa, University of Florida, USA
Zhigang Shen, University of Florida, USA
Haiyan Xie, University of Florida, USA

About the Authors 363

Index 374

Preface

Value-driven products and services today are delivered through complimentary competence sharing between different organisations. This is more of a commonality than an exception in industries such as construction, where the typical modus of operandi is that of the virtual enterprise. Within such inter-enterprise collaborative environments, the management of the valuable experience and knowledge gained is crucial for success. Easy-to-use information and communications technology (ICT) systems that work across organisations are mandatory for not only ensuring data and information integrity, but the smooth passage and flow of knowledge. This is essential, as a predecessor to the task to be accomplished by one organisation is the result of an undertaking by another organisation. ICT solutions to support the same are still in a relatively infant stage from the perspective of actual take-up and use in the industry. What's more, we are still coming to grips with what "knowledge management" can and cannot do for us.

Background

To the outside world, the media is meticulous in highlighting through reports of disasters and failings on large and small construction projects alike of a problem-plagued industry. People are reminded again and again of a highly inefficient, poorly managed industry. Success stories, rarely, if at all, make the headlines. Yet at the same time, we are a profitable industry that employs a very large workforce. If we were not successful, or profitable, we would be out of business.

The construction industry has long been characterised as a project-based business that delivers one-of-a-kind products (such as a building). No two buildings are alike, though they may appear similar from the outside. One only needs to look back at the construction of buildings such as the Petronas Twin Towers in Kuala Lumpur, Malaysia to understand this reality. Similar in look and feel, yes; in construction methods, more or so, yes; in problems faced, no. In most construction projects, it is no longer one skilled "master builder" who builds the building for us, but a set of organisations working together in parallel through sharing of complementary competencies to provide "value to the customer."

In the words of Albert Einstein, "The only source of knowledge is experience." This has been one of the earlier guiding principles of the construction industry. Skills were passed on from master to student, from one generation to the next; the construction industry of yester years was a pool of skilled craftsmen, artists in their own right. The birth of industrialisation, large-scale projects, and a large mind shift to profitability has over the years changed the very nature of the way the construction industry operates today. In the past, management focus was on the transfer of skills to maintain a set of skilled craftsmen; today the focus is on shortening construction times, minimising costs, and effectively utilising resources. As projects grew larger over time, emphasis was put on IT solutions to help manage them. These solutions were never really intended to support and maintain the traditional process of transferring skills.

There is a growing concern in the construction industry today that new recruits prefer to sit in front of a computer to "control" projects and "surf" for information, rather than spend some time on a construction site to master the skills of the trade. With a depleting skilled force (based on discussions with several lead contractors, estimates range from 40% to up to 70% of the skilled workforce retiring within the next 15-20 years), the urge to capture their experience and skills cannot be over-emphasised. Complex IT solutions were initially seen as the solution; opinions are now however divided over the issue. While terms such as "knowledge" and "knowledge management" were in the past seen by many in the construction industry as hype and a means for management consultants to make more money, such attitudes are now changing. Knowledge, or as some call it, "experience" management may be the solution to counter balance the effect of the skilled craftsmen who will be retiring soon.

Knowledge Management

The construction industry has generally been reluctant to pick up on new management paradigms. Many see these as complex "buzz words" or "re-packaging" of well-known principles that offer little added business value. The same has been the case for knowledge management. This can be attributed to one main reason. We are generally told of the "how" behind new management paradigms. Many in the construction industry insist on first knowing the "what" and "why" before approaching the question of "how." However, when presented in a form that is understandable to the industry, there is a sudden brewing of enthusiasm and a willingness to see what can be achieved with knowledge management.

During a recent discussion with the managing director of a building contractor on the subject of intellectual capital, the managing director immediately dismissed the word as being yet another management "buzzword." According to him, words such as these did not have any effect on the bottom line, the balance sheet. When told that by this reasoning, if his mind were to be replaced by that of a "horse," there too would be no difference on the bottom line. There was a very long silent pause. He snapped his fingers and said, "…now I know what you mean. I have suddenly realised how valuable the experience and insight of my employees really are. So is this what knowledge management is about, managing this experience and insight? What can I do to ensure that I can maximise on 'intellect sharing' in my company?" Towards the end of the very long and interesting discussion, the managing director suddenly asked, "In financial management, we talk of ROI as return on investment. So in knowledge management, do we see ROI as return on 'intellect' or return on 'intelligence'?"

Most researchers like to discern two main types of knowledge, explicit and tacit. A well-accepted definition of explicit knowledge is: knowledge that can be expressed in words and easily communicated and shared in the form of hard data, scientific formulae, codified procedures, or universal principles (Nonaka & Takeuchi, 1995). It is this form of knowledge that may be programmed in the form of IF, THEN, AND, OR rules. Tacit knowledge on the other hand, though, is difficult to grasp from a "programming" perspective. It is accepted by most as being highly personal and difficult to formalise. Subjective insights, intuitions, and hunches form the basis of this form of knowledge (Nonaka & Takeuchi, 1995). Within the construction industry, this form of knowledge is seen as the most important one, one that is difficult to write down on paper, and one that is typically transferred from peer to peer through face-to-face interaction (Kazi, Hannus, & Chareonngam, 1999). It pertains to not only sharing the intuition or hunch, but also sharing the reasoning ("ba") behind that intuition or hunch. The Japanese word "ba" (Nonaka, Toyama, & Konno, 2001) is used by many in the construction industry today to describe this experience and intuition.

There is general agreement that in the simplest of terms, knowledge management is that body of knowledge that deals with the management of both personal and organisational knowledge (Nonaka & Takeuchi, 1995; APQC, 1996; Davenport & De Long, 1997; Skyrme, 1997; Davenport & Prusak, 1998). Around this general understanding, different processes and mechanisms have been introduced to serve as frameworks or guidelines into how knowledge can be managed. There is no golden rule or method. Each organisation is unique, both in terms of its business processes and its culture. Of these, it is the latter that is usually the most influential element in designing an effective knowledge management strategy and follow-up plan for implementation. As an example, Kazi, Puttonen, Sulkusalmi, Välikangas, and Hannus (2001) used a combination of approaches. First of all, the framework developed by APQC (1996) showing a combination of processes and process enablers was translated to a set of "questions." For processes, questions were asked as to how knowledge is identified, collected, organised, shared, adapted to existing products and processes, used, and new knowledge is created from existing knowledge. In terms of enablers, questions were asked as to how technology supports the processes; the influence organisational culture has on the implementation, how knowledge is measured and what impact it has, and finally, how the strategy and leadership of an organisation supports knowledge creation and sharing. These were then mapped on to a "palm tree," used as a visual metaphor to illustrate how scattered knowledge "seeds" evolved into knowledge "coconuts." This was just one example used in the case of a company providing engineering, procurement, and construction services to the energy sector. You will find several such examples within the chapters of this book.

Once an organisation has appreciated the need for knowledge management by understanding what it is, what it can do for the organisation, and mainly why it is needed, some basic "how" questions emerge: How can I capture and then share knowledge? How can I represent knowledge? How can I convert my tacit into explicit and vice-versa? Knowledge capture and sharing in a form that is later usable and shareable with many is often difficult. While a certain degree of explicit knowledge may be captured through human-to-computer interaction, the bulk of tacit knowledge is mainly captured (or more precisely shared) through human-to-human interaction. The important message being that, "if we know what we know," then we can be a highly effective workforce (Sieloff, 1999). Murray (1999) identified conceptual indexing, conceptual mapping, hypertext (hypermedia), information modelling, and semantic networks as some means for knowledge representation. Social network analysis is another means being used to visually represent knowledge through links and interrelationships between individuals in a typical knowledge exchange chain (using who knows who, and who goes to whom for information on a specific topic). In terms of knowledge conversion, the most popular model in use today is the SECI model (Nonaka & Takeuchi, 1995; Skyrme, 1997; Nonaka et al., 2001). It advocates socialisation as a mechanism for tacit-tacit, externalisation for tacit-explicit, combination for explicit-tacit, and internalisation for explicit-tacit knowledge conversion/transfer.

Challenges and Emerging Solutions

Experience is the intellectual wealth of both individuals and organisations. There is a need to handle this experience (knowledge) properly so as not to reinvent the wheel on every new project, and not repeat past mistakes again and again (Stewart, 1997). This is ever more so important in project-based businesses. When people move from project to project and work in different teams (at times concurrently), valuable experience gained can be lost if not recorded or shared properly. This has in fact been one of the factors inhibiting progress in the construction industry. Success has many parents and failure is but an orphan. The construction industry is waking up to the fact that both, however, form integral parts of the intellectual capital of organisations and individuals. Means and mechanisms to capture and share this wealth are being urgently sought, as most of the skilled workforce (tacit knowledge base) is headed for retirement in the very near future.

We are always in pursuit of the right information at the right time. The reality is that we rarely find it. Either it is residing somewhere in an IT system, or worst of all was never really recorded. When a project ends, so do on countless occasions the valuable lessons and experiences gained. Today, many of us are overwhelmed with information and just do not have the time to share what we know. What little time we do have, we spend searching for that ever so important needle of information that is buried in our information haystack (document archive). This is no different from what many of us in the construction industry go through on a daily basis in mission-critical projects.

The Internet has been hailed by many as a free, open, information goldmine. Yet, we are overwhelmed with content, some useful, and most not, all scattered in different places. Proper organisation of these documents and information is but a first step. Each individual may have a different mindset when organising information. As an example, just look at the data folders of different people in an organisation. Each individual has his/her own way of categorising and organising information. A key challenge is to "sort the wheat from the chaff, and then sieve out the real drops of wisdom." This would mean not only providing the right information to the right person at the right time, but also ensuring the validity of the information, its source, relevant experts, and the provider of the information. On many occasions while the person who writes a document has one thing in mind, the reader grasps another. So it is always essential to be able to trace back to the source of the document or information to ensure its correct understanding and validity.

There is unfortunately no fixed recipe or tool for knowledge sharing. It all depends on the organisation and the people involved. Most successful organisations have adapted certain tools to their own needs. Foremost however, they have ensured that people mix and talk together. Without the willingness to acquire and share knowledge, even if the knowledge exists, few will know where it resides and even fewer will be able to use it.

People in the construction industry are very social people and proud of their experiences and achievements in problem solving. As you may know, each day on a construction site brings new problems that at times require on-the-spot solutions. People are proud to be a part of the process that solved the problem. They love to share this experience by narrating stories to peers, colleagues, and at times, to anyone who will listen. So a simple solution is to get people working on different construction sites together for a social event like a dinner and let them socialise. The main thing that connects them is their work, and that is what most will end up talking about and telling of what they did, and when. Once the evening is over, exchange names, phone numbers, and emails. Repeat this with the same people, but in different groups over a period of time. When these people encounter a problem, they will be able to recollect the stories they heard and will use this as a basis for either solving the problem, or getting in touch with the person who told them the story.

There are many interesting knowledge management tools on the market. Some are rebranded document management systems, while others focus on search functionality, and some support collaborative work in team environments. The construction industry in general is realising that a tool on its own is not enough. There is a need to instil a knowledge-sharing culture within the organisation that fosters social communication and interaction. Once an organisation can get its employees talking and sharing experiences face-to-face, it can later also encourage them to do more of this sharing using software tools.

There is no "*the* tool" for knowledge management, let alone one for the construction industry. Experience shows that most tools need to be built and customised to the needs and culture of an organisation. No two organisations are alike, so even if one uses the same tool in each, some customisation is needed. This of course is on top of the basic integration with the organisation's other legacy systems.

The construction industry has long been dominated by a project-based delivery paradigm to deliver unique products and services. Though two adjacent buildings may look the same, each has a characteristic of its own when it comes to constructing it. Valuable experience is gained throughout the process. In fact each day brings new surprises and lessons. These, however, may be lost if they are not properly recorded and shared widely with peers. This sharing is possible through a harmonious blend of both social and technical (supported by information communications technologies) interactions (Davenport, 1994). Construction organisations such as YIT Construction Ltd. in Finland for example, are trying to address this challenge by making both good and bad practices available to its employees and business partners during key construction process phases. At the same time, they are also identifying critical "intervention points" during these phases where "social processes" are purposely induced to allow for better tacit knowledge capture and sharing.

This Book and Its Contents

This book presents a portfolio of various concepts, methods, models, and tools supported by real-life case studies from various corners of the globe providing insights into the management of knowledge in the construction industry. Untangling the hype from the reality, practical means of implementing knowledge management in the construction industry through various mechanisms and tools are demonstrated. Case studies ranging from the implementation of decision support systems, external provision of knowledge management services, learning histories, innovation, human resources management, and so forth provide enriching practical experiences in knowledge management implementation in the construction industry.

This book is organised into 18 chapters. Each chapter lists a set of "Learning Objectives" that then are conveyed to the reader through the content of the chapter. At the end of each chapter, you will find a set of "Practical Tips and Lessons Learned," summarising the key lessons the chapter had to offer. Each chapter furthermore provides a wealth of sources for further reference on knowledge management in general and knowledge management in the construction industry in particular. A brief description of each chapter follows.

In Chapter 1, de Kretser and Wilkinson present a case study from New Zealand focusing on the development of project-generated knowledge management systems, and the benefits that construction organisations can reap from such developments. They discuss how wasted, lost, and inefficient use of knowledge leads to inefficiencies and reworking of past problems. We learn in this chapter that it is the people that use and produce the knowledge that are in the best position to help develop project-generated knowledge management systems, with senior managers and company knowledge managers providing structure, facilitation, and support.

In Chapter 2, Dainty, Qin, and Carrillo present a case study on HRM strategies for promoting knowledge sharing within construction project organisations from the perspective of a large construction company based in Hong Kong. Focusing on the use of innovative solutions to encourage knowledge sharing within the company, they demonstrate a set of techniques based on a sophisticated combination of HRM approaches aimed at breaking down barriers to open communication and reciprocal knowledge exchange. We learn that while different approaches could facilitate knowledge sharing and management within a construction project environment, to be effective they must be underpinned by a supportive culture, effective communication structures, and appropriate HRM practices.

In Chapter 3, Smyth explores the external provision of knowledge management services for projects through a case study from the United Kingdom. He argues that while knowledge management is largely considered from the perspective of internal generation, the external provision will increase due to the diversity of problems and scarce resources that most construction companies today have. We learn that project managers and teams must be equipped in advance with sources of knowledge for problem solving both internally and externally from their firms, with external provision of knowledge management expected to increase in the coming years.

In Chapter 4, Pasher and Horsky take us through the experience of developing the world's first intellectual capital report for a construction company, Danya Cebus Ltd. from Israel. They point to learning, appreciating, and capitalising on the intellectual capital of firms, and guide us through a phase-by-phase approach on how to develop an intellectual capital report. Their experience demonstrates that through intellectual capital reports, shareholders are able to better understand and appreciate an organisation's hidden competences and values. We learn how to create a vision and core values, and communicate them to serve as a compass for creative energies.

In Chapter 5, Fong and Wong take us through a case study in Hong Kong on the capturing and reusing of building maintenance knowledge. Their study shows that owing to the unique nature of building maintenance, building surveyors need sufficient knowledge and experience to facilitate their decision-making process, and therefore proper capture and reuse of knowledge is required to reduce the risk of "reinventing the wheel." We learn that knowledge usually resides in the minds of the individuals, and it is through sharing and socialisation that this knowledge is elicited.

In Chapter 6, Peansupap and Walker share their experience on the community of practice concept in a case study of three major Australian contractors. Communities of practice are analysed and presented from the perspective of the individual, work groups, and the organisation. The authors demonstrate how communities of practice can be used as an efficient diffusion mechanism for information and communications technologies within and across companies. We learn that communities of practice already exist in organisations. They need to be tapped into as useful and powerful instruments for knowledge creation and sharing.

In Chapter 7, Jewell and Walker provide insights from a large UK construction organisation case study where communities of practice have been supported through use of a software tool and management approach that encourages their spread across the organisation. They provide valuable insights from several years' reflection upon the tool's use and application, and highlight both drivers and barriers to its deployment. We learn that dynamic profiling, for example, can be used as an effective means to enable people to be more effectively connected with each other through their expressed interests.

In Chapter 8, Orange, Onions, Burke, and Colledge present the results of an action research project on facilitating organisational learning within the construction industry. They describe the development of the COLA (Cross Organisational Learning Approach) process and its use. We learn from the experience of two major construction clients, a privatised utility company, a major leisure services provider, a large construction company, and two leading construction consultancy firms, that reflective practice benefits the organisation and, when seen as a social process, contributes to organisational learning.

In Chapter 9, Ng presents experiences from the use of knowledge management methods in the higher education and professional development of members of the construction industry. He sets a case for lifelong learning and continuing professional development as essential ingredients to ensuring that graduates and workers remain competent and maintain a competitive advantage. We learn that "learning to learn" is an essential skill for successful knowledge sharing in the construction industry.

In Chapter 10, Dikmen, Birgonul, and Ataoglu explore organisational learning ability as a performance driver in construction through a survey of 85 Turkish construction contractors. Through empirical evidence, they demonstrate that as organisational learning ability increases, firm performance also increases. We learn that the measurement of organisational learning ability within an organisation necessitates an evaluation of both its knowledge management abilities and the appropriateness of its culture.

In Chapter 11, Jefferies, Eng, and Zenke present a case study on the application of the "learning history" approach to organisational learning in the Tasmanian State Public Health Sector, Australia. The approach relies on the learning history of an organisation as a basis for evaluating the organisation's methods of acquiring and managing knowledge, and thereafter its learning culture. We learn that a learning history tool adds value far beyond producing documentation, in that the process itself, particularly reflective interviews, it gives stakeholders the opportunity to discuss their experiences and therefore enables a form of knowledge capture.

In Chapter 12, Wild revisits early studies on socio-technical research in construction conducted by the Tavistock Institute of Human Relations in the early sixties. Using this as a reflective basis, he discusses the extent which the tacit order, instability, and diffuseness of construction may, practically, undermine knowledge management initiatives. A set of critical questions are presented, challenging the current discourse of knowledge management research in construction. We learn that the diffuseness of construction requires a significant tacit order, and that the solution to the problems of the construction industry may be simply to manage projects differently.

In Chapter 13, Walker, Maqsood, and Finegan introduce "the knowledge advantage" (K-Adv) approach that describes a model in which knowledge leadership provides the setting in which an organisation's knowledge vision can be developed and activated. Through a set of examples, they demonstrate some strategic as well as practical tactical tools that companies can use to develop their competitive advantage. We learn that the challenge is for the construction industry to embrace its commitment to a style and form of leadership that not only empowers its knowledge workforce, but also positively supports and underpins them with appropriate infrastructure, the means to unleash innovation and creativity.

In Chapter 14, Whelton, Pennanen, and Ballard use a case study to explore the co-production of project needs and requirements by client-specialist groups. They describe the significance of project definition as a project delivery phase and outline a project definition model for knowledge creation. Through the case study, it is demonstrated that project definition is a complex adaptive process, and that knowledge about project purpose emerges from group collaboration. We learn that at times, organisational boundaries often limit the creative process. There is a need, therefore, to search both internally and externally for solutions to stakeholder needs.

In Chapter 15, Abdou, Radaideh, and Lewis provide an overview of decision support systems and their application in the construction industry. They present a set of examples from current research demonstrating the use of different types of decision support systems in the construction industry. We learn that decision support systems, while useful, are there to support decision makers, but not to replace them. Talents such as creativity, imagination, or intuition should not be ignored in the course of making a decision.

In Chapter 16, Tserng and Lin demonstrate the use of a knowledge management portal built on the basis of knowledge maps for activity-based knowledge management in construction. Various life cycle stages of knowledge management from knowledge acquisition, extraction, storage, sharing, and updating are demonstrated, followed by a description as to how these are used to construct a knowledge map. We learn that knowledge maps are an efficient visual tool for presentation of the availability and source location of knowledge within a construction project.

In Chapter 17, Chen, Kong, Li, and Xu present an integrative knowledge management system for environmental-conscious construction based on a comprehensive integration of environmental management techniques and tools in construction. An experimental case study is used to demonstrate its use and applicability for the construction sector. We learn of the need for better integration of environmental management tools to support better decision making in various construction phases.

In Chapter 18, O'Brien, Issa, Shen, and Xie present a vision for future project information technologies in the form of configurable environments. These configurable environments combine data integration capabilities with visualisation and analysis tools in a format customised to the needs of specific users and projects. The need for and technical aspects of the configurable environments approach is presented in the context of an illustrative case study of construction space management. We learn that as information technologies evolve, they will shift the balance of job tasks from information gathering to analysis. Firms should consider grooming younger employees for such changes.

This book is not a primer on knowledge management, nor is it a primer or introductory text on the construction industry. It is in fact a collection of global perspectives, practical experiences, and research in the area of knowledge management in the construction industry. While revisiting the advent of socio-technical thinking in the construction sector from the early sixties, it goes well beyond the current state of the art in exploring various visions for knowledge management in the construction industry. Encapsulated within are recent research undertakings supplemented with real-life cases and lessons learned from various corners of the globe.

References

APQC. (1996). *Knowledge management: Consortium benchmarking study*. Houston, TX: American Productivity and Quality Center.

Davenport, T.H. (1994). Saving IT's soul: Human-centered information management. *Harvard Business Review*, (March-April), 119-131.

Davenport, T.H., & Prusak, L. (1998). *Working knowledge: How organizations manage what they know*. Boston: Harvard Business School Press.

Kazi, A.S., Hannus, M., & Charoenngam, C. (1999). An exploration of knowledge management for construction. In M. Hannus, M. Salonen, & A.S. Kazi (Eds.), *Concurrent engineering in construction, challenges for the new millennium.* Espoo: CIB Publication 236.

Kazi, A.S., Puttonen, J., Sulkusalmi, M., Välikangas, P., & Hannus, M. (2001). Knowledge creation and management: The case of Fortum Engineering Ltd. In E. Coakes, D. Willis, & S. Clarke (Eds.), *Knowledge management in the sociotechnical world: The graffiti continues.* London: Springer-Verlag.

Murray, P.C. (1999). New language for new leverage: The terminology of knowledge management. *KM Metazine*, Knowledge Transfer International. Retrieved April 2004 from www.ktic.com/ topic6/13_TERM4.htm.

Nonaka, I., & Takeuch, H. (1995). *The knowledge creating company.* Oxford: Oxford University Press.

Nonaka, I., Toyama, R., & Konno, N. (2001). SECI, ba and leadership: A unified model of dynamic knowledge creation. In I. Nonaka & D. Teece (Eds.), *Managing industrial knowledge—creation, transfer and utilisation.* London: Sage.

Sieloff, C.G. (1999). If only HP knows what HP knows: The roots of knowledge management at Hewlett-Packard. *Journal of Knowledge Management,* 3(1), 47-53.

Skyrme, D.J. (1997). From information to knowledge management: Are you prepared? *Information Age,* 1(20), 16-18.

Stewart, T.A. (1997). *Intellectual capital: The new wealth of organisations*. New York: Currency/Doubleday.

Acknowledgments

My personal journey with knowledge management in the construction industry through a socio-technical perspective began well over a decade ago. I am indebted to my mentor, Dr. Chotchai Charoenngam of the Asian Institute of Technology for reminding me not to take for granted the simple fact that the human is the most critical element in any undertaking, and not the technology. Long hours of discussions in the form of stories and experiences shared have formed my main basis for passing on the message further to numerous construction companies. I am furthermore grateful to my work colleagues at the Technical Research Centre of Finland (VTT) for sharing with me their research and practical experiences. Through all their experiences, I was taught to communicate in a language that not only the industry understands, but is willing to listen to.

I have been actively engaged over the past several years in helping several construction companies realise that knowledge management is not yet another management buzzword, but something that can deliver. In particular, I am grateful to YIT Construction Ltd. and Fortum Engineering Ltd., both of Finland, for allowing me to be a part of their exciting journeys in knowledge management. It was these journeys that led to the urge for a need of a book on the subject matter—a book that would not be about knowledge management or the construction industry in general, but one that would showcase examples of experiences and research investigations from different corners of the globe.

I was both honoured and thrilled when Mehdi Khosrow-Pour from Idea Group Inc. invited me to edit a book on the subject matter of this book. From here on, an exciting journey started. I am grateful to the publishing team at Idea Group Inc. for their professionalism and enthusiasm throughout the pre-publication process. Many thanks are due to Jan Travers for always being accessible in times of need, following up progress on a regular basis, and providing support throughout the process. Jennifer Sundstrom has been very helpful in helping to maintain the focus of the content of this book, through her at times difficult questions and inquiries. Thanks are also due to Amanda Appicello and Michele Rossi, for being an integral part of the process. This book would not have been possible without Mehdi, Jan, and the rest of the team at Idea Group Inc. Many thanks.

The planned content for this book would never have been realised had it not been for the willingness and enthusiasm of all contributing authors to share their experiences and wisdom. Many also took the time to suggest terms of inclusion in the glossary which you will find at the end of this book. They showed an amazing amount of patience in receiving and accepting the feedback of the reviewers, and my continuous probing to ensure that not only did they deliver the best of what they had, but to also deliver it on time, and in the correct format. I would like to thank all authors for their patience, insights, and foremost excellent contributions. I am also grateful to all the reviewers who reviewed the chapters. Some of the authors of this book in fact took on the additional role of acting as reviewers for chapters written by other authors. The reviews were critical, constructive, and comprehensive. I am sure all authors will join me in thanking the reviewers for helping improve the quality of the content of the chapters in this book.

No journey, adventure, or project is possible without the support of one's loved ones. I am grateful to both my parents for allowing me from a very young age to explore what this world has to offer. They have shared their experiences with me, from which I have learned, and have been an integral part of my journey throughout life itself. My life partner, Sajida, has been a source of love, support, and comfort throughout this and numerous other projects I have chosen to be part of. She has stood by my side and ensured that once I start something, I complete it. My wonderful children, Haani and Shumail, remind me each and every day that one learns with each step one takes, and that there is no better teacher than personal experience itself (if you burn your finger once, you will learn from experience never to again put your finger inside a cup of hot coffee). I am grateful to my whole family for being an integral part of this book project through their unending support and patience.

Abdul Samad (Sami) Kazi
Espoo, Finland
April 2004

Section I:

Knowledge Management in Construction Practice

Chapter I

Strategies for Managing Project Generated Knowledge: A New Zealand Case Study

Steve de Kretser
Sinclair Knight Merz Ltd, New Zealand

Suzanne Wilkinson
The University of Auckland, New Zealand

Abstract

If consulting construction companies made more effective use of project-generated knowledge, the financial health of the company would increase. This chapter discusses how wasted, lost, and ineffective use of knowledge leads to inefficiencies and reworking of past problems. The chapter focuses on the development of project-generated knowledge management systems and the benefits that construction companies can reap from such developments. Through a detailed case study, the chapter shows where the main problem areas occur in managing knowledge and proposes possible solutions to these problems. The chapter concludes that it is the people that use and produce the knowledge that are in the best position to help develop such a project-generated knowledge management system, with senior managers and company knowledge managers providing structure, facilitation, and support. With such collaboration and involvement of all staff in a company, failures of such systems will be reduced and the company as a whole will benefit.

Introduction

This chapter examines how a consulting engineering company can manage the knowledge acquired in the course of executing an engineering project with the aim of obtaining maximum future company-wide benefit from that knowledge. In the field of consulting engineering, specialist knowledge is the firm's primary product, even when manifested as design documentation and other physical, codified forms. There will be real benefits to a firm if it can ensure that important knowledge gained by team members on each project is identified and made available to others in the firm to learn from and use on future projects. The larger and more geographically diverse the firm, the more difficult this process becomes. A review of current practices within the case study company identified shortcomings in the extent and success of project-generated knowledge management. Based on this review, and drawing on current knowledge management literature, a general strategy was proposed for the initial and ongoing development of a management system for project-generated knowledge. Certain specific practices and system elements were identified for further development and implementation as part of the knowledge management system. A key conclusion from this research is that it is the producers and consumers of knowledge who are typically best placed to identify appropriate management methodologies, particularly at the "detail" level, with the central "corporate" role being primarily one of support, education, facilitation, and coordination.

Learning Objectives

After studying this chapter, the reader will:

1. Appreciate the importance of a management system for capturing project-generated knowledge.
2. Understand the typical problems that arise for consulting construction companies managing project-generated knowledge, and how to overcome these problems.
3. Learn some basic principles for developing a knowledge management system for capturing project-generated knowledge.

Background

In any organisation, there is a certain component of the work carried out that can be termed "knowledge work." Many researchers (Gurteen, 1998; Martin, 2000; Kamara, Anumba, & Carrillo, 2002; Kululanga & McCaffer, 2003; Koch, 2003) have provided discussions and definitions of the terms "knowledge," "knowledge work," and "knowledge management." For an engineering consultancy, "knowledge work" refers to work that relies on the accumulated knowledge, experience, and judgment of the individual, rather than the ability to be trained to learn and execute simple procedures. Garvin (1993) provides distinctions between tangible assets and intangible (knowledge) assets. The knowledge that is required to carry out those work elements can be considered to be an asset of the company—its "intellectual capital" (Garvin, 1993).

In consulting engineering, knowledge work makes up the major component of a firm's effort, output, and profitability. The primary product is intellectual—the firm's specialist knowledge—rather than physical. Any physical output, such as design drawings, documents, and the like, are simply a representation of the knowledge that is being sold. This is not only the technical or professional knowledge acquired by the firm's staff members during, and subsequent to, tertiary education, but also their knowledge of clients, business sectors, organisational procedures, other industry personnel, and so forth.

With knowledge considered as an asset in consulting engineering, then it is clear that loss of knowledge from a firm, or its inefficient use, will have an impact on the firm's economic performance (Kamara et al., 2002). A firm's knowledge can be lost either through staff turnover or simply a loss of organisational memory of what once was known. Furthermore, the firm may possess certain knowledge, but either: (i) not be aware that it does so, (ii) not be able to locate the specific source, or (iii) not disseminate the knowledge or make it available to all in the wider organisation who would benefit from it. Martin (2000) reports on the problems of inefficient knowledge management for a company and how this can impact on its operation and profitability. For example, he reports the problems include: reinventing the wheel, knowledge loss, knowledge hoarding, poor decisions, and the inability to learn. Added to this for consulting engineering can be the problems of rework to correct errors, time (and money) spent searching for held knowledge, lost or missed business opportunities, and financial liability for errors.

All knowledge is initially held by individuals, either as tacit knowledge or as personal explicit knowledge (Nonaka & Takeuchi, 1995). One key barrier to overcome in the process of knowledge management is the lack of communication of such relevant personal knowledge to others. This phenomenon has been called "knowledge hoarding" (Martin, 2000). Knowledge hoarding can be intentional, due to an individual's desire to maximise their own importance or power, or to protect their position in the firm (Martin, 2000). It may also be inadvertent, if there is no awareness or perception of others' potential interest in, or need for, certain knowledge. A third possibility is that the individual may recognise the value of their knowledge—and have a desire to share it, yet be frustrated in their efforts to communicate it to others—either due to lack of convenient means to do so (forums, systems, procedures) or lack of management support in the form of time, budget, or prioritisation. Koch (2003) suggests that a knowledge policy should recognise changing culture and be developed to foster a culture that acknowledges knowledge sharing rather than knowledge as individual power. If knowledge management problems are not addressed, they can reduce the competitiveness of a business (Sarvary, 1999).

Knowledge Management Theory

Kululanga and McCaffer (2003, p. 347) quoting Huber (1991) state that: "Knowledge management is a process of acquiring, creating, sharing, utilising, and storing intellectual assets and other stimuli from the internal and external business environments that facilitates an organisation to perform successfully." Knowledge management in the consulting engineering industry has existed in some form for many years—in technical libraries, codes and standards, standard procedures and specifications, and in mentoring of junior staff. However, the issue of knowledge management has assumed greater importance in recent times (Argote, 1999) due to factors such as: the increasing pace of technological change; the growth in the quantity of information available; the globalisation of companies and their spheres of activity; increasing competition for work and maturation of many industry sectors; and increasing client sophistication and expectations. Consulting companies in engineering and construction industries are actively seeking ways in which they can better capture and use the knowledge generated by their employees (Martin, 2000; Gupta & McDaniel, 2002). Gupta and McDaniel (2002) propose five components for effective management of knowledge. These are:

1. Harvesting (i.e., recognising and obtaining the knowledge from the individuals or groups).
2. Filtering (i.e., using knowledge that will produce a competitive advantage and discarding other knowledge).
3. Configuration (i.e., putting the knowledge in an readily understandable and useful way by organising and storing it).
4. Dissemination (i.e., communicating the knowledge or facilitating its use).
5. Application (i.e., applying the knowledge in everyday business activities).

For companies to develop a knowledge management system, it is important to understand current terminology and the theory of how knowledge is acquired. Davenport and Prusak (1998) describe the differences and relationships between the terms 'data', 'information', and 'knowledge' as follows: Data are the basic building blocks: discrete, objective facts about events. Being objective they only say "what"—there is no "how" or "why." Information is data that has been processed to add a degree of meaning (Parikh, 2001). The data may have been "contextualised," "categorised," "calculated," "corrected," or "condensed" in the process of becoming information (Davenport & Prusak, 1998). "Knowledge is a fluid mix of framed experience, values, contextual information, and expert insight that provides a framework for evaluating and incorporating new experiences and information. It originates and is applied in the minds of knowers. In organisations, it often becomes embedded not only in documents or repositories, but also in organisational routines, process, practices, and norms" (Davenport & Prusak, 1998, p. 49). The transformation of information into knowledge takes place through processes such as "comparison," "consequences," "connections," and "conversation." Because knowledge is such a complex and "fluid mix," it is extremely difficult to capture in words, to categorise, and to understand completely. The differentiation of knowledge into "tacit" or "explicit" is discussed by Allee (1997, p. 45), Roos and colleagues (1998, p. 16) and Nonaka and Takeuchi (1995, p. 63). Summarising from these sources, it can be said that tacit knowledge is the knowledge that is held inside an individual. It is personal and context-specific, derived from the holder's experiences and insights. Explicit knowledge is that which is codified in written or some other "hard" form (Nonaka & Takeuchi, 1995). It is explicit knowledge that can be transferred between people through documents, images, and other deliberate communication processes. Roos et al. (1998, p. 16) and Allee (1997, p. 45) describe the processes of conversion and interaction between these different forms of knowledge. Tacit-tacit interactions create knowledge through conversations, reflections, and so forth, which Allee describes as a "socialisation" process. Tacit-explicit interaction involves codification of some of an individual's tacit knowledge so that it is accessible to others (Allee's "externalisation"). Explicit-tacit transfer requires access to, and learning from, codified documentation, such as through reading ("internalisation" per Allee). Allee (1997) refers to explicit-explicit interaction as "combination"—that is, some processing, organisation, or arrangement of codified knowledge elements. However, Roos et al. (1998) consider that this fourth form of knowledge transfer cannot really exist, that it is merely "data" (or "information") transfer, since to be considered knowledge there must be some form of human interpretation in the process. In other words, "combination" of explicit knowledge actually involves an explicit-tacit-explicit sequence, to account for the human presence in the process. The process of knowledge development and creation is not simply one of making tacit knowledge explicit (i.e., attempting to codify it in some way). That is only one step of the sequence. Rather, as described by Roos et al. (1998) and Allee (1997), to have a truly effective process, there needs to be a "knowledge spiral" as per Nonaka and Takeuchi (1995) of knowledge creation moving back and forth through all four (or three) of the above conversion processes. This is accomplished by the communication and sharing of individuals' knowledge. The key to knowledge management, then, is to facilitate and encourage this sharing and communication of knowledge—either verbally or through written means (Binney, 2001). Roos et al. (1998, p. 17) identify that technical systems, such as "information technology," can help this process, but cannot, by themselves, ensure it.

Developing a Knowledge Management Strategy

Depending on the company and sector type, different strategic approaches can be used in adopting a knowledge management strategy (Haggie & Kingston, 2003). Hansen, Nohria, and Tierney (1999) propose ways in which companies can develop knowledge management strategies depending on the type of service or products they sell. They suggest that companies selling standardised products suit processes that standardise and codify knowledge (codification approach); but for those that produce customised solutions to unique problems (as in engineering consulting businesses), the information technology should be used to help people communicate (personalisation approach). Companies need to opt for one approach and use the other approach in support (Hansen et al., 2000). However, Stewart (1997) points out that the potential for creating knowledge databases of what he terms "structural capital" (i.e., codification approach) are almost limitless and vary enormously from company to company. This can be useful for consultancies. Stewart (1997) identifies three items that he considers are particularly worth the effort:

1. A "corporate Yellow Pages" (or "knowledge map"), also advocated by Davenport and Prusak (1998), that facilitates connecting inquirers with the relevant sources of specialist knowledge. The larger and more diverse the company, the more valuable such a system would become. It is much easier to identify the people who hold certain knowledge than it is to attempt to codify the knowledge that they hold, and tacit-tacit interaction is still a perfectly valid mechanism for knowledge transfer.
2. A system of identifying "lessons learned" from past projects, so they are available for consideration in future, similar projects. Although knowledge work is substantially custom work, there are normally sufficient elements of similarity to benefit from such documented lessons learned.
3. A system to facilitate recording and communication of knowledge and information about a company's customers and competitors.

Malhotra (2002) discusses the failure of knowledge management systems and proposes that they be designed to adapt to changing business characteristics. Davenport and Prusak (1998, p. 45) state that there "are many pitfalls and limitations in using information technology for knowledge work, trying to force fluid knowledge into rigid data structures, for example, or focusing too much on the system and not enough on content." However, "networks and desktop computers...can dramatically improve knowledge market efficiency [by providing] an infrastructure for moving knowledge and information about knowledge as well as for building virtual knowledge marketplaces" (Davenport & Prusak, 1998, p. 46). The role of information technology is ideally one of supporting, facilitating, and enhancing the natural exchange of knowledge for engineering consultancies (i.e., the personalisation approach as per Hansen et al., 1999). There is no doubt that information technology can provide great benefits in the transfer, storage, and accessibility of knowledge, particularly on a large, international scale. However, it will only be successful if the primary focus of the system development is on the principles of knowledge management rather than on the underlying technology itself. Allee (1997) discusses the "Knowledge Technologies Interface"—that being the overlap between the "The Organisation," "Knowledge-at-Large," and "Supporting Technologies." The target is the triple overlap between all three areas (Allee, 1997).

Knowledge Management Systems in a Project-Driven Environment

Since engineering and construction consulting businesses tend to be "project driven," knowledge management systems need to be designed to collect, disseminate, and use, for the benefit of the entire company, project-generated knowledge. For project-generated knowledge, Conroy and Soltan (1998) define three "knowledge bases" to contain the knowledge that is used and created in the execution of a project. These are:

1. The Organisation Knowledge Base, which contains the data and information specific to the organisation and wider environment in which the project is being executed.
2. The Project Management Knowledge Base, containing knowledge of the theory and application of project management.
3. The Project-Specific Knowledge Base, which contains the project-specific knowledge acquired from the user at the outset and developed over the project life cycle.

Conroy and Soltan (1998) suggest that during the course of the execution of a project, there will be a creation of knowledge in the Project-Specific Knowledge Base, but there will also be an evolution of the initial knowledge held in the Organisation and Project Management Knowledge bases. For example, this could occur in areas such as familiarity with the particular client; the equipment and technologies used on the project; and the project management systems, techniques, and software. The new knowledge in all three knowledge bases is initially held only by those individuals who are members of the project team. The goal, therefore, is for that knowledge, or at least the most important and relevant elements of it, to be identified, captured, and fed back into, or made available to, the wider knowledge pool of the organisation, along the lines Gupta and McDaniel (2002) propose. In this way, the knowledge can be used to the benefit of future work across the firm, rather than only on future projects involving those specific project team members. In considering the business of consulting engineering, the project-generated knowledge that is of interest for capture can be divided into three general categories (Conroy & Soltan, 1998):

1. Technical—relating to the techniques, technologies, work-processes, statutory requirements, costs, and so forth involved in the production of discipline-specific elements of the project. New knowledge needs to be fed back into and managed by the discipline departments of the organisation.
2. Project Management—relating to the techniques and technologies for managing the execution of projects. This knowledge is of benefit to all project managers and others involved in the wider aspects of project management, including the company quality system.
3. Project Related—knowledge of the client and the historical aspects of the project which are of use for future marketing purposes either in winning jobs with the same client or in improving the firm's "curricula vitae."

Thus by following such guidelines for an integrated knowledge management system, capturing project-generated knowledge is feasible.

A New Zealand Case Study

Through a detailed case study of an engineering consulting company, this research investigates the application of knowledge management, reviews the firm's current systems and processes, and proposes a more comprehensive strategy for management of project-generated knowledge. The aim is to focus on the initial key steps of understanding the problems and proposing a general strategy. The company in this case is a large, multi-disciplinary engineering consultancy that is based in Australia, but has offices around the world, including in New Zealand. For this company, and other such consulting firms, almost all work is executed as projects that have defined beginnings and ends. This company functions as a matrix organisational structure, whereby technical staff are members of discipline-based departments (e.g., mechanical or civil engineering), reporting to a disciple manager, but are also normally assigned at any time to one or more project teams, in which capacity they also report to the project manager. Upon completion of a project, its team members return to the discipline department ready for assignment to other projects.

In such an organisational structure, a significant proportion of the new knowledge generation in the company will occur in the project environment—staff will acquire knowledge in the course of, and as a direct result of working on, a specific project. There is thus a substantial incentive to identify, store, and communicate the key knowledge, experience, and feedback gained on each project carried out by the firm to ensure that the whole firm can obtain the maximum future benefit from it as part of a continuous improvement process. Since each discipline department is responsible for the technical standards of its members' work output, it is these departments that must manage the knowledge elements related to their field of expertise. However, since project managers can be drawn from any area of the company, the management of the project management knowledge base is a somewhat different exercise, and one that has significant links to the firm's quality system procedures. This knowledge management process is distinct from the separate requirement to properly manage project-specific information during the course of each project—that being essentially a project management function.

The approach of focusing the case study analysis on the general principles, strategy, and common knowledge management system elements means that the findings will be relevant to most construction consulting organisations. The following section focuses on the main problems identified during an examination of how the New Zealand branch of the company managed its project-generated knowledge, together with proposals on how a company can solve such problems.

A Company Knowledge Management Problem Identified

One of the main problems for the company was that, while there were some systems already in operation, there was evidence that these were not functioning as ideally intended and that there was further scope for improvement. Such examples include the commonly found repeating design problems. Design problems encountered in an area of a project can often be similar to those on a project completed several years earlier. At the company in this case, one such incident was when both projects were in the same market sector and were run from the same office. Site engineers from the earlier project had identified the problems in a "semi-official" post-project report, but this appears not to have been circulated to all the designers, or to have been readily accessible by the company employees who would benefit from the knowledge. If this can happen in the same office, the chances of this repeatedly happening throughout the large, multi-location company were high. One solution to this problem is to introduce a global information sharing system that allows all project-based documents to be shared across multiple offices, with an inbuilt coding system to facilitate understanding of updates and how these changes subsequently affected the project.

Another avenue for companies, particularly in technical knowledge management, is to develop standards, such as drawings, specifications, design methodologies, calculations, spreadsheets, checklists, and so forth. Traditionally these kinds of documents, which are used on most projects: (i) are frequently developed by editing comparable documents from a previous project, (ii) are normally directed only at the current project, and (iii) are usually kept only in the project files, including archived "as-builts." The problems with this usual approach include: the new document draws primarily on the knowledge contained in the document from the previous project from which it was derived. It may well miss out on certain knowledge that was omitted from that base document, or was identified on some other project; there is inconsistency in document format, style, and content between different projects; or there is wasted effort in reinventing the wheel on one project when the knowledge may already reside elsewhere. To avoid this, companies can identify specific areas relevant to a discipline for which standard (non-project-specific) documents would be of benefit; it is then possible to collate the accumulated knowledge of many projects in one accessible repository. The in-house standard documents can be developed alongside a fee-paying project, so that the maximum benefit is obtained. The extra cost to the organisation to make the documents 'standardised' for use on future projects is then relatively small. Since these documents are 'living', they must be updated to reflect the evolving knowledge of the organisation. A relevant specialist, or a nominated party, can be the means of accomplishing this, based on feedback from the entire organisation.

This approach is seen to be of particular benefit in capturing project-generated knowledge, particularly feedback from the construction and commissioning phases of projects—at this stage the project-specific drawings are typically updated to 'as-built' status, but other design-phase documentation is frequently not. For example, the installation specification used on a project is not normally retrospectively revised based on a deficiency identified during on-site installation or commissioning. Having a "living" document outside of the project-specific arena allows this to happen much more easily.

Identifying Knowledge Holders and Capturing Feedback

One of the main difficulties with large companies is that of finding out who's who in the larger organisation—who one's peers are and who is likely to hold knowledge from particular past projects or experience of interest. On the most basic level, even organisation charts can be hard to find. It seems to be common practice for people to send e-mail messages that often contain no indication of their position/discipline, the projects they are working on, or even which office they are based in. In order to try to locate experience elsewhere in the firm, the only readily available mechanism often resorted to is sending e-mail messages to all professional staff asking, "Does anyone know anything about...?" While reasonably effective at eliciting responses, such messages can tend to clutter up the e-mail system and serve as distractions to the majority of recipients who are not knowledgeable in the topic in question. Given the rapid and continuing growth of this case study company in recent years through mergers and acquisitions, the problem of knowing who's who is certain to continue. Since the basic requirement of knowledge management is for people to communicate with each other, in order that they openly share their knowledge, a first step is to facilitate and encourage such communication. Several company-wide elements of a knowledge management system are recommended to help address this problem—some are commonly used, although others should be considered to maximise the benefit of knowledge generated:

- an online database of all staff, their areas of expertise, and so forth—commonly known as a corporate "Yellow Pages";
- increased establishment and use of technets and newsgroups throughout the organisation, especially to share project-specific knowledge and experiences;
- a database identifying key staff who worked on each project; and
- an intranet that provides a resource for identifying areas of staff expertise and experience. This database (or several linked sub-databases) should contain the following kinds of information: the staff member's name and basic contact details—office, e-mail, phone and fax, their position and discipline; a hot-link to that person's CV; an experience database, using guided keywords, to track their areas of expertise (by role and by market sector); a database listing clients for whom they have worked.

For most companies, this database would be relatively straightforward to establish and to update, and staff should be encouraged to use it as a "first call" for any technical information about company personnel they require. It is expected that staff would be responsible for ensuring their own details are current, as is the case with CV's, and that staff would be required to regularly check and update details.

Capturing feedback through existing technets and newsgroups is one mechanism of obtaining project-generated feedback. This company's experience was that these were fairly effective but depended heavily on a volunteer coordinator for each group and the level of usage by relevant staff members. Discipline- or interest-based groups are seen as an extremely appropriate means of communicating information and knowledge around the company—particularly between offices. Therefore, their establishment and use needs to be actively encouraged. If a group that one would expect to exist does not, then that apparent lack needs to be followed up. This effort is best done at a divisional or departmental level, under the coordination of a company-wide knowledge manager. The development of sub-systems for managing technical, or specialist, knowledge is best placed in the hands of those who create or use that knowledge. This is on the basis that they are best placed to determine an appropriate mechanism for the identification, codification, storage, and access of this knowledge. However, unless information is communicated and disseminated across the company, the knowledge and experience gained can be lost, or at most only communicated to a few perceived interested parties. Hence the need for a knowledge manager to facilitate the flow of information through various company channels.

Managing Different Knowledge Types

Managing general and specialised knowledge is one of the problems faced by construction consulting companies. Many companies, this one included, tend to follow a relatively 'generalist' approach to work assignments among technical professionals, whereby staff members will end up working on a huge range of projects and tasks within their particular discipline. While this is considered an appropriate approach in a business where no two jobs are ever the same, and where the range of jobs undertaken is too wide to justify employment of teams of specifically qualified individuals, there is still room to introduce elements of specialisation. Requiring many staff across the organisation to possess or acquire certain knowledge, which they may need at some stage, but not continually, represents an over-investment in that knowledge. It is more efficient to promote one (or more) individuals to specialise in selected areas, and to communicate that specialisation to the other members of the organisation (through mechanisms discussed above). Certainly the ability to tap into a wider pool of expertise should be one of the significant benefits of belonging to a larger organisation. Identification of potential areas for development of specialisation is best done at lower levels of the organisation and coordinated between offices across the company. These need not be major areas of technical expertise, but even more general aspects of the regular business of a discipline. For example, mechanical engineers could be nominated to specialise in areas such as pumps, control valves, or pipe support systems. This aspect of knowledge management obviously ties in with issues of employee training and development. Should such a specialist leave the company, there is certainly a resulting loss of knowledge, however the extent of this loss will be minimised if the individual has been actively sharing their knowledge with their peers in the normal course of carrying out their work, and has codified aspects of their knowledge through drawings and documentation which they have produced or contributed to while in that role. Having such identified specialists in the company allows them to be used as a resource for the entire firm in areas such as reviewing specific relevant sections of project documentation. It ensures a consistent and correct output from each specific knowledge area, and also provides a readily identifiable target for project-related feedback in their area of expertise. The specialist should be responsible for updating standard documents to reflect the evolving knowledge of the organisation.

Communicating Captured Knowledge

Capturing and using lessons learned more effectively is one of the ways in which the management of project-generated knowledge can financially benefit the company. The production of "lessons learned" reports is seen as beneficial to the organisation, provided they reach the appropriate parties. Furthermore, the dissemination of the knowledge contained in these reports should not have to wait until the completion of a project. It is recommended that during the life cycle of a project, reports be treated as a "knowledge bulletin," which is used any time there is a need for written communication of some aspect of technical knowledge to the wider organisation. The content should include a "lesson learned," identify "what went well," suggest an improvement opportunity, and so forth. If companies were to use such project-generated knowledge bulletins, they should be in standard form, or template, which would include key project details in a specified format to facilitate its tracing and storage. The feedback content is best left as "free format," adopting the approach of trying not to force the knowledge into a rigid system, and allowing the author's anecdotal style to best convey the details. In the normal course of a project, or whenever else necessary, a bulletin would be prepared and issued—to a technet, newsgroup, or technical specialist, as appropriate. When a project is completed, all bulletins prepared during its course and as part of the post-project review under cover of a "project close" form, and be fed the project documentation under cover of a modified project close and validation form, and fed into the company-wide information sharing system. Not only will this approach facilitate the communication of knowledge at an early stage in a project's life, but it will help prevent the knowledge from being forgotten, as could happen if the feedback process were deferred until the completion of the project. Guided keywords will assist database searching by knowledge seekers. As part of the project close-out, any separate soft-copy files can be linked and indexed to the main form in the database. Communication of the content of the form would be by several mechanisms such as: direct hard or soft copy circulation to relevant local staff; electronic circulation to relevant staff in other offices via technets and newsgroups; and storage in a database for future search access. It would also naturally be stored in the project files, but without any real expectation of that ever being a primary search location once the project has been archived.

In addition to using the more obvious keywords such as project title and client, the content of each "lesson learned" and "what went well" needs to be identified by the discipline/area(s) involved—for example, project management, marketing/proposals, technical (specific discipline), contract administration, QA—and additional keywords for specific details covered (e.g., relevant technical terms). This information can be entered quite rapidly, using a system of guided keywords, and will greatly facilitate future database searching. These bulletins, and other project information, such as post-project audit, review, and feedback information, must cater for the wide range of project sizes and types that most construction companies handle and thus must differentiate between mandatory minimum entry requirements and suggested/optional elements.

Coordination, Control, and Use of Knowledge

Coordination, control, and effective use of the knowledge generated are essential if a system for using project-generated knowledge is established. One of the key problems is how to get people involved. The role of a knowledge manager can greatly ease this problem as they can facilitate the generation and flow of knowledge generated through the company. For effective knowledge management, an implementation and control plan needs to be prepared and specific responsibilities assigned. Discussion and training sessions around the company's offices should aim to promote the initiative, conveying the strategy and general techniques. A survey should be undertaken of what is currently being done and how effective it is—in particular identifying "best practice/lessons learned" techniques and approaches which can then be spread throughout the company. Technical managers and discipline-based groups (departments and divisions) should be encouraged to identify and discuss opportunities in their fields. They must come up with plans for managing their own knowledge, which would then be communicated to the knowledge manager for comment and coordination. The various company-wide activities, such as the staff experience database, should be initiated following the initial round of discussion/training sessions. Identified strategies for knowledge management can be reviewed against available technologies, such as intranet, extranet, project sharing databases, and so forth, in order to identify any improvements. For most companies, it is recommended that they start with identified "big gain" items—those with the most potential benefits and least chance of failure. Although individual or "local" elements of the knowledge management system should not be developed without considering the larger, system-wide issues, appropriate application of the basic knowledge management principles identified should permit development of effective elements to occur. Certainly it is neither appropriate nor beneficial to try to determine a "complete" detailed system strategy before proceeding with any implementation steps.

Discussion

For engineering and construction companies, it is important to establish a strategy for managing the project-generated knowledge. The key questions to be addressed are:

- What is the system is aiming to achieve?
- What are the underlying general requirements of the system and each of its component elements?
- What are the requirements from the organisation to allow the system to occur and function effectively?

The term "system" is not intended to indicate a rigidly defined set of procedures and software programs; rather, "system" refers to a looser set of tools and elements developed through a consistent methodology and coordinated across the company, with the common underlying goal of knowledge management. The aim of this management system for project-generated knowledge is to provide maximum benefit to the entire company from the knowledge gained in the process of executing projects. The identification and exchange of this acquired knowledge can be by many means such as: codifying and storing it for later retrieval by the same or different staff; codifying it and disseminating it to other relevant individuals; or making other relevant individuals aware of who holds particular knowledge.

The key is to avoid knowledge hoarding, either deliberate or inadvertent, but instead to promote and facilitate the communication and sharing of this knowledge through one of the processes identified. For such an initiative to succeed, it must receive explicit and continued support from senior management. In particular, management throughout the company must be constantly aware of the issues relating to knowledge management and be actively looking for opportunities for improving the knowledge flow around the company. They must create a culture of open knowledge communication in the company. Furthermore, companies should assign a "company knowledge manager." This function is not to implement all of the detailed elements of a knowledge management system, nor is it to lead the IT development across the company. Rather it is to oversee, guide, coordinate, and facilitate the implementation of all of the various and diverse elements of the system across the company. Consideration of the issue of managing project-generated knowledge in the light of the basic principles of knowledge management leads to the conclusion that the most effective "system" for managing this knowledge will not consist of a comprehensive, detailed set of databases, forms, and procedures developed by a single person or committee. Instead, it will be a whole range of initiatives generally identified and implemented at a more "localised" level, facilitated and coordinated by a knowledge manager, assisted by a number of company-wide system components, and promoted by company management. Rather than trying to force knowledge transfer, it will be more effective and efficient to actively encourage it; facilitate it through strategy, frameworks, guidelines, and training; then let it happen, and monitor its progress and development. A strategy and key requirements of a system to manage project-generated knowledge will substantially consist of sub-elements created at a level local to the producers and consumers of the specific knowledge. There are, however, several areas where elements of the system can and should be company wide.

Conclusion

It is clear that the effective use of project-generated knowledge will increase the financial health of a consulting construction company. Wasted, lost, and inefficient use of knowledge, through various means discussed in this chapter, will lead to inefficiencies and reworking of past problems. The discussions in this chapter have focussed on the development of project-generated knowledge management systems and the benefits that construction companies can reap from such developments. In conclusion, it is the people that use and produce the knowledge that are in the best position to help develop such a project-generated knowledge management system, with senior managers and company knowledge managers providing structure, facilitation, and support. With such collaboration and involvement of all staff in a company, failures of such systems will be reduced and the company as a whole will benefit.

Practical Tips and Lessons Learned

From the literature review and the case study, it is possible to outline some basic principles for companies wishing to develop a knowledge management system for capturing project-generated knowledge:

- The total system and system elements need to be easy to use by all parties such as contributor, receiver, or searcher of knowledge.
- Creators and users of specific knowledge are in the best position to deal with the most appropriate mechanisms for its management, so companies should avoid adopting and using a system that has been imposed from "outside."
- The appropriate mix of tacit knowledge (e.g., discipline specialists) and explicit knowledge (e.g., standard documents) repositories must be selected, based on the nature of the company's work and its organisational structure.
- The system and its sub-elements need to cater for a wide range of project types, sizes, and disciplines.
- The progressive evolution of knowledge areas may lead to significant changes in approach. Hence flexibility and customisability are important factors.
- It must be developed with an overall "systems perspective," rather than simply being a group of ad-hoc pieces; hence the need for a knowledge manager.
- It should draw on "best practice" and "lessons learned" knowledge management examples from throughout the organisation.
- The system should make use of technology as a tool, but not be unnecessarily bound or complicated by it.
- Where practical, knowledge should be identified and/or disseminated as soon as possible, rather than waiting until project completion.
- There must be sufficient direction to the approach to ensure a degree of consistency and commonality, such as a company-wide standardisation of keywords to be used for inputting and outputting database information.
- People must be put in touch with each other and encouraged to communicate.

These basic principles will allow companies to develop a project-generated knowledge management strategy that best suits their company focus and needs.

References

Allee, V. (1997). *The knowledge evolution: Expanding organizational intelligence.* Newton, MA: Butterworth-Heinemann.

Argote, L., (1999). *Organizational learning: Creating, retaining and transferring knowledge.* Boston: Kluwer Academic.

Binney, D. (2001). The knowledge management spectrum—understanding the KM landscape. *Journal of Knowledge Management, 5*(1), 33-42.

Conroy, G., & Soltan, H. (1998). ConSERV, as a continual audit concept to provide traceability and accountability over the project life cycle. *International Journal of Project Management, 16*(3), 185-197.

Davenport, T.H., & Prusak, L. (1998). *Working knowledge: How organizations manage what they know.* Boston: Harvard Business School Press.

Garvin, D.A. (1998), Building a learning organization. *Harvard Business Review on Knowledge Management.* Boston: Harvard Business School Press.

Gupta, A., & McDaniel, J. (2002). Creating competitive advantage by effectively managing knowledge: A framework for knowledge management. *Journal of Knowledge Management Practice, 3.* Retrieved November 2003 from www.tlainc.com.

Gurteen, D. (1998). Knowledge management and creativity. *Journal of Knowledge Management, 2*(1), 25-37.

Haggie, K., & Kingston, J. (2003). Choosing your knowledge management strategy. *Journal of Knowledge Management Practice, 4.* Retrieved November 2003 from www.tlainc.com.

Hansen, M.T., Nohria, N., & Tierney, T. (1999). What's your strategy for managing knowledge? *Harvard Business Review, 77*(2), 106-117.

Kamara, J.M., Anumba, C.J., & Carrillo, P. (2002). A CLEVER approach to selecting a knowledge management strategy. *International Journal of Project Management, 20*(3), 205-211.

Koch, C. (2003). Knowledge management in consulting engineering—joining IT and human resources to support the production of knowledge. *Engineering, Construction and Architectural Management, 10*(6), 391-401.

Kululanga, G.K., & McCaffer, R. (2001). Measuring knowledge management for construction organisations. *Engineering, Construction and Architectural Management, 8*(5/6), 346-354.

Martin, B. (2000). Knowledge-based organizations: Emerging trends in local government in Australia. *Journal of Knowledge Management Practice, 2.* Retrieved November 2003 from www.tlainc.com.

Nonaka, I., & Takeuchi, H. (1995). *The knowledge creating company.* Oxford: Oxford University Press.

Parikh, M. (2001). Knowledge management framework for high-tech research and development. *Engineering Management, 13*(3), 27-33.

Roos, J. et al. (1998). *Intellectual capital: Navigating in the new business landscape.* New York: New York University Press.

Sarvary, M. (1999). Knowledge management and competition in the consulting industry. *California Management Review, 41*(2), 95-107.

Stewart, T.A. (1997). *Intellectual capital: The new wealth of organizations.* New York: Currency/Doubleday.

Chapter II

HRM Strategies for Promoting Knowledge Sharing within Construction Project Organisations: A Case Study

Andrew R.J. Dainty
Loughborough University, UK

Jidong Qin
Loughborough University, UK

Patricia M. Carrillo
Loughborough University, UK

Abstract

Construction projects present an extremely challenging context for exchanging knowledge, as information must flow across both project and professional interfaces. This chapter explores innovative solutions to encouraging knowledge sharing within a large construction company based in Hong Kong. The techniques used comprised a sophisticated combination of HRM approaches aimed at breaking down barriers to open communication and reciprocal knowledge exchange. These were supported by a bespoke intranet-based knowledge repository to allow access to knowledge as it was captured from employees' project experiences. The case study demonstrates that, whilst all of the approaches could facilitate knowledge sharing and management within the construction project environment, to be effective they must be underpinned by a supportive culture, effective communication structures, and appropriate HRM practices. A policy framework is proposed which attempts to align reward and appraisal systems, work processes, and organisation design in a way that encourages knowledge sharing in other large construction firms.

Introduction

Knowledge is a vital resource which lies at the heart of both organisational and project success (Nonaka & Takeuchi, 1995; Egbu, 1999). In the post-capitalist society, knowledge is not just another resource alongside the traditional factors of production (land, labour, and capital), but the only meaningful resource (Drucker, 1993). Unsurprisingly, there has been a great deal of attention on developing ways to manage knowledge more effectively in recent years, primarily through the development of new information and communications technologies (ICTs). However, this emphasis has largely been at the expense of efforts to explore the human dimensions of effective knowledge management (Scarbrough, 1999; Swan, Newell, & Robertson, 2000). Arguably, the overemphasis on technological solutions for managing knowledge within large organisations has contributed to the relatively high failure rate of knowledge management (KM) within many industries and organisations (Ambrosio, 2000; Carter & Scarbrough, 2001; McDermott & O'Dell, 2001). Thus, there is a practical need to integrate KM programmes with human resource management (HRM) policy, to ensure its effective contribution to the performance of the modern business (Blackler, 1995; Carter & Scarbrough, 2001; Swan et al., 2000).

One important hurdle that must be overcome by a KM strategy is how to encourage people to willingly share their knowledge for the betterment of project and organisational performance. Although recent studies have attempted to address this gap in the KM literature, few studies have explored how knowledge sharing could be improved within the construction industry's unique context. This has left construction organisations relatively uninformed as to how they should develop and manage their knowledge-sharing practices in such a way as to improve their performance. This chapter reports on research that set out to reveal the knowledge-sharing barriers that exist within the construction project environment, and the close interrelationship between KM strategies and HRM practices that underpin them. It explores the current approaches and problems inherent in knowledge sharing within a leading construction project organisation, and drawing upon the good practice identified, discovers potential HRM solutions that can be used to enhance knowledge sharing in other construction organisations. Current approaches used within the case study organisation to encourage employees to share knowledge were explored via in-depth interviews and focus groups involving both managers and employees. These provide rich insights into the structural and cultural framework necessary for construction managers and professionals to share their knowledge for the betterment of the organisation.

Learning Objectives

1. To distinguish between tacit and explicit knowledge in the context of construction organisations.
2. To appreciate theoretical perspectives on how employees can be encouraged to share and benefit from a knowledge-sharing culture.
3. To develop a sophisticated and mutually supportive range of HRM mechanisms for supporting the knowledge management function within large and complex construction organisations.

Background

The Need for Tacit Knowledge Sharing Within Construction Organisations

Much of the current work in the area of knowledge has emerged from the initial work of philosopher Michael Polanyi (Kane, 2003). Polanyi first made the distinction between tacit and explicit knowledge (see Polanyi, 1958, 1966). He describes these as two different types of knowledge rather than two forms of the same thing (Shapiro, 1999). Tacit knowledge can be defined as things we know, but *cannot* say (i.e., it is locked in the mind). Conversely, explicit knowledge can be expressed in words or numbers and shared in the form of data, scientific formula, specifications, and manuals (Civi, 2000). This kind of knowledge is readily transferable within the business community and should accessible for most companies regardless of their cooperative activities. It can be codified and stored in databases where it can be accessed and used easily by anyone in the company. Explicit knowledge, however, represents only the metaphorical 'tip of the iceberg' of the entire body of knowledge. When Polanyi (1966, p. 4) says, "We can know more than we can tell," he is implicitly suggesting that there are difficulties inherent in the communication of tacit knowledge, or even that there are aspects we as individuals cannot convey to others (Kane, 2003). Tacit knowledge is therefore not easily visible and expressible, but is highly personal and therefore difficult to communicate or share with others (Haldin-Herrgard, 2000; Smith, 2001).

Nonaka expands the concept of tacit knowledge and states that tacit knowledge includes cognitive and technical elements (see Nonaka, 1994; Nonaka & Takeuchi, 1995). The cognitive element centres on mental models in which humans form working models of the world by creating and manipulating analogies in their minds. These include schemas, beliefs, and viewpoints, which help individuals perceive and define their world. The technical element of tacit knowledge includes concrete know-how, crafts, and skills. For example, a manager may develop a wealth of expertise after years of experience, but s/he is often unable to articulate the technical and philosophical principles behind it. In contrast to Polanyi, Nonaka also sees a close relationship between tacit and explicit knowledge. In Nonaka's view, tacit knowledge and explicit knowledge are not totally separate, but mutually complementary entities. They interchange into each other in the creative activities of human beings. Human knowledge is created and expanded through social interaction between tacit and explicit knowledge through conversion. Nonaka postulates four modes of knowledge conversion: socialisation (from tacit to tacit), externalisation (from tacit knowledge to explicit knowledge), combination (from explicit to explicit), and internalisation (from explicit to tacit). Nonaka states that the key to knowledge creation lies in the mobilisation and conversion of tacit knowledge.

According to Haldin-Herrgard (2000), the main difficulties in sharing tacit knowledge are related to perception and language. It is not so much that people have difficulty expressing and articulating what they know, but that they may not be conscious of what it is that they know, or the interconnection between their tacit and explicit knowledge. The second difficulty in sharing knowledge is the time it takes for the internalisation of tacit knowledge. In today's business world, time is such a scarce resource that it is rarely set aside for the sharing of this knowledge to take place. Furthermore, the need for rationality and logic in business matters often relegates the use of tacit knowledge to a subordinate position in comparison to more tangible and easily transferable explicit knowledge.

Organisational Culture in a Knowledge-Sharing Environment

A knowledge-enterprising culture is arguably one of the most important conditions leading to the success of a KM project. Indeed, many authors have identified culture as the most significant barrier to sharing knowledge (Ruggles, 1998; McDermott & O'Dell, 2001; Moore & Dainty, 2001; Holton, 2001). The prevailing organisational culture effectively shapes the attitudes of employees towards knowledge sharing (Carter & Scarbrough, 2001). However, managed effectively, a supportive culture can be a positive benefit to knowledge sharing. In an organisation with a culture supportive of knowledge sharing, people share ideas and insights because they see it as natural, rather than because they are forced to do it (McDermott & O'Dell, 2001).

De Long (2000) identifies four ways in which culture influences the behaviours central to knowledge creation, sharing, and use. Firstly, it shapes assumptions about what knowledge is and which knowledge is worth managing. Secondly, it defines the relationships between individual and organisational knowledge, determining who is expected to control specific knowledge, as well as who must share it and who can own it. Thirdly, it creates the context for social interaction that determines how knowledge will be used in particular situations. Fourthly, it shapes the processes by which new knowledge, with its accompanying uncertainties, are created, legitimised, and distributed within organisations. Understanding the influence of knowledge in this regard is the first critical step in developing a strategy and specific interventions to align the firm's culture with the KM strategy.

According to McDermott and O'Dell (2001), the most obvious place to begin understanding an organisation's culture is to explore the espoused values, philosophy, and mission of the organisation. This is often encapsulated within the organisational mission statement—an explicit articulation of the strategic direction and operating goals of the organisation. These statements will almost certainly say something about the culture, even if they simply represent the aspirations of the firm. At a deeper level, most organisations have an unspoken set of core values that guide both what people do and how they make sense of each other's actions. These often comprise simple precepts, but effectively form the core values of the organisation. These values are often most clearly manifested in what it sees as its rewards. These rewards can be obvious, such as in the form of pay/financial benefits or training and promotional opportunities. Alternatively, recognition can come in simpler forms of reward such as verbal praise for a job well done. Regardless of their nature, rewards reinforce and transmit the culture by providing tangible evidence of what the organisation does or does not value. Changes to the reward system, such as introducing a new performance management system, often lead to changes in culture (Harman & Brelade, 2000). This, in turn, will affect the way in which people view the role of knowledge sharing within the organisation.

Encouraging Knowledge Sharing Within Organisations

Organisations can use a variety of HRM mechanisms and techniques to encourage a knowledge-sharing culture. Some of these form straightforward reward strategies, whereas others represent the sophisticated manipulation of the way in which people work in order to create the desired working environment. Desirable aspects of the organisational culture are discussed below:

- **Reciprocity.** Knowledge is seen as a personal intellectual asset (Carter & Scarbrough, 2001) and so people do not want to give it away unless there are benefits in doing so. The reciprocal act of receiving something in response to giving something else has been coined 'reciprocity' and is central to the concept of knowledge sharing (Ellis, 2001). This could be a formal reward from the organisation in recognition for an individual sharing his or her knowledge, or less tangible rewards such as motivational praise or new knowledge returned from a colleague. Using recognition as a reward is often considered as one of the most powerful motivators by knowledge workers (Storey & Quintas, 2001).
- **A Shared Vision.** The development of a shared vision provides the focus and energy for knowledge sharing. It encourages and energises people to uncover the organisation's view of the future, and should provide meanings and value for everyone (Pan & Scarbrough, 1998). Indeed, there are few examples of great companies that have achieved success without incorporating a vision (Griego, Geroy, & Wright, 2000). This demands a learning system that allows employees to achieve the vision (Campbell & Cairns, 1997; Robinson, Carrillo, Anumba, & Al-Ghassani, 2001).
- **Trust.** This is a key element for knowledge sharing, as employees may not be willing to share their work-related knowledge if they believe that hoarding knowledge will assist in furthering their careers, or if they feel ill-treated at work (Storey & Quintas,

2001). Moreover, if employees feel that their employment relationship is not fair, if they feel that work-related promises or commitments have not been kept, and if they do not trust their employer to keep promises or commitments in the future, then they are less likely to feel disposed to share knowledge at work (Patch, Guest, Davey, & Kidd, 2000). Ellis (2001) argues that it is difficult to discuss knowledge sharing or knowledge management without trust, as most people will not risk sharing what they know without it. It is fundamental to developing a culture of openness and collective ownership which underpin successful learning environments and encourage knowledge acquisition and sharing (Smith, 2001). Arguably, trust develops through frequent and meaningful interaction, where individuals learn to feel comfortable and open in sharing their individual insights and concerns. This demands an atmosphere where ideas and assumptions can be challenged without fear or risk of repercussion, and where diversity of opinion is valued over commonality or compliance.

- **Teamwork.** This is important for knowledge sharing, because the barriers between team members can be broken by creating a sense of fraternity and bonding (Huang, 1997). Rather than require each person to contribute individually, the company can assign them to small groups and give them collective responsibility for knowledge sharing. If they are asked to meet a few times a year and contribute three pieces of knowledge every time they meet, they will start sharing what they know and even develop new knowledge (Ellis, 2000). As pointed out by Ingram and Desombre (1999), teamworking can be characterised by friendliness or camaraderie. For some respondents they surveyed, this friendliness is an important part of their job satisfaction and has been extended outside of normal working hours.

HRM and Encouraging a Knowledge Sharing Culture

Hansen, Nohria, and Tiemey (1999) identify some important interrelationships between HRM and KM. In essence they assert that there are fundamentally two KM strategies: codification and personalisation. Codification is where "knowledge is carefully codified and stored in databases where it can be accessed and used readily by anyone in the company." Conversely, personalisation is where "knowledge is closely tied to the person who developed it and is shared mainly through direct person-to-person contacts" (Hansen et al., 1999, p. 107). The two knowledge management strategies require different incentive systems. In the codification model, managers need to develop a system that encourages people to record and submit information to the electronic repository. In contrast, companies following the personalisation approach need to reward people for sharing knowledge directly with others.

Arguably, codification has traditionally represented the dominant paradigm for construction organisations in attempting to develop effective KM approaches. However, the failure rate of KM initiatives is as high as 70% when they have relied too heavily on technology-led approaches (Ambrosio, 2000). This is because IT-driven KM initiatives place greater emphasis on exploitation of knowledge rather than on its exploration.

Indeed, the lack of concern with people in the KM literature is reminiscent of early approaches to Business Process Reengineering (BPR), which ignored the centrality of people to the acceptance of change programmes (Swan et al., 2000).

Developing personalised approaches demands the support of appropriate HRM mechanisms to ensure a culture within which people freely communicate and exchange information in return for reciprocal gestures from their colleagues. The HRM function has a central role to play in the development of such an open knowledge-sharing environment within large organisations, because it can help to create the cultural framework necessary for communities of practice to develop within them. The challenge for the construction industry, however, is to identify the HRM approaches which support knowledge sharing within its dynamic workplace environment.

Construction projects, being characterised by short-term involvement and specialist functional roles, present an extremely challenging context for exchanging knowledge, as information must flow across both project and professional interfaces (Loosemore, Dainty, & Lingard, 2003; Emmitt & Gorse, 2003). In order to explore ways of ensuring the exchange of knowledge between employees in a way that promotes inter-project learning, this chapter now explores how this can be achieved by exploring the techniques used within a large construction organisation.

Case Study Findings

The research strategy adopted for this investigation was to explore knowledge-sharing interventions through an in-depth case study of a large, Hong Kong-based construction organisation. Following the financial crisis in 1997, the whole of the Asian economy went into decline, and after the terrorist acts of September 2001, the Hong Kong economy has been particularly hard hit. It suffered a sharp slowdown in 2001, with GDP growth decelerating to only 0.1% (Anson et al., 2002). This had led to problems for many construction firms. Total expenditure in building and construction have continued to shrink, and competition has become more intense. Foreign and domestic companies are treated equally in terms of business involvement, which has led to fierce international competition for the work that is available. These circumstances have forced many Hong Kong-based contractors to take action to adapt their policies and management styles to the new economic climate.

The case study company turns over almost 10b HK$ and directly employs more than 3,200 employees, making it one of the largest construction organisations operating in Hong Kong. More than 2,000 of these are site-based employees working remotely from its offices. Although many employees are qualified, over 2,200 have no degree-level qualifications. However, more than 2,500 have over 10 years experience of working within the industry, and therefore offer significant knowledge potential. With top managers eager to improve performance by restructuring the organisation and introducing new management strategies, developing an effective knowledge management strategy is regarded as critical for securing competitive advantage. For an organisation as large and complex as the case study company, the knowledge management challenges are considerable and demand sophisticated mechanisms to cope with the current challenging business environment.

A mix of data collection methods was used to explore the efficacy of the knowledge-sharing approaches used from the perspective of both managers and employees. The data was collected through open-ended interviews and small focus groups with 20 respondents selected from different levels within the organisation. These were senior managers ($n = 5$), specialist HRM staff ($n = 4$), and site-based staff ($n = 11$). This allowed the prevailing culture and learning environment to be established, the mechanisms used to create this environment and importantly, the practical impact of the HRM policies to be evaluated. The main findings of the case study are presented below in three parts. Firstly, the barriers to knowledge sharing as perceived by the respondents are discussed; secondly, the organisation's practices for encouraging knowledge sharing are described; and thirdly, the HRM implications of these mechanisms are evaluated.

Barriers to Developing a Knowledge-Sharing Culture

The organisation was acutely aware of the importance of KM and had invested a great deal of time and resources in improving their systems and approaches to ensure that they improved their performance through an embedded KM strategy. The participants highlighted the three principal barriers to the creation of a knowledge-sharing culture that they felt the organisation must overcome through its KM strategy:

- an unsupportive culture,
- poor communication structures, and
- time constraints.

These are discussed below.

- **Unsupportive Culture.** All of the interviewees agreed that hoarding knowledge may help to maintain their own promotion prospects within the workplace. Employees expressed concerns over losing their chance of promotion or even their jobs if they failed to maintain a degree of speciality knowledge which differed from that of their colleagues. The specialist staff all agreed that the hoarding of knowledge was detrimental to knowledge sharing, but that it was inevitable given that performance management systems tended to be individually based. Under such circumstances, people who shared their knowledge received little benefit and recognition.
- **Poor Communication.** The lack of effective mechanisms and processes to encourage open communication was raised as a serious issue within the interviews. Although the company was acknowledged to encourage knowledge sharing in a generic sense, such open communication required facilitation from specialists with a remit to maintain open communication channels. Without this, the day-to-day

pressures of working within the construction project environment and the fragmented nature of the industry were likely to prevent open knowledge exchange. In this regard, the structure of the industry was seen as a serious hindrance to engendering a knowledge-sharing culture.

- **Time Constraints.** Most of the site-based employees complained that their working responsibilities allowed little time to engage in knowledge-sharing activities. The lack of time for reflection and learning upon completion of projects was cited as a particular problem by those interviewed. The organisation's senior managers had also found their organisational structure too lean to exploit knowledge-sharing opportunities to the full.

Organisational Knowledge-Sharing Practices

Over the past three years, the organisation had attempted to overcome these barriers to knowledge sharing by investing in a range of knowledge-transfer techniques. These included workshops, seminars, routine meetings, site visits, summary reports, project award scheme, intranet discussion groups, and an e-library. These are discussed in turn below.

- **Informal Knowledge Workshops.** Monthly workshops covering a wide range of topics were held by the organisation to which a large number of operations staff were invited. These very informal sessions aimed to encourage employees to discuss relevant problems and innovative solutions openly amongst themselves within a social network. The workshops were often attached to social events to encourage the informality of the knowledge exchange. These events had proven popular enough for them to be run outside of normal working hours and still attract a sufficient number of participants.
- **Knowledge Exchange Seminars.** A seminar programme had been used within the case study company for some time. They were seen as an effective mechanism for encouraging people to share good practice and create new ideas in response to common problems arising on projects. In particular, they afforded an opportunity for knowledge sharing between projects and for experts on particular issues to showcase their expertise amongst their colleagues and peers. Many participants had relished the opportunity to showcase the results of the efforts to improve processes and enhance the performance of their organisation.
- **Departmental Meetings.** Every department conducted monthly meetings to which all employees were required to attend. These meetings were problem-centred in nature and designed to encourage face-to-face interactions between staff nominally based on different projects. Although effective in bringing like-minded staff together, these were seen as not being as effective as the seminars, as they tended to be intra-disciplinary in nature and, therefore, did not stimulate knowledge exchange between the functional specialists employed by the company. However, they had been effective in helping to shape communities of practice within the

organisation, and so had been used as a supplementary activity in support of the seminar programme.

- **Site Visit Programme.** The organisation ensured that upon completion, a cohort of staff who had not previously had any involvement with the management team visited every project. The aim was to transfer lessons learned during the project to others within the organisation whilst their experiences remained fresh in their minds. More staff visited projects that had experienced particular problems, or that had performed exceptionally, in order to maximise the learning opportunities. The managers of particularly successful projects were also asked to give a presentation on their achievements to their colleagues. Site visits were considered a particularly efficient method of knowledge sharing, particularly as staff liked to showcase their successes and innovative practices at the end of their involvement with projects
- **Summary Reports.** Every project team was required to submit a standardised summary report of their project. The report covers special issues, such as successful experiences or lessons they have learned in the project. Senior managers believed it was an effective way for the employees to learn from the past. Although well intentioned, in this case the project record documents were fairly superficial, merely recording basic performance data and project outcomes. The nature of problems occurring during projects and particular solutions achieved were mostly omitted.
- **Project Award Scheme.** The organisation had launched a project award scheme in 2001. The aim was to foster knowledge-sharing behaviour by rewarding innovation in project planning and preparation. The scheme was operated by inviting project teams to submit proposals for how they would develop and manage a project that the organisation had been awarded through the competitive bidding process. Five project teams are then short-listed to present their proposals to senior managers from relevant company departments and to each other. After careful consideration, the panel made a decision as to which team the project should be awarded to, with all of the participants having benefited from the open knowledge exchange afforded by the presentations. The teams being awarded the project would have the opportunity to benefit from a share of the profits if their innovative solutions led to the successful completion of the works. There was also a requirement for the winning team to offer feedback of their experiences as part of the other mechanisms listed above.
- **Coaching and Mentoring.** These were used to encourage senior experienced employees to work alongside their junior colleagues in order to offer support and advice to them. This face-to-face training was regarded as one of the most effective ways of encouraging learning and was appreciated by younger, less experienced employees. Those taking part in the scheme were rewarded financially for the time they put into training their colleagues, in order that it became a recognised activity within the organisation.
- **Intranet and E-Library.** A local area network (LAN) system had been set up to facilitate communication between sites and the head office. This included online bulletin boards where people could highlight and discuss with their colleagues

problems and issues arising within projects. It also provided an electronic repository for new and innovative ideas to be housed and accessed by others within the organisation. Allied to this was a searchable e-learning system, which comprised an electronic library of resources and other information collated from the mechanisms discussed above. These systems offered technological support for the organisation's KM activities.

Discussion

The barriers to knowledge sharing acknowledged by the case study organisation and their responses suggest that they have a comprehensive understanding of the challenges involved in developing an open learning culture. Indeed, their efforts to facilitate mutually supportive communities of practice suggest a carefully measured response to the challenges posed by the challenging environment in which construction firms tend to operate. The organisation's HRM response to the challenges imposed by the nature of the industry include strategies to: (a) recognise and reward knowledge sharing within the organisation; (b) train and develop staff in order that new knowledge can be embedded throughout the organisation; and (c) structural solutions for facilitating the collection and storage of such knowledge for the benefit of every employee. These are discussed below in relation to the wider literature surrounding the relationship between HRM practices and the KM within larger organisations.

Recognising and Rewarding Knowledge Sharing

Wasko and Faraj (2000) suggest that knowledge is a private good that is exchanged in the expectation of a commensurable return. Thus, people rarely give away something without expecting anything in return. Hall (2003) also argues that knowledge is a private commodity and it is up to the owner to decide whether to share it or not. Thus, to entice people to share their knowledge as part of a social exchange transaction, they need to be persuaded it is worth doing so. Social exchange theory suggests that participants in peer-to-peer community networks expect mutual reciprocity that justifies their expense in terms of time and energy spent sharing their knowledge. Group knowledge-sharing processes will result in poor outcomes if members are dissatisfied with them. This presents a serious challenge for organisations attempting to foster open communications and knowledge-transfer opportunities, particularly where an individual's involvement with colleagues has a temporal dimension, such as within the construction project environment.

To achieve successful knowledge sharing, companies need to convince people to reject the old-school thinking that they are being measured by what they know and do individually. Such thinking only perpetuates knowledge hoarding and the development of knowledge 'silos' from which little value-adding transfer takes place. One way in which this can be achieved is to build knowledge sharing into daily work processes such as performance appraisal (Robertson & Hammersley, 2000; McDermott & O'Dell, 2001). Reward and appraisal systems can encourage actors to participate in knowledge-sharing activities so knowledge sharing can be seen to be valued by the organisation. They can reinforce and convey the desired culture of an organisation by providing tangible evidence of what it values. Changes to the reward and appraisal system are therefore closely linked to changes in culture. To convey that the company values knowledge sharing, it should therefore align performance assessment and compensation to encourage a sharing environment (Davenport & Prusak, 1998). Appraising and rewarding knowledge sharing raises its profile and ingrains it as a core organisational value. It highlights the things the company feels important and demonstrates that the time and energy people spend on knowledge sharing are highly valued (McDermot & O'Dell, 2001). The efforts of the organisation to reward those individuals and teams who share their knowledge in a spirit of collaboration and innovation are therefore congruent with creating the type of learning organisation known to thrive in the contemporary business environment. The concurrent emphasis on rewarding team performance rather than just the excellence of the individual is also supportive in this context.

Developing Effective Training and Development Systems

According to Harman and Brelade (2000), the most obvious link between KM and HRM is in the field of training and development. Training itself is a knowledge-sharing process, as well as a primary human resource development mechanism. It is one of the key roles through which HR can be involved in and contribute to effective knowledge management within an organisation. Civi (2000) suggests that training is the best way to start to induce a culture within organisations seeking to engender a more open collaborative environment.

Several elements of the case study company's approach are directly supportive to the ethos of supporting knowledge sharing through human resource development activities. Firstly, they have instituted a continuous professional development programme, through their coaching and mentoring schemes, which facilitates lifelong learning through knowledge sharing. Secondly, they have established mechanisms for knowledge workers to interact on a face-to-face basis (such as through their seminars, workshops, meetings, and site visits). This formalises knowledge exchange activities as a defined element of their employees' role within the organisation. Thirdly, the organisation is actively seeking to learn lessons from its experiences. Following each project, the organisation seeks to explore what actually occurred and then identify what can be done differently to improve future performance. Collectively, these activities can be seen to have provided a robust training and development approach to support their KM aspirations.

Changing Organisational Structures and Work Content

In contemporary organisations, the structure and work content need to be designed to suit the needs of knowledge workers in a way that facilitates knowledge creation and sharing (Storey & Quintas, 2001). Furthermore, the creation of knowledge teams and centres are fundamental to the success of an overall KM strategy (Civi, 2000). Gross (2001) suggests that organisations should establish knowledge teams responsible for searching out new knowledge, compiling the information in accessible knowledge repositories, and sharing it with those who require it. The case study organisation had gone through considerable efforts to attempt to create knowledge centres and teams, redesign working practices, and embed knowledge-sharing activities as part of their job roles. The project reporting mechanisms, seminar programmes, and mentoring activities helped to ensure that employees viewed knowledge-sharing activities as an integral aspect of their job roles. In addition, the company had time set aside for employees to create and share knowledge as part of their functional activities. This had raised the profile of knowledge-sharing activities within the organisation.

Conclusion

Knowledge sharing should be at the centre of any effective KM program. Without mechanisms for the workforce to transfer what they know, KM cannot succeed. Few empirical studies have explored how knowledge is shared in the construction project environment and how the barriers that prevent this from occurring effectively can be overcome. The aim of the chapter was therefore to investigate potential HRM solutions that are demonstrably effective in promoting knowledge sharing within the construction project environment.

Whilst many KM approaches could facilitate knowledge sharing and management within the construction project environment, they need to be underpinned by a supportive culture, a robust communication strategy, and appropriate HRM practices if they are to be effective. The case study organisation developed an HRM policy framework that had managed to align reward and appraisal systems, work processes, and organisation/job design in a way that encouraged knowledge sharing as a natural function of employees' job roles. This demonstrates the importance of placing people at the centre of an organisational knowledge strategy.

The findings of this study indicate that the case organisation regards knowledge and knowledge sharing as essential for its sustained development over the next few years. Having a relatively large amount of knowledge workers, senior managers believed knowledge sharing to be crucial to the development of the company, in terms of increasing the value of their knowledge asset and improving the overall performance. This is because the continual updating of their knowledge should improve their competitive position. The current approaches to knowledge sharing within the case study organisation were wide ranging and included seminars, workshops, routine meetings, site visits, summary reports, coaching and mentoring systems, profit-sharing schemes, and intranet technologies. Used together, they had overcome many of the barriers to knowledge sharing such as the unsupportive culture, poor communication, and time constraints. They had developed a cultural environment in which knowledge sharing was a central aspect of employees' job roles.

Although the findings are based on a single case study of a Hong Kong-based contractor, the consistency between what was discovered to work effectively within the organisation and the findings of previous research reinforces the importance of culture if knowledge sharing is to impact upon performance. The role of HRM is thus to foster such an organisational culture through changes to existing organisational systems and the processes that underpin it. Some of these may be formal systems that convey clear messages about what the organisation values, whereas others are less formal, but they also send a clear message about the behaviour expected of employees.

Practical Tips and Lessons Learned

- Knowledge sharing is the cornerstone of any KM program. Without mechanisms for the workforce to transfer what they know, KM cannot succeed.
- Employees must be recognised and rewarded for sharing knowledge within a culture characterised by reciprocity and mutual commitment.
- Construction organisations must integrate and align their knowledge management policies with their broader business strategies and priorities.

References

Ambrosio, J. (2000). Knowledge management mistakes. *Computerworld, 34*(27), 44-45.

Anson, M., Chan, E.H., Chiang, Y.H., Choy, H.T., Hui, C.M., & Sharma, S. (2002). Hong Kong country report. *Proceedings of the 8th ASIACONSTRUCT Conference*, 5-7 November, Beijing.

Blackler, F. (1995). Knowledge, knowledge work and organisations: An overview and interpretation. *Organisation Studies, 16*(6), 1021-1046.

Campbell, T., & Cairns, H. (1994). Developing and measuring the learning organization: From buzz words to behaviours. *Industrial and Commercial Training, 26*(7), 10-15.

Carter, C., & Scarbrough, H. (2001). Towards a second generation of KM? The people management challenge. *Education + Training, 43*(4/5), 215-224.

Civi, E. (2000). Knowledge management as a competitive asset: A review. *Marketing Intelligence & Planning, 18*(4), 166-174.

Davenport, T., & Prusak, L. (1998). *Working knowledge: How organizations manage what they know*. Boston: Harvard Business Review Press.

De Long, F. (2000). Diagnosing cultural barriers to knowledge management. *Academy of Management Executive, 14*(4), 113-127.

Drucker, P.F. (1993). *Post-capitalist society*. Oxford: Butterworth Heinemann.

Egbu, C.O. (1999). The role of knowledge management and innovation in improving construction competitiveness. *Building Technology and Management Journal, 25,* 1-10.

Ellis, K. (2001). Dare to share. *Training, 38*(2), 74-80.

Emmitt, S., & Gorse, C. (2003). *Construction communication*. Oxford: Blackwell.

Griego, O., Geroy, G., & Wright, P. (2000). Predictors of learning organizations: A human resource development practitioner's perspective. *The Learning Organization, 7*(1), 5-12.

Gross, A. (2001). Knowledge sharing—the crux of quality. *Annual Quality Congress Proceedings* (pp. 452-457), Milwaukee, Wisconsin, USA.

Haldin-Herrgard, T. (2000). Difficulties in diffusion of tacit knowledge in organizations. *Journal of Intellectual Capital, 1*(4), 357-369.

Hall, H. (2003). Borrowed theory: Applying exchange theories in information science research. *Library and Information Science Research, 25,* 287-306.

Hansen, T., Nohria, N., & Tiemey, T. (1999). What's your strategy for managing knowledge? *Harvard Business Review,* (March-April), 106-116.

Harman, C., & Brelade, S. (2000). *Knowledge management and the role of HR: Securing competitive advantage in the knowledge economy*. London: Prentice-Hall.

Holton, J. (2001). Building trust and collaboration in a virtual team. *Team Performance Management: An International Journal, 7*(3/4), 36-47.

Huang, K. (1997). Capitalizing collective knowledge for winning, execution and teamwork. *Journal of Knowledge Management, 1*(2), 149-156.

Ingram, H., & Desombre, T. (1999). Teamwork: Comparing academic and practitioners' perceptions. *Team Performance Management: An International Journal, 5*(1), 16-22.

Kane, H. (2003). Reframing the knowledge debate with a little help from the Greeks. *Electronic Journal of Knowledge Management, 1*(1), 33-38.

Loosemore, M., Dainty, A.R.J., & Lingard, H. (2003). *Human resource management in construction projects: Strategic and operational approaches*. London: Spon Press.

McDermott, R., & O'Dell, C. (2001). Overcoming cultural barriers to sharing knowledge. *Journal of Knowledge Management, 5*(1), 76-85.

Moore, D., & Dainty, A. (2001). Intra-team boundaries as inhibitors of performance improvement in UK design and build projects: A call for change. *Construction Management and Economics, 19,* 559-562.

Nonaka, I. (1994). A Dynamic theory of organizational knowledge creation. *Organization Science, 5,* 14-37.

Nonaka, I., & Takeuchi, H. (1995). *The knowledge creating company*. Oxford: Oxford University Press.

Pan, S., & Scarbrough, H. (1998). A socio-technical view of knowledge sharing at Buckman Laboratories. *Journal of Knowledge Management, 2*(1), 55-66.

Patch, A., Guest, D., Davey, K., & Kidd, J. (2000). What will encourage employees to acquire and share knowledge at work? *British Psychological Society 2000 Occupational Psychology Conference*, Brighton.

Polanyi, M. (1958). *Personal knowledge*. Chicago: University of Chicago.

Polanyi, M. (1966). *The tacit dimension*. London: Routledge & Kegan Paul.

Robertson, M., & Hammersley, G. (2000). Knowledge management practices within a knowledge-intensive firm: The significance of the people management dimension. *Journal of European Industrial Training, 24*(2/3/4), 241-253.

Robinson, H.S., Carrillo, P.M., Anumba, C.J., & Al-Ghassani, A.M. (2001). Perceptions and barriers in implementing knowledge management strategies in large construction organizations. *Proceedings of the RICS Research Foundation Construction and Building Research Conference* (Volume 2, pp. 451-460), 3-5 September, Glasgow Caledonian University.

Ruggles, R. (1998). The state of the notion: KNOWLEDGE management in practice. *California Management Review, 40*(3), 80-89.

Scarborough, H. (1999). System error. *People Management,* (8 April), 68-74.

Schindler, M., & Eppler, M. (2003). Harvesting project knowledge: A review of project learning methods and success factors. *International Journal of Project Management, 21,* 219-228.

Shapiro, G. (1999). *Inter-project knowledge capture and transfer: An overview of definitions, tools and practices.* Working Paper, CENTRIM, University of Brighton, UK.

Smith, E. (2001). The role of tacit and explicit knowledge in the workplace. *Journal of Knowledge Management, 5*(4), 311-321.

Storey, J., & Quintas, P. (2001). Knowledge management and HRM. In J. Storey (Ed.), *Human resource management: A critical text* (2nd edition). London: Thomson Learning.

Swan, J., Newell, S., & Robertson, M. (2000). Knowledge management—when will people management enter the debate? *Proceedings of the 33rd Hawaii International Conference on System Sciences* (Volume 3). 4-7 January, Maui, Hawaii.

Wasko, M., & Faraj, S. (2000). "It is what one does": Why people participate and help others in electronic communities of practice. *Journal of Strategic Information Systems, 9,* 155-173.

Chapter III

Managing the External Provision of "Knowledge Management" Services for Projects

Hedley Smyth
University College London, UK

Abstract

Knowledge management has largely been considered from the perspective of internal generation. It is posed that external provision will increase due to the diversity of problems and scarce resources, the arising issues being explored through a case study. The aim is to explore this process for construction projects. It will be shown that the successful delivery of knowledge services for the construction industry depends upon the way in which the provider conducts its marketing and the operation of gatekeepers in the project organisation. These issues are explored in relation to the main concepts and theories, applying and evaluating these through the case study approach.

Introduction

This chapter explores the issues for knowledge management where the knowledge and the application are derived in one organisation and used in another—the *external provision* of information and knowledge management services. The aim of the chapter is to explore this process for construction projects. It will be shown that there are particular issues to be addressed in the successful delivery of knowledge services for the construction industry, specifically the role of *marketing* and the operation of *gatekeepers* where the project is the main means of delivery. These issues will therefore be explored within the project working environment, bringing to bear the main concepts and theories, and applying and evaluating these through a case study approach.

Learning Objectives

This chapter provides the opportunity for:

1. Analysing the issues involved in the external provision of KM.
2. Gaining an understanding of the gatekeeping role of the knowledge manager in the user firm.
3. Securing a knowledge of the marketing requirements of the supplier in external knowledge provision.

Background

Nonaka and Takeuchi (1995), following Polanyi (1966), have classified knowledge as being explicit or tacit. It has been argued that specific strategies and processes are needed to make tacit knowledge explicit and to manage that knowledge to the benefit of the organisation (Teece, Pisano, & Sheun, 1997). The process of developing competencies and capacities to achieve these requirements has been divided into four areas of organisational activity (Leonard-Barton, 1995):

- Shared creative problem solving
- Integrating and implementing new methods and adopting new tools
- Formal and informal experimentation
- Learning outside of the organisation

This chapter focuses upon the last of the four areas—the external provision of learning. For external provision from one organisation to another, the knowledge has to be made explicit and managed in such a way that it can be marketed and sold either as a knowledge management service package or a bespoke service to meet specific needs. This is an important and under-researched area in industry and in the whole sphere of knowledge management. The external provision of knowledge management services for problem solving is likely to increase. The strategy for an organisation demands choices, which precludes options, hence the allocation of resources in support (see, for example, Mintzberg et al., 2002; Johnson & Scholes, 2002). KM may not be a strategic area for investment, but even where it is, there are two primary reasons why external KM provision will increase:

1. The specific processes selected by any one organisation to realise knowledge management and organisational learning objectives inherently limit the range of knowledge that can be captured, managed, and successfully applied.
2. There is a high cost to the generation of knowledge and learning, so efficiency of generation is likely to be maximised by those specialising in certain areas of expertise and bodies of knowledge.

Strategies and specific processes provide direction for a firm, which provides the context for knowledge management and organisational learning pursuit both in overall emphasis and the specific areas in which it can be maximised. Other aspects of learning and knowledge management will therefore take second place or be excluded. Hughes and Kao (2001) also demonstrate that there can be advantages to acquiring external information rather than developing it in-house in an integrated way (cf., Shavell, 1994). These include speed and cost of provision, as well as potentially effective solutions mobilised from a different resource base.

Cost, the second issue, will relate to the above in that budgets and spending will be focused upon the areas of strategic preference. The areas where less effort has been focused will have higher costs as the starting point and will be weaker or at a comparative disadvantage. The internal costs will be high. Thus, when the cost of internal generation becomes high, the cost of purchasing a solution in the market is more favourable. The external production and transaction costs are lower than the internal production costs (Williamson, 1985).

There is the market for information and knowledge. The provision process in the sense of the knowledge generation and application is demonstrated in Figure 1. The market exists where there is asymmetry of information between internal sources (including those that can be obtained 'free') and the external provision of information (Milgrom & Roberts, 1992). Although the quality of information is uncertain in advance (Akerlof, 1970), and therefore the transaction costs may prove to increase in order to search out the quality and assess risk (Williamson, 1985), the external provision may prove the best option. Shavell (1994), however, argues that partial disclosure of valuable information provides an incentive for buyers to acquire remaining information. In these ways the market for knowledge is similar to the market for information. Drucker (1995) was amongst the first to view knowledge as a capital asset or resource to be utilised, and therefore intellectual capital becomes recognised as a scarce commodity. Drucker (1989, 1993) identified the *knowledge worker* with ways of thinking that transcends industries. Groák (1994) coined the concept *demand chain* to indicate how technology and innovation for construction projects can be drawn from a range of industries. In this context, the external knowledge

Figure 1. Information and knowledge generation and application from an external provider

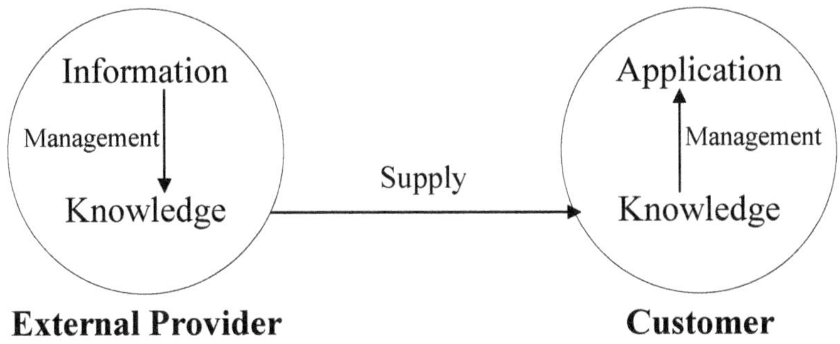

can then be injected from outside the construction enterprise, whether in design or contracting, into the other three activities, mostly the first concerning problem solving in a project working environment.

In construction and many other project working environments, teams are multi-organisational (Cherns & Bryant, 1983), giving rise to two additional costs:

1. The creation and realisation of knowledge at the level of the project and capturing; management and diffusion of that knowledge across the organisation and other projects (Smyth, 2004).
2. The transfer of knowledge and learning across organisational boundaries at the project level.

External sources of knowledge are therefore a valuable supplement, if complementary and additional to internal sources. *Project-specific knowledge* is therefore a notion or concept for which external provision is likely to be a practical solution to meeting the practical need—a socio-technical solution for construction. In summary, such sources will be accessed where the value added is greater than the marginal cost to the service or product, or where the organisation secures a competitive advantage relative to the costs.

The Knowledge Management Market

What is the knowledge service, product, or commodity? Egbu and colleagues (2003a) have drawn attention to the paucity in definition and articulation of the tools for knowledge management (KM). The tools can be divided into:

- KM techniques
- KM technologies

The tools of techniques and technologies both imply a service, which can be dynamic and tailor-made, created and applied internally, or sourced from an external provider. External provision can be sourced as a pre-packaged service or *bought off the shelf,* hence reconstituted to appear as a *product.* Such products can be used by a number of clients; the higher the rate of usage, the more the product becomes a commodity, and the value, hence price, diminishes over time. Bespoke services will be relatively, if not totally unique, hence are of potentially high value and price. KM techniques require strategies for development and realisation (Egbu et al., 2003a), yet the external provision of KM services requires a particular form of strategy—marketing principles (Smyth, 2000). The particular problems the KM-marketing interface must address are:

- Identifying the KM needs of consultants and contractors managing projects in order to generate KM service packages and bespoke solutions.
- Marketing the bespoke services and packages to consultant and contractor organisations to create awareness in advance of their perceived need for the KM services.
- Getting past those responsible for information, areas of expertise, and knowledge—*gatekeepers*—in the organisations to the real end user requiring the project-specific knowledge.

This chapter explores these issues for KM provision in construction through a case study. The case study, Concrete Information Limited (CIL), has its origins in the forerunner to and within the British Cement Association as its former library and information service. The British Cement Association has operated an extensive library and information service for many years. This has been restructured into a knowledge-based service. There are two foci. The first is pre-specification, which is the historic focus for information, which is now being developed around market data. This switch in 'product' area is in its infancy; therefore, this chapter mainly develops the case around the post-specification segment of CIL's business. The pre-specification market segment is composed of a number of areas, The Concrete Centre being a key client. TCC was launched in September 2003 as a new initiative to help specifiers in the choice of structural materials in design. The second and recent focus is on KM service provision post-specification for the cement and concrete sectors. CIL is thus an external organisation responsible for supplying knowledge within a complex field of business-to-business social networks, a large proportion of which are and will increasingly be aimed at problem solving within projects. This case arises from a two-year research project on CIL, during which CIL has been restructured from an information provider based upon a public-sector approach to a commercial KM provider. It is providing information and knowledge to a range of end users, and especially to serve project-specific needs as a key market segment. It is jointly owned by the British Cement Association and the Concrete Society, both trade bodies representing concrete and cement interests.

Conceptual Issues in the External Provision of KM

The main objective is to address the three problems cited: identifying the KM needs of consultants and contractors managing projects, marketing the bespoke services and packages, and reaching the real end user. This is achieved by engaging with the literature and by exploring the practical outworking through the case study.

Identifying KM Needs

Using the classification identified in recent research by Egbu et al. (2003b), a profile of services can be generated. Their classification is:

1. Main triggers of knowledge production (and in this case it includes need):
   - Problem solving
   - Innovation
   - Managing change
2. Main sources of knowledge production:
   - Expert individuals
   - Team and supply chain
   - Routines
   - Repositories
   - Communities of practice
   - Knowledge gatekeepers

In the case of CIL, the primary trigger is *problem solving* in developing the design details and production information or site-based problems on a project. The main source in this context is the *repository*, CIL, which is external, and therefore knowledge has to be accessed as a free advice service or purchased as a service. CIL therefore operates as part of a *community of practice* (Egbu et al., 2003a), that is, part of a network of different skill bases, histories, and experience that combine at points to realise shared goals (see Ruggles, 1997; cf., Egbu et al., 2003a). These can be located within a diversified organisation, but in a project context typically include a range of organisations working to realise mutual or compatible goals in completing a project.

The knowledge needs of experts and managers of knowledge, often assigned to the role of technical director and/or knowledge manager in large contracting organisations or consulting groups, will be rather different from a contract's director or a site manager, whose requirements are project-specific knowledge. The technical director or knowledge manager will serve design and production functions, and thus will apply knowledge to reduce uncertainty further down the project process and ensure quality for the finished structure. The KM need is primarily problem-solving innovation according to the classification above. Project-specific knowledge will be entirely focused upon problem solving. These will be issues that have not been recognised upstream or were totally unexpected and require an immediate solution, not only in their own right, but also in the context of completing the project on time and within budget without compromising quality. These needs at the project level are difficult to predict for an external provider, just as they are unforeseen internally. Therefore these needs have to be met from the repository of information available, the external provider creating generic packages of services. The provider derives the packages from experience of past patterns of demand, feedback, and market research. Other services will be bespoke of necessity.

The next issue is to make the market aware of the service and to sell them. This is the marketing function.

Marketing Bespoke Services and KM Packages

Organisational learning and knowledge management are perceived as competencies for the creation of effectively serving clients and improving internal efficiencies. This perspective of KM concerns strategy. At a tactical level, few people in practice talk about knowledge management as a starting point. They talk of *performance*. Knowledge management follows as a means to enhance performance. Harnessing KM is a means to an end, and is not the driver (Construction Productivity Network 2003). The consequence is that the awareness of the need for KM is secondary within an organisation, performance being primary. Where knowledge is provided from an external source, the perceived need for KM becomes one step removed and thus is tertiary in priority. Within a project environment, especially where the project is spatially removed from the organisational hub, the need for KM becomes a further step removed or quaternary in priority. This is of crucial significance. The final step is where the role of the gatekeeper within the organisation becomes critical and will be returned to below. At this stage, the main aim is to look at the overall picture created by this 'distance' and the role of marketing in bridging this gap.

Marketing is concerned with how to sell services, and the implementation is the sales process. Marketing is about managing the market; that is, creating and retaining a customer (see Kotler, Armstrong, Saunders, & Wong, 1996; Levitt, 1983; and in the construction context, Smyth, 2000). An initial issue for marketing is how to create awareness of the service and then how to promote the content of the service, both in package and bespoke forms. It has been stated that the quality of information is uncertain in advance (Akerlof, 1970); therefore, the transaction costs may increase in order to search out the quality and assess risk (Williamson, 1985). The marketing strategy and the promotion effort are designed to overcome these obstacles that arise from the external provision of information and KM services. The obstacles to overcome are greater in this case:

- Lack of advance need for KM services without increased awareness—project-specific knowledge is mobilised to solve problems as they arise.

- Distance created by external provision—a barrier to being aware of availability.
- Uncertainty of information quality in advance—location in a separate organisation, coupled with private ownership, prevent the uncertainty being automatically overcome.
- Internal distance created by dislocated projects and (in)action of gatekeepers.

These latter two factors arise out of the way in which organisations are constituted and operate (Child & Heavens, 2001).

There are two primary marketing paradigms: the marketing mix or transaction approach and the relationship marketing approach (see, for example, Gummesson, 2001; Smyth, 2000). The marketing mix approach grew up in the post-war era of mass markets of the United States and thus has its origin in the consumer or over-the-counter (OTC) retailing market. The mix is made up of four ingredients, measured out according to the recipe required for the circumstances. The ingredients are called the 4Ps:

1. Product
2. Price
3. Place
4. Promotion

Setting the right price to a product design and specification are clearly key. The delivery channel, which embraces distribution and the retail outlets, are the elements of place. Promotion embraces the full range of branding, advertising, PR, endorsements, and so on. Put simply, if you get the mix right, the marketing campaign will prove successful. This approach works where a number of factors exist:

- The product or service can be specified in advance.
- The market is of sufficient size to justify the packaging and promotion of the services.

This approach is certainly not suited to meeting very specific problems at any stage of the project life, but especially on site. A bespoke service is required. Over the last 20 years, it has been argued that business-to-business marketing is more appropriately conducted on a relationship basis. The Nordic School developed the paradigm of *relationship marketing* (see, for example, Gummesson, 2001; Smyth, 2000) in recognition that this is how people originally sold their wares, that it had become more problematic as organisations grew, but that systems could be put in place to replicate what had been lost on a business-to-business basis, often articulated through some key personnel.

This paradigm has received much wider recognition, even in consumer markets through reward cards. In this way retailers can profile the spending patterns of groups of customers representing particular market segments. There is a level of certainty of demand patterns that can be derived from aggregated data. In a business-to-business service environment, that feedback comes from listening to and interacting with the customer. The issue with a project problem-solving context once the project is underway on site is that there may be a relationship with the technical director, but not those solving problems on the project itself. The technical director or equivalent essentially becomes the prime relationship, acting as a gatekeeper to others.

The Project End User and Gatekeepers

Gatekeeper is a sociological concept applied to managers of a system in relation to the allocation of scarce resources, and was investigated widely in the public domain (see, for example, Pahl, 1975, for an early influential example). There has been less attention within private enterprises to the allocation. As in the public domain, allocation of resources in private organisations provides a source of power for the *gatekeeper*. The *gatekeeper* is an important information source for KM within the customer organisation. In the KM context the gatekeeper is a facilitator of knowledge production, or for its flow from the external environment to the project. However, the gatekeeper can also be an obstacle to effective dissemination of knowledge within an organisation. Child and Heavens (2001) give two reasons:

- Historical embeddedness of processes that goes against the grain of KM creation and dissemination in terms of;
  - Social systems (see also Stinchcombe, 1965; Johnson, 1990)
  - Social behaviour (see also Granovetter, 1985; Leonard-Barton, 1995)
- Leadership and authority systems that do not legitimise or facilitate dissemination—Senge (1993) places leadership in a key position for initiating learning and hence KM.

Child and Heavens recognise that learning and knowledge management are not only a process of generation and processing, but are also outcomes, that is, they impact upon performance, which does not automatically flow from the inputs. The issue about external provision is that it flows at that point of potential disjuncture between the processing and application. In that sense demand is not driven by strategy from above, but is driven by needs at the operational level. Therefore three levels for generating learning and knowledge occur: senior management, operational, and external suppliers (see Figure 2).

From this a model of learning and KM can be constructed that encompasses internal and external sources, which demonstrates the interface both in terms of how firms are organised in hierarchy and the horizontal management functions (see Figure 3).

The boundary of the firm must simultaneously be retained whilst being penetrable. Using a military analogy, the management issue is to know when the drawbridge of the castle should be up or down. In an economic and management environment, where the ability to generate a range of learning and knowledge to be effective and efficient in performance

Figure 2. Three relationship channels for effective organisational learning (adapted from Child & Heavens, 2001, p. 314)

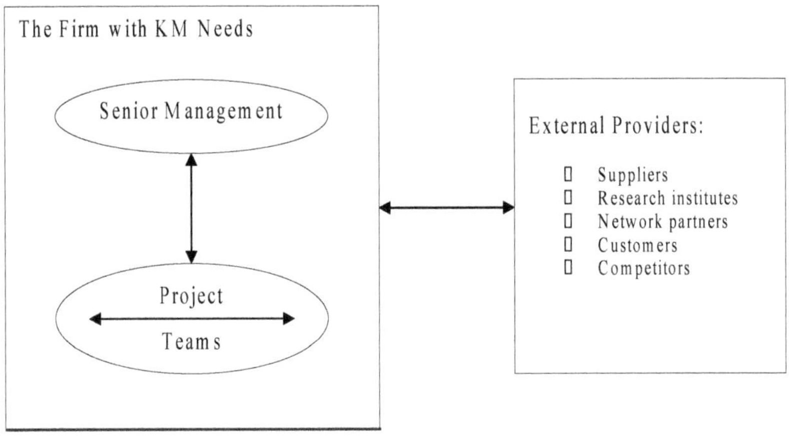

Figure 3. The interface between internal and external learning and KM

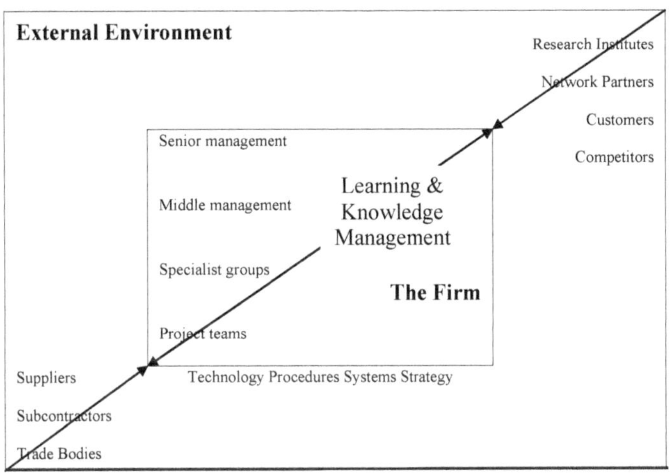

is constrained by management strategy and costs, external sources are becoming and will continue to become of greater significance. Boundaries therefore need to be penetrable—the drawbridge must be down so that key players in the firm are open to external ideas from the stage of receiving information that creates awareness to the stage when a problem surfaces that needs resolving using external information and KM services, hence the possibility of the key player connecting the awareness of the external source with the source to create the solution (cf., Tushman & Katz, 1980).

The key player becomes the gatekeeper of the knowledge source and some of the content, especially service packages. The role of the gatekeeper is:

- Being aware of the external sources and services
- Accessing information and knowledge
- Processing information into knowledge and integrating knowledge to match the culture, systems, and technical needs of the firm
- Disseminating the right knowledge to the right problems at the right time

In practice, there are problems with the effective functioning of those who are responsible for technical information and knowledge management—the *gatekeepers*:

- Those who primarily have an inward focus will have a tendency to have low levels of awareness of services and therefore will be ineffective at identifying the relevance of external knowledge to fit current and future internal needs.
- Those who primarily have an external focus will have a tendency to be less aware of internal needs and will have more difficulty processing knowledge to fit the internal language, systems, and strategy.
- Knowledge is power and some gatekeepers may use their role to reinforce their position, profile, or political stance in the firm, hence trading off effectiveness in the best interests of the firm in favour of personal power.

The Case of the Cement and Concrete Sector in the UK

Case Study Method

The methodology is ethnographic, whereby participation and observation occurred within the primary actors (see Dunnette, 1976). The primary actors are CIL, a joint venture between the British Concrete Association (BCA) and the Concrete Society (CS). There has also been an element of action research, as participation has led to influencing the shape of service provision. This started initially with a consultancy contract and then developed through the UK government Teaching Company Scheme, now called Knowledge Transfer Partnerships.

The data is therefore mixed in sources, using observation, formal documentation, and notes derived from participatory activities. The evidence is primarily qualitative and provides greater capacity to address "how" questions in a dynamic and complex context. The method has permitted close contact with the subject over a long timeframe. It assures data validity through both: (1) an iterative process of evaluation over an extended time in conducting the research, and (2) reflection among the actors in examining the findings, thus offering the opportunity to address misunderstandings or misinterpretation in the data analysis. It is inappropriate due to confidentiality to reference all internal documentation. This is provided where possible.

Context

The British Cement Association (BCA) is a key trade organisation for the sector. It provided specialist technical information to the concrete and cement industries through its former library and information services. This service was characterised as being the provider of *information*, the aim being to provide a service that would help to promote the use of concrete and cement in construction and the design of projects. The approach was a quasi-public sector one. Recipients were typically technical directors and heads of specialist fields, who were responsible for dissemination within their organisations. Some were occasional users, some regular users and subscribed as members, receiving documents, standards, books, and other publications on a preferential basis. Other income was received through some sales and direct support from the BCA.

Restructuring

This service has been restructured, exhibiting some classic features of 'marketisation', to independently serve the knowledge management market. Serving this market requires not only the provision of information for others to process, but also the processing of information into knowledge. The service is now delivered through Concrete Information Limited (CIL).

This change has gone hand in hand with some restructuring of the information services. In essence, the sector launched The Concrete Centre (TCC) in September 2003 to parallel the successful initiatives of the steel sector. TCC is supported by the industry trade organisations and by leading producers. Its remit is to focus upon the specifier market. This was the traditional operating area of CIL. BCA continues direct support to CIL and has also brought in the Concrete Society, having transferred 50% of its shareholding in CIL to this other leading trade association.

Market Reconfiguration: Awareness, Marketing, and Gatekeepers

Under the new configuration CIL is working towards being a commercial entity in its own right, income derived from open market sales and through membership. One of the key member clients is TCC; therefore a relationship marketing approach is most appropriate, and it is anticipated that the links will strengthen over time. TCC has a pre-specification focus and therefore is one important CIL client in this area. CIL will serve the regional representatives of TCC, and through them build relationships with their customers where concrete is specified and innovation or problem-solving services are required. TCC regional representatives are acting as facilitators rather than gatekeepers. They are more akin to sluice gates, that is, encouraging the rate of information flow, than gatekeepers, who tend to keep back knowledge that is not perceived to be in the best interests to release. The particularities for these circumstances are shown in Figure 4.

Whilst the model may infer that the relationship is three-way due to the automatic dispersal or dissemination up and down the client organisation, it was always dependent upon the internal functioning of the client organisation. That has changed into a new situation. An additional primary market target is the project level, and the technical director, knowledge manager, or people with the responsibility cannot be assumed to be motivated nor committed to making those who may need the service aware of the possibilities, nor be in an informed position to give advice to call upon the services of CIL. In terms of Figure 3, the target has moved from the top right-hand segment—the Research Institutes and Network role of supplying the senior management within the firm—to the bottom-right segment—the Suppliers and Trade Body relating to the project of the firm—where it was always partially located, but is now more concentrated, except for its role in relation to TCC.

While clients, some introduced through TCC, may be tracked and the contact maintained, many other projects may not require the involvement at CIL at the initial stages. So generally, the prime point of contact is no longer solely nor mainly the technical director or the knowledge manager for the entire consultant, contracting, or supplier organisation to those implementing the project.

This raises a crucial issue about reaching the end user at the project level. The two main and related marketing issues are:

Figure 4. Relationship channels for CIL

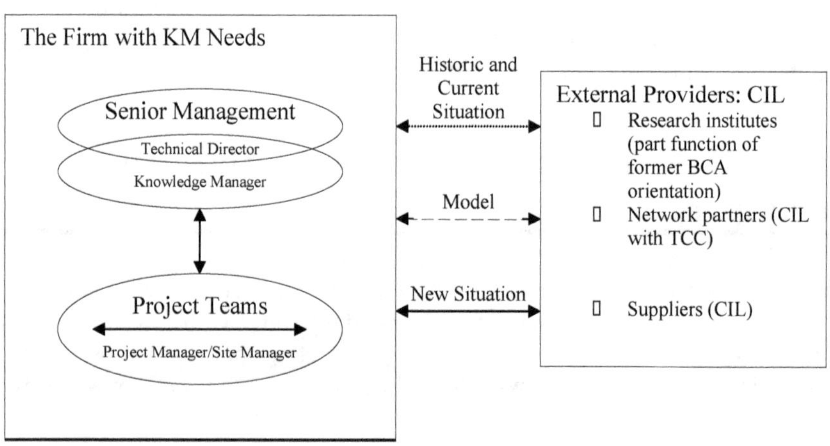

1. Paradigm choice
2. Creating awareness

Theoretically, a relationship marketing approach would be the ideal choice in this context, as the needs are likely to be very different for each project. However, the project staff confronts a number of issues that have been identified earlier in the chapter which act as filters:

- Uncertainty about the need for information and knowledge in the future (Milgrom & Roberts, 1992)
- Uncertainty of the quality of information and knowledge available (Akerlof, 1970)
- Uncertainty of the availability of information and knowledge
  - Role of gatekeepers
  - Effectiveness of awareness creation from external provider

CIL is facing all these issues and thus the relationship approach is difficult to establish, particularly at the front end of the market changes. Reputation in the new market will help and the project network will create referrals in due course. Until such time, CIL is dependent upon relationship opportunities introduced through TCC and a marketing mix approach. The mix is generated from several sources:

- Market feedback and intuitive feel for service packages based upon experience
- Short-term market research and longer term monitoring of demand patterns

Price is a sensitive issue. The construction industry is reticent to invest by economic predisposition. The previous regime led to a 'free' or minimum charge service. The service is being developed to reflect the high value with a corresponding price tag (Oxon Consulting, 2001). The true cost of production, transactions, and renewing the resource has to be recovered to perpetuate availability. The remix of the services, tailored to needs, will help the transition. Hence the 'product', in this case the service package, is changing. The delivery channel is essentially the same, but both more proactive and more targeted towards the end user segment concerned with projects. Promotion is reliant upon mailouts targeted both across the market and towards niches.

It is recognised that there are limits to this type of marketing mix promotion. The market is not so large that it can withstand too many and sustained direct marketing campaigns of this nature. However, it creates awareness, generates cash flow from a low cost base, and hence is an important baseline for growing more sophisticated and targeted marketing in the longer run. Repositioning in the market and creating awareness amongst a new customer base is a dynamic process, which requires a changing mix (Concrete Information Limited, 2003). The industry network is strong. The former network was based around the roles of technical directors, experts, and whoever performed these information gathering and disseminating roles. This new market is a slightly different network, a network of end users looking for project-specific knowledge to solve problems. Gaining access is the key, as market research indicates that the traditional gatekeepers are proving to be a barrier—rather than facilitators—to reaching the new market. The motives for this are unclear—pure stress and volume of work, wishing to use information as a tool of power within the organisation, or some combination. It is probably that gatekeepers are working in different ways in different organisations—some using their role as political kudos, some as a power position, some simply overloaded to search out and disseminate the information.

Successful access to the market will help to build relationships at the individual level. Multiple needs will help convert some relationships into members. There used to be one key membership package. Conversion is being facilitated by a series of different membership packages designed for different needs and closeness of relationship. It is anticipated that firms will identify the appropriate entry level and subsequently purchase bespoke services. The rationale is that through membership, issues of uncertainty are overcome concerning quality and availability of information.

Some direct enquiries are being made. These unsolicited enquiries have a long history. Most of these have been one-off requests for information or have been treated as such. Those that were not one-off tended to be from existing members or individuals well known in the industry network. What has changed in the market is that there are a number of enquiries that are being made which are more complex. In essence the demand is for knowledge, not information alone. This is positive market data for the changing patterns of demand and the need for external KM services. In addition, direct enquiries are being treated as potentially longer term, with the opportunity to establish or develop relationships. In all these cases where one-off information services are not required, the relationship approach is being used to define and develop bespoke KM services.

Primary Results

The restructured library and information services, formerly of the British Cement Association, have been re-branded Concrete Information Ltd. (CIL) and placed under joint venture ownership between the Concrete Society and the BCA. The restructuring has been aimed at providing the context for new service provision, summarised as:

- From information services to information and knowledge, especially for the project-specific KM market, that is, at the post-specification project end of the construction business
- From a low-cost service to a commercially sound service, priced to reflect the historic high value and value-added bespoke and service packages
- From a predominantly (and frequently reactive) marketing mix approach, towards securing new business to a combined highly proactive marketing mix and relationship marketing approach to new business

- From targeting technical directors or equivalent to also gaining access to end users on a project's post-specification

While there was some initial internal inertia and adjustment within CIL in the face and wake of restructuring, the new approach has been embraced and actively developed, momentum building up over the duration of change. CIL has faced constraints to the implementation of a commercial approach due to:

- The need for additional market data of client needs
- The reactive nature of demand within client organisations
- The inability of client organisations to be able to adequately identify their future needs

The run up to the launch of TCC provided a surrogate framework for CIL in identifying packages that mirror the services offered pre-specification in the post-specification market. CIL now offers levels of membership associated with different service packages (in addition to pay-as-you-go services). Membership levels are important, as services frequently have to be paid out of project budgets, at least initially.

However, the main constraint is external. Gatekeepers within the client organisation are creating a major obstacle to awareness creation of the services. Developing Internet and e-based solutions will provide wider access, hence awareness to the marketplace; however, targeted contact has to be established. The combination of a marketing mix and relationship marketing approach has been developed through a series of short campaigns to gain access and build new networks in the market.

Conclusion

The case study has been used to explore some of the issues raised in the literature. The results are of significance. The external provision of KM services to customers requires that KM theory on developing services needs to be engaged with marketing paradigms. This determines the approach to selling in the market. Theoretically, and even intuitively, the relationship approach would appear most suited. Both KM and relationship marketing are competency based and both therefore typically are located at the interface of human capacity and systems. However, this case has shown that this is not always necessary or appropriate. The larger the market, the more potential for a marketing mix approach, but also gaining market access to create awareness is a major issue and the marketing mix approach is efficient. This is likely to be the case in more start-up cases or cases of restructuring provision, as established networks and relations will not exist or will have been 'broken'.

It has also been demonstrated that the role of people within organisations is key. This is the case in terms of relationship building generally, and critical where gatekeepers are evident. They can be a very positive force—the internal marketers of the external service. Gatekeepers can also be a negative force, constraining the ability to create awareness and reach the end user. While CIL presents a case with some unique features, and thus it is not expected that gatekeepers will present an issue for some external providers, it is anticipated from this case evidence that there will be similar circumstances in place in many customer organisations for external providers.

Therefore the management of knowledge in the context of external provision becomes a potential nexus of *knowledge-relationship-gatekeeper management*. The areas are interlinked conceptually and empirically. The theoretical issues, such as those giving rise to uncertainty and a lack of awareness, asymmetry of the quality of information in advance can be overcome by addressing marketing and gatekeeper management in this context.

The impact also serves to highlight the need for further research to explore the external provision of KM services. The potential for market growth of external services needs further exploration. Further research at the KM-marketing interface is needed in terms of the application of KM tools, that is, techniques and technologies. In addition, these need to be explored within the paradigms of the marketing mix and relationship marketing. Specific research is needed into whether the experiences of the cement and concrete industry and CIL in particular are replicated elsewhere.

In terms of project managers and industry, this chapter is itself part of the awareness-creation process and thus makes an impact in its own right. In terms of the content, project managers are provided with lessons and guidance on accessing KM from external sources on proactive and reactive bases. External providers are provided with lessons and guidance on the creation of KM services through to the delivery of these services to the end user within the client organisation.

Practical Tips and Lessons Learned

This chapter has provided pointers and insight into the following areas:

- Traditional information providers and consultants with strengths in particular areas of learning and knowledge can explore opportunities for externally providing KM services to others.
- Careful consideration should be given to the role of marketing, especially weighing the marketing mix and relationship marketing approaches against costs and opportunities.
- KM user firms must carefully manage the knowledge managers in order that the best interests of the firm are served and that project needs are clearly identified in time.
- Project managers and teams must be equipped in advance with sources of knowledge for problem solving both internally and externally from their firms, being aware that external provision of KM is expected to increase in the coming years.

References

Akerlof, G. (1970). The market for lemons: Qualitative uncertainty and the market mechanism. *Quarterly Journal of Economics, 84,* 488-500.

Cherns, A.B., & Bryant, D.T. (1983). Studying the client's role in project management. *Construction Management and Economics, 1,* 177-184.

Child, J., & Heavens, S.J. (2001). The social constitution of organizations and its implications for organizational learning. In M. Dierkes, A.B. Antal, J. Child, & I. Nonaka (Eds.), *Handbook of organizational learning and knowledge.* Oxford: Oxford University Press.

Concrete Information Limited. (2003). *Marketing plan.* Crowthorne.

Construction Productivity Network. (2003). *Knowledge management: Construction industry experience.* Presentation by Bob Robinson, BAE Systems, Members Report E3127, CIRIA, Bracknell.

Drucker, P.F. (1989). *The new realities.* London: Butterworth-Heinemann.

Drucker, P.F. (1993). *Post-capitalist society.* London: Butterworth-Heinemann.

Drucker, P.F. (1995). The information executives truly need. *Harvard Business Review,* (January-February), 54-62.

Dunnette, M.D. (Ed.). (1976). *Handbook of industrial and organisational psychology.* Chicago: Rand & McNally College Publishing.

Egbu, C., Kurul, E., Quintas, P., Hutchinson, V., Anumba, C., & Al-Ghassani, A. (2003a). *Techniques and technologies for knowledge management, work package 3.* Interim Report, Knowledge Management for Sustainable Construction Competitiveness Partners in Innovation Project, Commissioned by the Department of Trade and Industry. Retrieved from www.knowledgemanagement.uk.net.

Egbu, C., Kurul, E., Quintas, P., Hutchinson, V., Anumba, C., & Ruikar, K. (2003b). *Knowledge production, resources and capabilities in the construction industry, work package 1.* Final Report, Knowledge Management for Sustainable Construction Competitiveness Partners in Innovation Project, Commissioned by the Department of Trade and Industry. Retrieved from www.knowledgemanagement.uk.net.

Granovetter, M. (1985). Economic action and social structure: The problem of embeddedness. *American Journal of Sociology, 91,* 481-510.

Groák, S. (1994). Is construction an industry? Notes towards a greater analytical emphasis on external linkages. *Construction Management and Economics, 12,* 287-293.

Gummesson, E. (2001). *Total relationship marketing.* Oxford: Butterworth-Heinemann.

Hughes, J.S., & Kao, J. (2001). Vertical information and proprietary information transfers. *Journal of Economics and Management Strategy, 10*(2), 277-299.

Johnson, G. (1990). Managing strategic change: The role of symbolic action. *British Journal of Management, 1,* 183-200.

Johnson, G., & Scholes, K. (2002). *Exploring corporate strategy* (6th edition). London: Financial Times-Prentice Hall.

Kotler, P., Armstrong, G., Saunders, J., & Wong, V. (1996). *Principles of marketing.* London: Prentice-Hall.

Leonard-Barton, D. (1995). *Wellsprings of knowledge.* Boston: Harvard Business School Press.

Levitt, T. (1983). *The marketing imagination.* New York: The Free Press/MacMillan.

Milgrom, P., & Roberts, J. (1992). *Economics, organization and management.* Englewood Cliffs, NJ: Prentice-Hall.

Mintzberg, H. et al. (2002). *The strategy process* (4th edition). Englewood Cliffs, NJ: Prentice-Hall.

Nonaka, I., & Takeuchi, H. (1995). *The knowledge-creating company.* Oxford: Oxford University Press.

Oxon Consulting. (2001). *BCA CCI business plan approach.* Unpublished Report.

Pahl, R. (1975). *Whose city?* London: Penguin Books.

Polanyi, M. (1966). *The tacit dimension.* London: Routledge & Kegan Paul.

Ruggles, R. (1997). *Knowledge tools: Using technology to manage knowledge better.* Working Paper, Ernst & Young.

Senge, P. (1993). *The fifth discipline.* London: Random House.

Shavell, S. (1994). Acquisition and disclosure of information prior to sale. *Rand Journal of Economics, 25*(1), 20-36.

Smyth, H.J. (2000). Marketing and selling construction services. Oxford: Blackwell Science.

Smyth, H.J. (2004). Competencies for improving construction performance: Theories and practice for developing capacity. *International Journal for Construction Management, 4*(1), 41-56.

Stinchcombe, A.L. (1965). Social structure and organizations. In J.G. March (Ed.), *Handbook of organizations.* Chicago: Rand McNally.

Teece, D., Pisano, G., & Sheun, A. (1997). Dynamic capabilities and strategic management. *Strategic Management Journal, 18*(7), 509-533.

Tushman, M.L., & Katz, R. (1980). External communication and project performance: An investigation into the role of gatekeepers. *Management Science, 26,* 1071-1085.

Williamson, O.E. (1985). *The economic institutions of capitalism.* New York: The Free Press.

Chapter IV

The Intellectual Capital Report of Danya Cebus: Smart Construction

Edna Pasher
Edna Pasher Ph.D. & Associates, Israel

Gil Horsky
Edna Pasher Ph.D. & Associates, Israel

Abstract

This chapter is a "case study" portraying the use of a unique method to explore the hidden values and intangible assets of an Israeli construction company, Danya Cebus Ltd., as a basis for its future growth. The chapter describes in detail the process of producing the intellectual capital (IC) report of Danya Cebus, the first IC report produced by a construction company in the world. The report was the output of an in-depth strategic process which was guided professionally by the Israeli management consulting firm, Edna Pasher PhD & Associates. The authors hope that more companies will adapt the method of creating an IC report as described through the Danya Cebus case study. A periodical display of the organisation's competences will increase the market value of the company and diminish stock fluctuation in the market routed in reasons that are irrelevant to the organisation's capacity. Moreover, an IC report can serve as a central management tool, which enables a systematic examination of the organisation's competences.

Introduction

The market value of successful firms traded on the stock market is high—sometimes higher than many times the value if their assets as shown in their financial reports. Take, for example, the market value of Microsoft, which is approximately 15 times greater than its book value. The disparity between these two values reflects the market's valuation regarding the worth of the hidden values. These values, which are not expressed in the financial reports, combine to create the organisation's intellectual capital. The intellectual capital comprises the knowledge, wisdom, capability, and expertise that provide the organisation with a competitive advantage over other organisations and visualizes its potential for future growth.

When an investor chooses between buying IBM shares or Microsoft shares, the financial reports will not provide him with the full picture of the companies' value. Professor Baruch Lev (2001) of the NYU School of Business conducted research in which he analysed the correlation between the stock prices and the value of the companies as shown in their books. He found that the correlation weakened over the years. According to Professor Baruch Lev, this phenomenon confirms decreasing relevance of financial reports as indicators of a company's market value.

The financial report, which is approximately 500 years old, is still used for tax purposes and historical documentation, among other things. It provides us with a picture of the past. Nevertheless, it does not provide a reliable basis for assessing an organisation's potential for future growth.

The financial balance sheet is based on the measurement of assets by means of their historic cost, assuming that the cost of purchasing the asset reflects its actual value. The model is of no relevance to us when we consider intangible assets, let alone knowledge. The financial reports provide no indication of intellectual assets except for goodwill. This is due to difficulty in measuring intangible assets.

Intellectual capital provides a basis for creating a sustainable competitive advantage of the organisation in the "Age of Knowledge." Intellectual capital has much greater value and intensity than any of the tangible assets, which include natural resources, large factories, equipment, and land—all of which provided the organisation with a competitive edge in the past. Today, intellectual capital is the organisation's most valuable resource. Yet, it is neither managed well nor exploited efficiently. The source of an organisation's growth and future affluence lies in its correct management of intellectual capital in order to produce a high yield from its valuable assets.

One of the tools for assessing the intellectual capital in an organisation is a report of its hidden values. This balance sheet, which complements the regular balance sheet, helps us define and visualize the hidden values of an organisation. The intellectual capital report enables the organisation and other stakeholders to obtain an integrated, comprehensive view of all the organisation's assets, to learn about its potential for future growth, to make more reliable assessment, and to navigate the organisation towards the realisation of its goal and vision.

Danya Cebus's intellectual capital report (1999) includes the identification and presentation of the company's core competencies, key success factors, hidden and intangible values that provide the company's competitive advantage, and growth potential. Measurement of the intellectual capital of Danya Cebus was based upon a model evolved and applied by Leif Edvinsson, Former Vice President for Intellectual Capital at **Skandia,** a Swedish global financial service company.

Learning Objectives

1. Learning the essence of intellectual capital
2. Learning how to conduct an intellectual capital report based on the Skandia model
3. Learning how to design strategic processes within organisations

What is the Skandia Model for Measuring Intellectual Capital?

The Skandia model provides a balanced and holistic picture of both financial capital and intellectual capital. According to the model, there are four areas of focus with regard to intellectual capital: customer capital, human capital, process capital, and renewal and development capital. Theses areas of focus are used as the basis for assessing the intellectual capital within a competitive environment.

The model for measuring intellectual capital uses "the house" as a metaphor for the organisation. financial capital constitutes the roof of the house, and reflects the organisation's history and achievements of the past, which do not necessarily enlighten us in terms of future achievements. The supporting columns are process capital and market capital, and they are the areas upon which the present operations of the organisation are based. Renewal and development capital, which is found in the centre of the house, interacts with all the different focal points. Human capital, which is also found in the centre of the house, interacts with all the different focal points. Human capital is the heart of the organisation, i.e., the capabilities, expertise, and wisdom of the people. It is the role of the organisation to assist, guide, and support its people towards realisation of their vision strategic goals.

The organisation can also be described as a tree. The various means that express the tangible assets—the annual reports, catalogues, protocols, and others—are compared to the leaves and branches of the tree. The wise investor who desires the fruit of the tree will examine its roots in order to learn about its future. The roots of the tree and the foundation of the house reflect the renewal and development capital of the organisation, which is the source for growth and affluence.

Figure 1. Skandia model

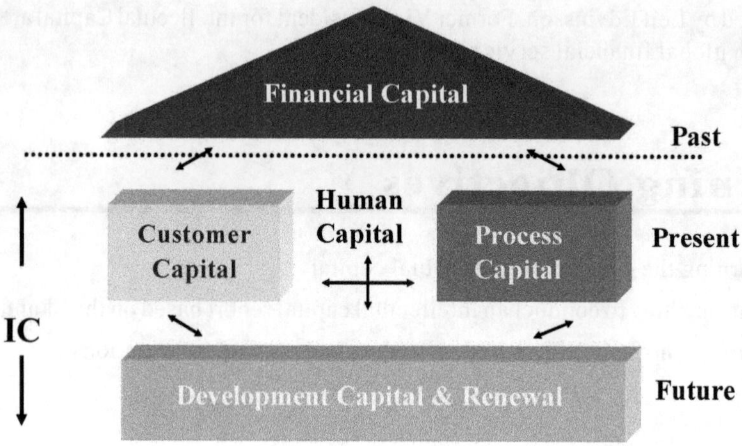

What are the Diverse Focal Areas of Capital?

The value chain according to Leif Edvinsson's model expresses the various components of market value on the basis of the following model (Edvinsson & Malone, 1997):

Market Value = Financial Capital + Intellectual Capital

Human Capital

Human capital includes all the individual capabilities, the knowledge, skill set, wisdom, expertise, and the abilities of the employees and managers to realize organisational tasks and goals. This focus also includes the values, the culture, and the philosophy of the organisation. Human capital is the property of the individuals, and not of the organisation, and therefore must be capable of capturing the dynamics of an intelligent organisation in a changing competitive environment. For example: Did the company's employees acquire any new professional skills? Does the organisation effectively utilize these new skills?

Process Capital

Capitalizing on the flow of knowledge requires structural intellectual assets, i.e., information systems, hardware, software, databases, laboratories, patents, trademarks, organisational structure, and management focus—and everything else of organisational capability which sustains and amplifies the output of the human capital. One of the important challenges of companies is turning the human capital into process capital. According to Hubert Saint Onge, a leading theory developer of intellectual capital, the relationship between human and structural capital is a "double-arrow dynamic." In Saint-Onge's words, "Human capital is what builds structural capital, but the better your structural capital, the better your human capital is." The human capital is a limited asset and is stipulated in the company's ability to become attractive to the people it needs. In fact, process capital is the capital that remains in the organisation after the employees go home. Unlike human capital, process capital can be owned and thereby traded

Customer Capital

Customer assets are assets that are embedded in the relationship with the organisation's customers, and enable the full utilisation of their knowledge. The once common idea that customers are essentially one-time targets, which can be forgotten once the sale is made, is irrelevant theses days. Today most companies realize the importance of providing customers with "total customer service"; even some, such as Debra M. Amidon (in press), are talking about "customer success," which means that companies should collaborate with their customers in a joint value creation process.

This capital component includes: customer loyalty, customer knowledge, satisfaction level among customers, branding, and more. Market share is, of course, part of this category. The number of clients a company has becomes in the "new economy" the most prominent measure in estimating the market-value of the companies.

Renewal and Development Capital

This is probably the most important category in exhibiting the organisation's competences. In a competitive market, companies, products, and services are fast duplicated more and more. The ability of the organisation to innovate and develop new knowledge in each of its core-competences is a central feature of the organisation's future growth, and therefore is an excellent measure for evaluating the company. Renewal and development assets include investments in research and development, patents, trademarks, new product development, usage of advanced technological tools, and the like.

But, unfortunately these are times when organisations are going into the bunker. They are busying themselves with primary survival. How? By cutting costs, of course. This is why it is so important to remember, particularly in these lean times, that long-term sustainability does not come from cutting expenses, but from creating new value. And the most important source for creating new value is innovation: new ideas, new solutions to unsolved problems, new products, and new services. We are even talking about new ways to be efficient and save.

We are living in a dynamic world, and chances are that it will only get more dynamic in the future. The bursting of the bubble has not changed the fact that what worked yesterday, namely in the industrial age, will probably not work tomorrow, in the knowledge age. As Gary Hamel has claimed, companies inherited an important set of virtues from the industrial era: diligence, efficiency, replication, and control. But these virtues are becoming less important in an age where the new required virtues are creativity, imagination, diversity, speed, openness, and flexibility (Hamel, 1999, 2000, 2001).

This is not a trivial insight. It is not obvious to most managers, just as it wasn't always so obvious to knowledge management experts. The first generation of knowledge management didn't catch on to this important point. The original driving force of the knowledge management pioneers was how to make better use of existing knowledge. In Peter Drucker's terms, we were after higher productivity of knowledge, a higher return on investment in knowledge (Drucker, 1993). The big news back then was that knowledge could be managed at all—like other assets. Everyone was excited about capturing knowledge, knowledge repositories, and the potential contributions of information technologies. We wanted to make knowledge available, avoid repeating mistakes, and prevent inventing what others had already thought of.

But managing the knowledge asset is not enough. You need to create new knowledge. You need to innovate in order to create value for the company. It turns out that knowledge assets age real fast. The danger in overly reusing knowledge is that it may become obsolete along the way. And then you find yourself left behind. Gradually it dawned on knowledge experts that reuse had its dangers and that companies need to also learn to unlearn, to forget what they know. It became increasingly evident that becoming free of the binds of previous modes of thought is more important than knowledge capturing, because it is this process that enables value creation through innovation.

If in the industrial age, organisations produced tangible products, in the knowledge age, knowledge-based organisations are knowledge factories. This is why Nonaka, one of the visionaries of knowledge management, titled his book, *The Knowledge Creating*

Figure 2. Skandia market value scheme

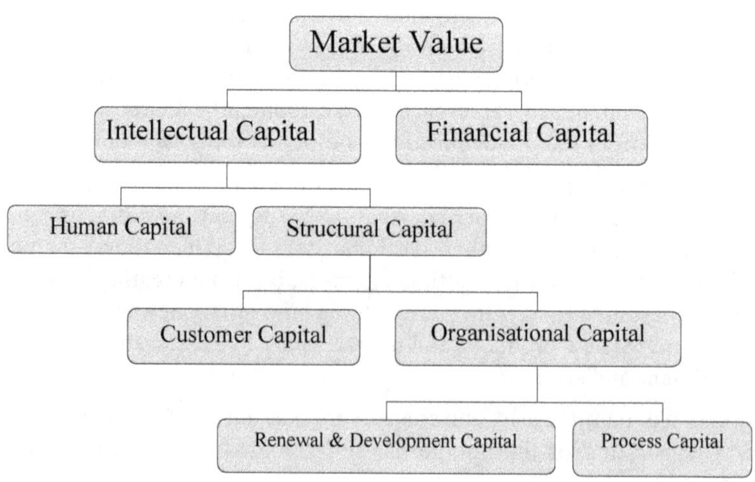

Company. Not *The Knowledge-Reusing Company,* not *The Knowledge-Capturing Company,* but *The Knowledge-Creating Company* (Nonaka & Takeuchi, 1995). That is what it is all about.

In the mid-1990s, at the height of the reengineering revolution, Gary Hamel and C.K. Prahalad informed Corporate America that it was headed in the wrong direction. In *Competing for the Future*—one of the most influential books on strategy of the 1990s—they argued that innovation, not efficiency, was the path to greatness (Hamel & Prahalad, 1994). Only by creating new businesses, and reinventing existing ones, could companies become dominant market leaders.

So we are focusing on knowledge creation. But what is it? It is something that creates value for the organisation. It is innovation. The NIMCube project, a European-funded project, has concluded that innovation is composed of three components: new use of existing knowledge, creation of new ideas, and exploitation of these ideas to create value for the organisation.

The NIMCube innovation formula:

$$Innovation = (New\text{-}use + Invention) \times Exploitation$$

The Process of Measuring Intellectual Assets is based on four phases:

- **Phase 1-Creating a vision and core values:** Since it is a tool that is due to predict the growing capacity of the company in the future, there ought to be a consolidation of the future picture of the company and its vision. Without a clear vision, its future potential cannot be measured. The focus while consolidating the vision must be in regard to the core competencies of the company. If you do not decide what the company should do in order to succeed, how would you check the company's level of knowledge, which is its future growth potential?
- **Phase 2 - Identifying the core competencies needed to realise the vision:** The decision regarding the organisation's core competences is a basic fundamental decision. It deals with many questions such as: Which types of knowledge will you develop within the organisation and which will you outsource? What are these competences, which form a relative continuous advantage in a competitive market and which will differentiate the company from existing and potential competitors? What are the scopes of knowledge, which are not easily entered, those which cannot be easily duplicated and therefore should become in-house core competences? What are the other scopes of knowledge, on the contrary, that should not be developed in-house and which can be acquired or outsourced relatively easily?
- **Phase 3 - Identifying the key success factors for each competence:** In order to measure the company's accomplishments and assets—its status in comparison with other companies—you will need to define the key success factors for each of the company's core competencies.

- **Phase 4 - Identifying the indicators for each key success factor.** After making your decision about what the company knows in order to succeed, you can check what the company knows in each of these areas of knowledge. Each key success factor must be linked with a list of indicators. These indicators depict the organisation's level of knowledge in that area.

The Process of Preparing the Danya Cebus Report

The initiative to conduct an intellectual report for the Israeli construction company, Danya Cebus, originated with the CEO's vision to visualize the hidden values and assets of the company in order to leverage its business activities. The need for the report originated from several reasons:

1. Forming a supplementary management tool for strategic planning
2. Defining an effective tool for assessing the company's value in preparation for the issuing of its stocks
3. Creating an innovative marketing channel for the company and its activities
4. Identifying and managing effectively the company's intangible assets and core competencies

Danya Cebus is a genuine new-generation construction company, with all the added values of a high-tech enterprise. Its activities are centred in the fields of residential construction, building commercial and office space, infrastructure development, and industrial design and manufacture, with projects in Israel and overseas.

Danya Cebus, a publicly owned company and subsidiary of Africa Israel Investments Ltd., has exhibited an extraordinary rate of consistent growth—an average of over 25% per annum in the last five years.

With a staff of more than 1,500 employees, Danya Cebus is currently one of Israel's largest construction companies. Generating annual revenue of $225 million, the company's financial might is inherent, with impressive year-to-year growth in profits and shareholders' equity. Over the past five years, it has been able to triple its building volumes—despite an increasingly unfavourable local market and fierce competition, and to exhibit an extraordinary rate of consistent growth—an average of over 25% per annum in the last five years. Today, boasting a higher market value than other top construction enterprises on the Tel Aviv Stock Exchange, Danya Cebus is among the most financially formidable construction companies in Israel.

To maintain this outstanding performance, the company relies upon the creative ideas and pioneering spirit of its exceptional staff to lead it along new professional paths and into new global territory, all the while adhering to its vision of sustaining a leadership role in construction project management and use of "smart construction" techniques. As part of the company's pioneering spirit, its management decided to produce an IC report, and to become the first construction company in the world to produce an IC report. The report was conducted with the professional guidance of the consulting firm Edna Pasher PhD & Associates (1999), an Israeli management consulting firm that has vast experience in professionally guiding companies to create their IC reports.

The Skandia model was chosen as the model for measuring the intellectual capital of Danya Cebus.

The first phase for the preparation of the Danya Cebus IC report was the formation of a steering committee headed by the CEO. The steering committee included various functions from all different departments of the company, including senior management, engineers, customer service representatives, and others. The purpose of the steering committee was to manage the process of preparing the report.

After forming the steering committee, there was a need to launch the project in all parts of the organisation, and to motivate and commit the employees to take an active role in the process. Therefore, a "Knowledge Café" session with more than 40 employees and managers took place in the company. Knowledge Café is a unique method that enables the conduct of effective brainstorming sessions with a large group of people. It is basically a forum in which employees from all over the organisation discuss strategic and management issues, as they sit in small groups over a cup of coffee.

This forum is knowledge-sharing junctions, which enables a wide discussion, but at the same time an intimate level of dialogue. Such a session usually begins with a presentation of one or more management dilemmas that start the conversation around the tables. After all the groups have discussed the topic around the tables, a representative from each table will present the outputs to the full audience. The aim of the Knowledge Café conducted in Danya Cebus was to present to all the participants the concept of the intellectual capital report, and the different phases of the planned project. In addition, all the participants were seated in small groups and were presented with questions that were relevant for the report preparation:

1. What is the vision of Danya Cebus and what are the core competencies which are derived from that vision?
2. What are the key success factors for the future development of Danya Cebus, and how can they be measured?

The outputs of the group discussions in the Knowledge Café session included the definition of Danya Cebus' vision and values:

Vision:

Danya Cebus is the leading Israeli construction company in quality, timetables, and service to its clients, while constantly expanding its activities to new markets.

Values:

- **Meticulous quality in construction.** Top quality, ensured by exacting, continuous quality assurance procedures, is a guiding principle in every Danya Cebus project—from the first blueprint to final tiling.

  The company's quality culture, endorsed by an ISO 9002 certificate, is based on a professional "do it right the first time" approach, which saves a great deal of construction time, non-quality costs, and customer dissatisfaction.

  Danya Cebus operates a professional quality assurance department, employing experts thoroughly versed in today's international quality assurance procedures and processes.

  The department is responsible for the continual monitoring of all projects under construction at any given time.

  Danya Cebus' comprehensive quality assurance system is supported by a regular training program, which incorporates both long-term quality awareness education and specialized professional updates. In addition, Danya Cebus promotes research and development activities—endeavouring to introduce new concepts and improve work processes, as it continues to build the future.

- **Professionalism in customer service.** Satisfied customers are the ultimate targets of all the company's complex and sophisticated operations. To achieve optimal results, Danya Cebus offers its future and present residents a responsive, flexible customer support network, consisting of two professional departments.

  The Residents' Modification Department provides purchasers with personal design services, tailoring their homes to their precise needs and lifestyles, and offering them full support and assistance until they are comfortably settled in the new residence. The fully computerized modification system monitors and records every alteration in planning and construction—for perfect implementation and detailed future reference.

  The Customer Service Department takes over once the residents have settled in their new home or office—offering a round-the-clock open service line and an experienced, get-up-and-go maintenance and repair crew. Always ready to help, the department provides the customers with quality solutions to any post-construction problems, for many years of guaranteed satisfaction.

  Together, the two customer support departments maintain close contact with Danya Cebus' satisfied clientele—the company's most important asset.

- **Safety comes first.** Danya Cebus believes in safety as a primary value in and of itself, and invests considerable resources—both human and financial—in preventing work-site accidents.

  Aiming to make safety an inherent, central component of its corporate culture, Danya Cebus implements a thorough, ongoing Safety Education Program that is intended for all managerial levels and sub-contractors.

Management and employees accept personal as well as collective safety responsibilities, in order to achieve the company's objective of an "incident- and injury-free work environment."

It is Danya Cebus' policy to strive to provide the highest safety standards on all our projects. No effort or money is spared in providing workers with the finest safety equipment, such as high-quality harnesses, cables, railings, and scaffolding, as well as full personal safety gear. On-site hazards are eliminated by an active and meticulous safety control system that continuously monitors site conditions, spotting and preventing any potential trouble.

- **Commitment to the quality of living in the community.** The company has contributed extensively to the country's rapid development over the past five decades. Today, the quality signature of Danya Cebus may be recognized throughout the land, at scores of elegant residential neighbourhoods, vast industrial and commercial complexes, and major public and national projects. The planning and building of residential neighbourhoods is the firm historical basis of the company's activities. In the past few years, Danya Cebus has built tens of thousands of residential units—from private homes and terraced housing clusters to high- rise apartment towers. With meticulous planning and precisely tailored environmental development, Danya Cebus created finely styled neighbourhoods incorporating modern community services and education facilities that offer residents a true 21st century quality of life.
- **State-of-the-art technology and construction methods.** Danya Cebus brings high technology into the construction business, with a comprehensive array of 21st century methods and tools—from advanced professional software to innovative construction techniques. Building for the future, Danya Cebus implements the smartest new-age construction techniques, optimising building procedures and ensuring top-quality construction, low costs, and the shortest possible time schedules.

Advanced methods include:

- Sophisticated mold-based construction, utilizing molds designed by some of the world's leading manufacturers.
- Flexible modular construction, based on the extensive variety of combinations enabled by the perfectly matched products of the Cebus Rimon in-house plant.
- Efficient, cost-effective industrialized construction, utilising top-quality prefabricated elements manufactured at Cebus Rimon.
- All processes at Danya Cebus are fully computerized, based on state-of-the-art software tools precisely tailored to the needs of the company's diverse activities and departments.

The rest of the meeting's outputs were later analysed and used as a basis for the next phases of the report preparation.

The third phase in the report's preparation process was an in-depth evaluation phase. The evaluation phase was conducted by interviewing key functions in the organisation, strategic partners, and clients. The interviews were mainly focused around the identification of Danya Cebus' core competencies and key success factors that will promote its future growth. In addition the evaluation phase included the analysis of existing organisational materials, such as working plans, protocols of meetings, and working procedure manuals.

The outputs of the evaluation phase were the definition of Danya Cebus' core competencies:

- Smart project management
- High quality service for the tenant—from purchase to maintenance
- Logistical, planning, and implementation competencies
- Strong and solid financial backing
- Clients' portfolio
- Employees' loyalty
- In-house industrial solutions
- Safety and quality assurance
- Usage of state-of-the-art IT solutions
- A learning organisation
- International alliances

In the next phase, several strategic workshops were conducted in the steering committee forum. The aim of these workshops was to finalize the list of the company's key success factors for future growth, and their measurement indicators. The heterogonous make up of the steering committee enabled each participant to suggest various indicators, based on his unique perspective, which is derived from his professional function. This enriched the workshop discussions and consequently led to better outputs.

The proposed indicators were evaluated in their accessibility, practicability, and the way they fit into the Skandia model. Based on that input, the final list of indicators was chosen for each of the four capital areas: human, process, customer, and renewal and development capital.

Finally, the relevant information and knowledge about the different indicators was collected and analysed from the company's existing databases. That enabled completion of the formulation of an in-depth IC report for Danya Cebus.

Conclusion

The Danya Cebus IC report is a pioneering effort in portraying a holistic visualisation of the company's knowledge and intellectual capital. The report represents in a focused way the areas in which promoting and managing their knowledge effectively will result in Danya Cebus' future growth and prosperity.

The report clearly indicates that Danya Cebus accomplished outstanding achievements in its field, despite the fact that the Israeli construction market is constantly decreasing and becoming more competitive. Currently, Danya Cebus is one of the leading construction companies in the Israeli market, and the report's results indicate that it will continually develop and grow in the future.

The report provided a unique tool for the company's investors to learn, in addition to its financial reports, about the company's competences and hidden values. This consequently led to successfully issuing the company's shares, which subsequently raised the value of the company's stocks. Additionally, the report initiated a solid infrastructure for the future development of the company, which assisted Danya Cebus in determining methods for its future businesses in order to expand its market share both in Israel and abroad. And finally, the company's management decided to use the report as a navigation tool that will direct the process of defining the company's policies and strategy. In order to evaluate the company's progress towards its defined vision and values, the management decided to update on a regular basis its intellectual capital report, while re-examining and defining its key success factors and indicators.

To conclude, companies that will adapt the method of creating an IC report as described through the Danya Cebus case study will be able to diminish their company's stock fluctuation in the market. Periodical display of the organisation's competences will increase the market value of the company and block falls, routed in reasons irrelevant to the organisation's capacity. In addition, a systematic and periodic examination of the organisation's competences offers a central management tool for developing these competences, which is an additional bonus to the company.

And finally, if a company is not traded in the stock market, but considering to issue its shares, an IC report can be a great tool in optimising the process.

Practical Tips and Lessons Learned

1. The first phase in the process of preparing an intellectual capital report is the formation of a heterogeneous steering committee headed by a senior executive.
2. Make sure you have diversity of input from all the employees to contribute to the process by conducting a Knowledge Café workshop.
3. Create a vision and core values, and communicate them to serve as a compass for creative energies.

4. Identify the core competencies needed to realise the vision, and define the key success factors for each of the company's core competencies.
5. Identify tangible indicators for each key success factor, and categorize them into one of the four capital areas: human, process, customer, and renewal and development capital.
6. Finally, collect and analyse the relevant information and knowledge about the different indicators from the company's existing databases, and formulate them into an in-depth intellectual report of your organisation.

References

Amidon, D.M. (In press). *Knowledge Innovation Strategy.*

Danya Cebus. (1999). *The intellectual capital report of Danya Cebus.* Israel.

Drucker, P. (1993). *Post capitalist society.* New York: HarperBusiness.

Dvir, R., Pasher, E., & Roth, N. (Eds.). (2002). *From knowledge to value, the innovation cube.* Israel: Edna Pasher PhD & Associates.

Edna Pasher PhD & Associates. (1999). *The intellectual capital report of the state of Israel, a look to the future: The hidden values of the dessert.* Israel

Edvinsson, L., & Malone, M.S. (1997). *The intellectual capital.* New York: Harper Collins.

Hamel, G. (1999). Bringing Silicon Valley inside. *Harvard Business Review,* (September–October).

Hamel, G. (2000). *Leading the revolution.* Boston: Harvard Business School Press.

Hamel, G. (2001). Companies need to synthesize industrial and post-industrial virtues. *CIO Magazine,* (August 15).

Hamel, G., & Prahalad, C.K. (1994). *Competing for the future.* Boston: Harvard Business School Press.

Lev, B. (2001). *Intangibles: Management, measurement, and reporting.* Washington, DC: Brookings Institute Press.

Nonaka, I., & Takeuchi, H. (1995). *The knowledge-creating company: How Japanese companies create the dynamics of innovation.* New York: Oxford University Press.

Chapter V

Capturing and Reusing Building Maintenance Knowledge: A Socio-Technical Perspective

Patrick S.W. Fong
Hong Kong Polytechnic University, Hong Kong

Ka-chi Wong
Raymond Chan Surveyors Ltd, Hong Kong

Abstract

Knowledge is regarded as the most important asset for sustainable success in today's knowledge-based economy, and it has become the main competitive tool for many businesses. Proper capture and reuse of knowledge reduces the risk of "reinventing the wheel." Building surveyors play a major role in building maintenance. Owing to the unique nature of building maintenance, building surveyors need sufficient knowledge and experience to facilitate their decision-making process. Apart from gaining this knowledge and experience by direct participation, learning from others is a sound alternative. However, insufficiencies have been found in the current practice on this aspect, and thus the aim of this chapter is to study whether a proposed Web-based prototype knowledge management system would be a feasible solution for capturing and reusing knowledge and experience in building maintenance. A questionnaire survey was conducted in this research to study the opinions of professional building surveyors on the capture and reuse of knowledge and experience in building maintenance, as well as the requirements of a proposed Web-based prototype system.

Introduction

With the start of the new century, our economy has increasingly become a knowledge-based economy. In such an economy, knowledge is regarded as the most important asset, with many people in the world believing that knowledge is more vital for sustainable success than assets such as capital and land (Kelleher & Levene, 2001).

A vast amount of knowledge is involved in many professions, and managing that knowledge is a critical task. As Rowley (1999) points out: "Knowledge has become the main competitive tool for many businesses" (p. 416). Building surveying, one of the key professions in the Hong Kong construction industry, is no exception.

Building surveyors act as "building doctors" to diagnose defects in buildings, and play a major role in the management and maintenance of buildings after they have been completed (Hong Kong Institute of Surveyors, 2002). In fact, building maintenance is a complex process requiring a vast amount of knowledge, both explicit and tacit, particularly in light of the unique nature of each project. Apart from acquiring such knowledge and experience by direct participation, building surveyors need to learn from others.

In the current situation, it is difficult for a building surveyor outside of a project team to gain the tacit knowledge required and understand the rationale for decisions that have been made, simply by referring to materials kept after the completion of the project, such as written reports and construction drawings. However, it is argued that the "capture, transfer, and reuse of the project knowledge are critical" (Kamara, Augenbro, Anumba, & Carrillo, 2002, p. 63), as properly reusing the knowledge and experience gained while working on a similar project will reduce the risk of "reinventing the wheel."

Learning Objectives

1. To study how knowledge and experience in building maintenance can be reused in a more effective and efficient way by applying the principles stated in knowledge management.
2. To investigate the opinions of professional building surveyors on the capture and reuse of knowledge and experience in building maintenance.
3. To study the requirements of a proposed Web-based prototype system for capturing, storing, transforming, and retrieving knowledge and experience.

Reusing Knowledge and Experience

As with the reusing of experiences in experience management, reusing knowledge is the ultimate goal of knowledge management. In order to reuse knowledge and experience from previous projects, it is necessary to recapture, transfer, and store the knowledge involved, as well as to devise a means for retrieving this knowledge.

Hicks, Culley, Allen, and Mullineux (2002) mention that reusing knowledge could be the repeated utilisation of the same information for the same or similar tasks, or the further use or extension of the same information for a different purpose. In their study of engineering design, they stated that reusing information to evaluate previous decision processes would require the capturing of additional information on the alternatives to the decision, and on the various considerations involved. Also, the aim of reusing knowledge is to make available the knowledge of individuals or organisations acquired from previous design activities for better-informed decision-making processes in future design activities.

Yeung and Holden (2000) emphasise that the reusing of knowledge goes beyond simply reusing the tangible outputs. As knowledge gained by people is dynamic and context-dependent, this means that it cannot be generalised. In reusing knowledge, it is important to employ a holistic method involving the sharing of knowledge. Thus, "knowledge reuse requires situated understanding and interpretation of the lesson learnt, organisational best practices and capabilities" (Yeung & Holden, 2000, p. 24/2).

In addition, tacit knowledge is the key component in the reusing of knowledge. This is because "tacit knowledge underpins the understanding, evaluation, and application of all knowledge," which "provides the clue allowing individuals to adapt the conceptual model in their minds and clarifies the contextual correlations between the original and the current situation" (Yeung & Holden, 2000, p. 24/7).

In considering the reusing of experience and knowledge, Bergmann (2002) suggests two basic scenarios. In one scenario, a person who has gained experience and knowledge in a previous problem-solving situation will act to provide the knowledge to another person, the user, who will in turn interpret the knowledge obtained in the context of the new problem-solving task. In another scenario, a group of people will share experience and knowledge for mutual benefit, via a common experience pool or experience bus.

Knowledge Required in Building Maintenance

Understanding the nature of building maintenance will facilitate the discussion on current practices in reusing knowledge and experience.

In general, maintenance is "the keeping, holding, sustaining, or preserving of a building and its services at an acceptable standard to enable its function" (Brett, 1997, p. 111). A more comprehensive definition is "building maintenance is the combination of all technical and associated administrative actions to retain an item or restore it to a state in which it can perform its required function" (Allan, 1993, p. 8).

Building maintenance is usually a short-term and discontinuous process (Zavadskas, Bejder, & Kaklauskas, 1998). Professional building surveyors play a major role in this process. They carry out inspection and condition surveys of existing buildings to diagnose defects, and they prepare schedules of dilapidation. They provide expert advice on remedial work, designs, and the performance of materials. In addition, they supervise subsequent construction work to bring the building to a state of good repair and into compliance with the relevant legal requirements (Hong Kong Institute of Surveyors, 2002). All of these activities in building maintenance involve the consideration of many alternative decisions (Horner, El-Haram, & Munns, 1997). To facilitate the making of decisions, it is necessary that the building surveyors involved have the relevant knowledge and experience.

Current Practices in Reusing Knowledge and Experience in the Construction Industry

Although the practice of capturing, transferring, and reusing knowledge gained from building maintenance projects in the building surveying field has not been mentioned separately in the literature, it can be determined from studies on practices in the construction industry, to which the field of building surveying actually belongs.

Egbu and Botterill (2002) state that the major challenge for construction organisations is to transfer knowledge and information across projects, and that it is common for specialist and technical knowledge to become lost from one project to the next.

A study by Kamara et al. (2002) on current practices in the construction industry regarding capturing, transferring, and reusing project knowledge shows that the industry is not very effective in capturing lessons learned from projects. The usual methods employed include reassigning people from one project to the next, and using standards and best practice guides. The reusing of knowledge is mainly done by people on an individual basis, and the risk is that such knowledge will be lost when people with the knowledge leave the organisation.

Wetherill, Rezgui, Lima, and Zaril (2002) also point out that there are limitations in the construction industry in capturing and retrieving knowledge gained by working on a project. Some of the reasons for the limitations include the fact that the intent behind the decision is often not recorded or documented, the lessons learned from projects are not well organised, and the construction knowledge resides in the minds of the participating personnel.

In addition, Wong (2001) states that knowledge gained in a construction project is usually stored in a set of documents, which is unorganised and difficult to find. Also, knowledge is usually stored in the minds of the people who have developed it, which makes access to it difficult when required.

Inadequacies of the Current Practices

The current practices of reusing knowledge and experience in the construction industry have been found to be inadequate, and similar to those mentioned by Disterer (2002) regarding IT projects.

The documentation of a project is stored without efforts being made to retain essential knowledge and experience for later reuse. Certain knowledge and experience becomes lost when the project is completed. The lessons learned in previous projects cannot therefore be reused, and the consequence is that "actions and decisions that caused problems and errors may be repeated" (Disterer, 2002, p. 513). Clearly, the current situation needs to be improved.

Data Collection

A questionnaire survey was used to investigate the opinions of professional building surveyors on the capturing and reusing of knowledge and experience in building maintenance; and to study the requirements of a proposed Web-based prototype system for capturing, storing, transforming, and retrieving knowledge and experience.

Table 1. The specific purpose of each part of the questionnaire

Part I: Background Information
- To confirm that the respondent has worked in the area of building maintenance - To find out the number of years the respondent has been involved in the area
Total number of questions: 2
Part II: Capturing and Reusing Knowledge and Experience
- To study opinions on the nature of knowledge in building maintenance - To study opinions on the sharing of knowledge and experience - To study opinions on current practices in reusing knowledge and experience
Total number of questions: 11

The advantage of using a questionnaire is that the range of participants is greater than in an interview. In this case, a questionnaire survey allowed the opinions of a larger number of professional building surveyors to be collected in a more limited period of time than would have been possible with interviews.

The questionnaire consisted of 13 questions, which were divided into two parts. Each part consisted of different numbers of questions. The questions in each part were designed to collect information on a specific area from the respondent, as summarised in Table 1.

The questions in this questionnaire were in a six-point scale format; this six-point scale ranged from one to six, representing "strongly agree" to "strongly disagree" respectively. The respondent was asked to give the rating he or she considered most appropriate for a particular question.

Sampling and Response Rate

Professional building surveyors working in two government departments, namely the Architectural Services Department (ASD) and the Housing Department (HD) were selected to be the potential respondents in this questionnaire survey.

According to the nature of their work, most of the professional building surveyors in the ASD and HD participate in the maintenance of government buildings and buildings built by the Housing Authority, respectively. Therefore, these professional building surveyors were expected to have the relevant experience in building maintenance and to be able to provide valuable information in the questionnaire survey.

Although there are professional building surveyors in the Buildings Department, they were not considered for the questionnaire survey because they mainly work in the area of building control, not building maintenance. Nor were professional building surveyors practicing in the private sector selected, because these surveyors work in various companies, from developers, surveying firms, and architectural firms to management companies. There is no systematic record containing information on surveyors in these firms, so it is difficult to collect their opinions by sending them questionnaires.

In view of this, only the professional building surveyors in the ASD and HD were targeted. A list of professional building surveyors in these two government departments was formulated by searching the Telephone Directory of the Government of the Hong Kong Special Administrative Region and Related Organisations, and checking with the list of members of the Hong Kong Institute of Surveyors. Random sampling was used to form the sample by referring to the list.

Fifty questionnaires were sent to the professional building surveyors in the sample. Of these, half work in the Architectural Services Department, and the other half in the Housing Department. Out of 50 questionnaires sent to the proposed respondents, 31 completed questionnaires were returned, for a response rate of 62%. Gillham (2000) states that "as a rule of thumb, a 30% return [of the questionnaire survey] has to be seen as fairly satisfactory, and more than 50% is good" (p. 14). Based on this assessment, the response rate in this questionnaire survey can be considered to be good.

Figure 1. Distribution of respondents (according to their years of experience in building maintenance)

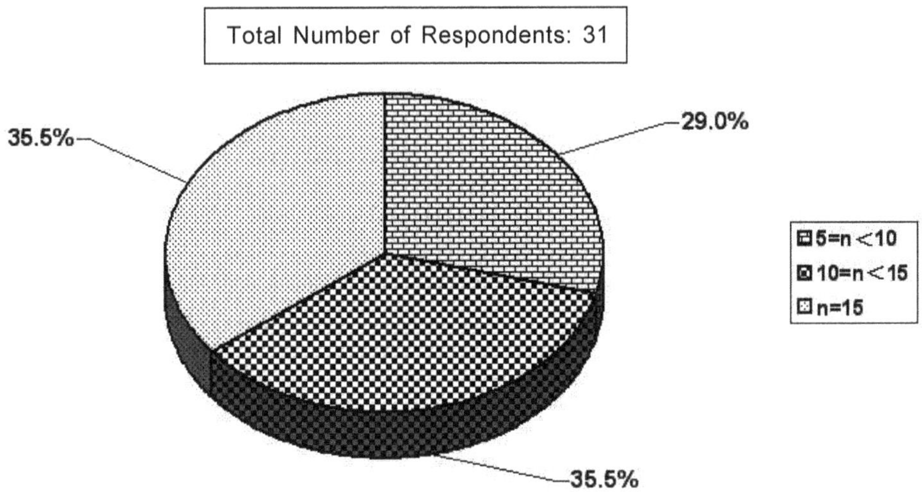

Survey Findings and Analysis

This section presents and analyses the results of the questionnaire survey, and provides a summary of the opinions of professional building surveyors regarding the capture and reuse of knowledge and experience in building maintenance.

Background Information

All of the returned questionnaires were completed by professional building surveyors who have worked in the area of building maintenance.

These 31 professional building surveyors have been involved in building maintenance and have varying years of experience. Eleven have been involved in the area for more than 15 years, and the same number have been involved for 10 to 15 years, while the remaining nine have been involved for five to 10 years (Figure 1). From the respondents' feedback, it can be seen that all respondents have substantial experience in building maintenance, making their replies to this survey useful.

Capture and Reuse of Knowledge and Experience

Understanding the nature of knowledge and experience in building maintenance, as well as the opinions of professional building surveyors regarding the capture and reuse of knowledge and experience, could enhance the development of the proposed Web-based prototype knowledge management system.

Nature of Knowledge and Experience in Building Maintenance

The results of the survey regarding the nature of knowledge and experience in building maintenance are summarised in Table 2. From the mean scores of all the four statements in Table 2, the results lay between "strongly agree" and "agree." We can interpret this as meaning that respondents view building maintenance as a knowledge-intensive discipline where experience gained from previous projects can enhance future practice. Though some building maintenance knowledge has been codified into explicit knowledge such as codes of practice, standards, or guidelines, practitioners find that they also need to rely on their own personal experience accumulated through years of practice. This is something that cannot be expressed: It is an embodied skill or something that can only be learned by socialisation, accrued by experience, or embedded. It is almost impossible to reproduce in a document or database. Examples include engineering judgment, 'rules of thumb', and so forth. To use Nonaka and Takeuchi's term (1995), it is referred to as tacit knowledge.

It is interesting to note that practitioners with longer working experience tend to agree more that their work requires lots of knowledge, which includes both explicit and tacit (Table 3). Less than 4% of all the respondents disagree that building maintenance requires a lot of knowledge. This may be due to the fact that they see their daily work as routine, not challenging enough and requiring little knowledge input.

Table 2. Nature of knowledge in building maintenance

Item	Mean	Standard deviation (S.D.)
• Building maintenance requires a lot of knowledge	1.65	0.84
• The knowledge required in building maintenance is not covered completely in codes of practice and other well-known standards and guidelines	1.97	1.05
• Some knowledge of building maintenance is built up from personal experience	1.71	1.13
• The knowledge gained from experience could enhance similar works in building maintenance	1.61	0.99

Key: in 6-point scale - 1: Strongly Agree - 6: Strongly Disagree

Table 3. Opinions on knowledge in building maintenance

Building maintenance requires a lot of knowledge				
Group of respondents with years of experience (n)	Summation of "Strongly Agree", "Agree" and "Slightly Agree"	"Strongly Agree"	"Agree"	"Slightly Agree"
$5 \leq n < 10$	88.89%	55.56%	33.33%	0%
$10 \leq n < 15$	100.00%	36.36%	54.55%	9.09%
$n \geq 15$	100.00%	54.55%	45.45%	0%
All	96.78%	48.39%	45.16%	3.23%

Tables 4 and 5 show that almost 94% of the respondents agree that the knowledge required in building maintenance is not completely covered in codes of practice and other well-known standards and guidelines, and that some knowledge is built up from personal experience. This shows that building surveyors practicing in building maintenance cannot simply obtain all their required knowledge by referring to published materials. Research by Fong (2003) reveals that a practitioner's social network can be an important vehicle for information and knowledge exchange, with people within the network generating knowledge through their experience or issues that they have come across previously. Thamhain (1991) investigated 220 project managers and concluded that 85% of all skill developments are derived from experience, that is, either from experiential learning or observing other peoples' practices. The remainder are derived from reading, seminars, and schooling. From their survey of 116 project managers of companies who are members of the European Construction Institute, McCaffer and Adham (1992) discovered that the source from which project managers gain the most knowledge and skills is on-the-job experience. Though the two studies are not directed at building maintenance professionals, one thing that project managers and building surveyors have in common is that both work in project-based industries where knowledge gains are useful for both future and concurrent projects.

In addition, most of the respondents agree that the knowledge gained from experience can enhance similar works in building maintenance (Table 6). The personnel involved in a project can be referred to as "knowledge carriers" whose knowledge learned from one project can be applied to others. By the same token, knowledge will be lost if the person retires or leaves the company. This is one of the key factors that motivates most

Table 4. Opinions on the sources of knowledge in building maintenance

The knowledge required in building maintenance is not covered completely in codes of practice and other well-known standards and guidelines				
Group of respondents with years of experience (n)	Summation of "Strongly Agree", "Agree" and "Slightly Agree"	"Strongly Agree"	"Agree"	"Slightly Agree"
5 ≤ n < 10	77.77%	22.22%	44.44%	11.11%
10 ≤ n < 15	100.00%	36.36%	63.64%	0.00%
n ≥ 15	99.99%	36.36%	45.45%	18.18%
All	93.55%	32.26%	51.61%	9.68%

Table 5. Personal experience helps build up knowledge

Some knowledge of building maintenance is built up from personal experience				
Group of respondents with years of experience (n)	Summation of "Strongly Agree", "Agree" and "Slightly Agree"	"Strongly Agree"	"Agree"	"Slightly Agree"
5 ≤ n < 10	88.89%	55.56%	22.22%	11.11%
10 ≤ n < 15	100.00%	63.64%	36.36%	0.00%
n ≥ 15	90.91%	54.55%	18.18%	18.18%
All	93.55%	58.06%	25.81%	9.68%

companies to capture or codify their employees' knowledge for future use/reuse: it is costly to train a new staff member. However, it must be noted that knowledge without context is futile. Hence, context plays an important role in producing knowledge. It is one of the key aspects that is considered when designing the prototype building maintenance knowledge management system.

Table 6. Knowledge and experience enhance similar works

Knowledge gained from experience can enhance similar works in building maintenance				
Group of respondents with years of experience (n)	Summation of "Strongly Agree", "Agree" and "Slightly Agree"	"Strongly Agree"	"Agree"	"Slightly Agree"
$5 \leq n < 10$	88.89%	55.56%	33.33%	0.00%
$10 \leq n < 15$	100.00%	54.55%	45.45%	0.00%
$n \geq 15$	100.00%	54.55%	36.36%	9.09%
All	96.78%	54.84%	38.71%	3.23%

Willingness to Share Knowledge and Experience in Building Maintenance

The results of the survey regarding the willingness of professional building surveyors to share knowledge and experience in building maintenance are summarised in Table 7. The majority of the respondents agree that sharing building maintenance knowledge and experience will benefit each other, as the knowledge required for each project is so diverse that no one will have all the knowledge required for the job. In fact, this is another objective of the prototype knowledge management system: to allow people to share and tap into the knowledge of others. In addition, from the results, people see knowledge gained as a personal asset which can enhance their employability in the job market or help them attain promotion within their companies. Unfortunately, when people see knowledge as having a price tag, this also creates a knowledge-hoarding culture in which they see "knowledge as power," or one perceives that sharing one's knowledge with one's colleagues will undermine one's position within the company, or worse still, the company may get rid of the one who shared. Generally, one of the key missions for most companies is to cultivate a knowledge-sharing culture in order for knowledge management to be successful. Another major barrier for companies to cross is trying to capture employees' knowledge into their database or repository.

Regarding the statement whether building surveyors in practice like to share their knowledge and experience in building maintenance with others, the mean score falls to between "agree" and "slightly agree." This shows that respondents generally feel that

Table 7. Willingness to share knowledge in building maintenance

Item	Mean	Standard deviation (S.D.)
• Knowledge gained from experience is a kind of personal asset	1.68	1.30
• The sharing of knowledge and experience benefits you and others	1.55	0.96
• Building surveyors in practice like to share their knowledge and experience in building maintenance with others	2.45	1.15

Key: in 6-point scale - 1: Strongly Agree - 6: Strongly Disagree

professionals are reluctant to share their professional know-how with others. This result also contradicts the other statement. On one hand, people know that sharing knowledge and experience can be mutually beneficial as no one has all the knowledge. On the other hand, they see knowledge as a personal asset and will guard it carefully. Currently, there is an increasing number of publications (e.g., Cohen & Prusak, 2001; Anand, Glick, & Manz, 2002) emphasising that social capital should be used to make knowledge management work.

Figure 2 illustrates the fact that most of the respondents think that knowledge gained from experience is a kind of personal asset, with a very small percentage disagreeing on this statement. This has motivated people to learn through different means in order to increase their value within their companies.

Figure 3 reveals that nearly all of the professional building surveyors who responded strongly agree or agree with the point of view that sharing knowledge and experience would benefit them and others. Since technologies, processes, methods, and tools are constantly changing, it is in this respect that some of the knowledge we have acquired in the past becomes obsolete or no longer useful in the current context. One good example to illustrate the change in technology is the way practitioners examine cracks in building structures. In the past, people relied very much on visual examination, but now professionals can deploy infra-red or ultrasound instruments to do the job for them.

It is important to note from Figure 4 that although the majority of professional building surveyors agree that building surveyors in practice like to share their knowledge and experiences in building maintenance with others, five building surveyors (accounting for 16% of the total) expressed a contrary opinion. This is perhaps a genuine reflection of their personal experience that not all building surveyors like to share their personal intangible assets with others, for fear of losing their competitive edge.

In general, Table 8 shows that professional building surveyors with fewer years of experience were more likely than those with more experience to agree with the proposition

Figure 2. The knowledge gained from experience is a kind of personal asset

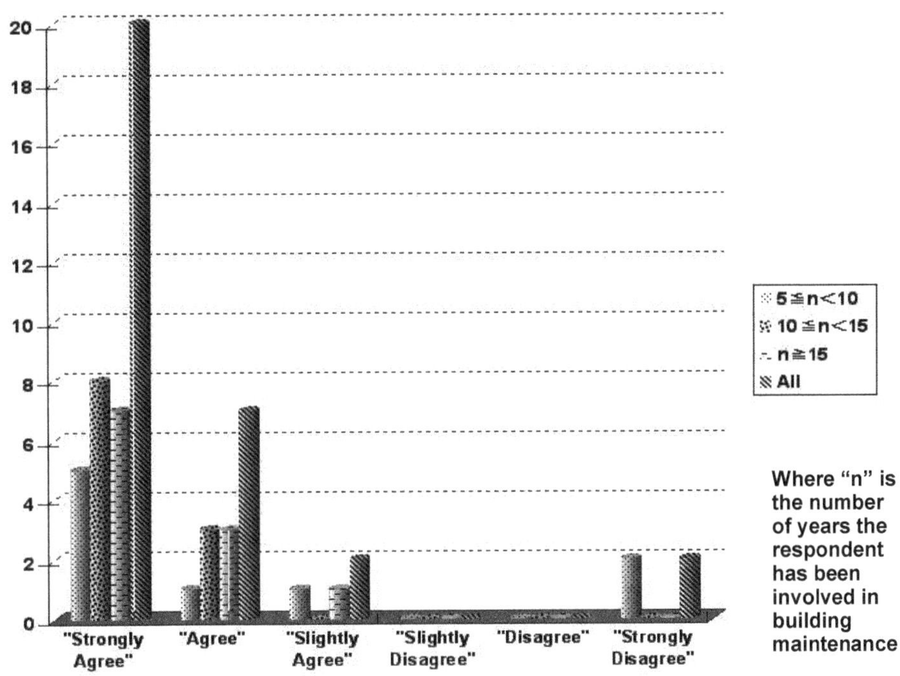

that building surveyors in practice like to share their knowledge and experience. This indicates that professionals with more accumulated experience think that people are more reluctant to share their knowledge with others. This can be explained by the fact that people see knowledge as an important personal asset for advancing within their organisations, as well as an asset to offer to their clients. This situation explains why they become "knowledge silos" within their organisations: Unless they are willing to share their knowledge and experience, their companies are taking the risk of losing that knowledge when they leave the firm or upon their retirement. In fact, there is a bigger chance that colleagues will repeat the same mistakes if the knowledge learned from previous projects is not shared.

Current Practice of Knowledge and Experience Management

The results of the survey regarding current practices in knowledge and experience management in building maintenance are summarised in Table 9. Regarding reassigning

Figure 3. The sharing of knowledge and experience would benefit you and others

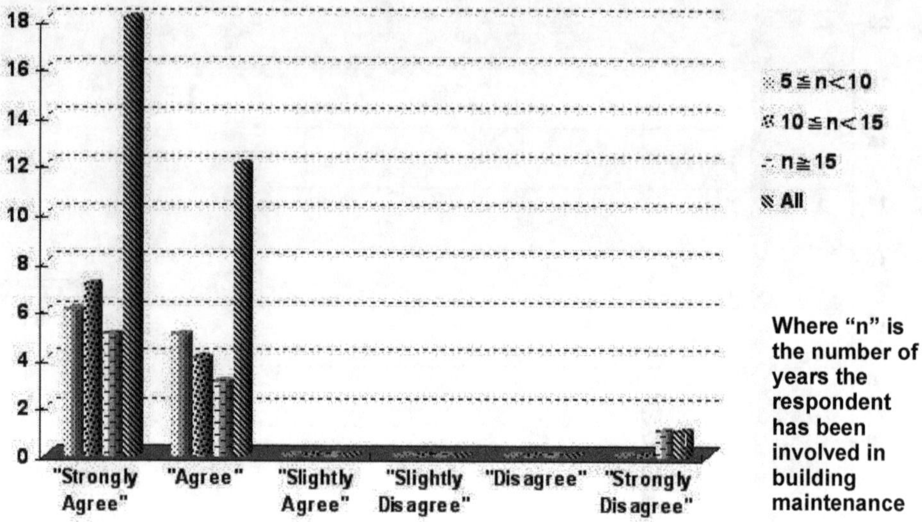

Figure 4. Building surveyors like to share their knowledge and experience

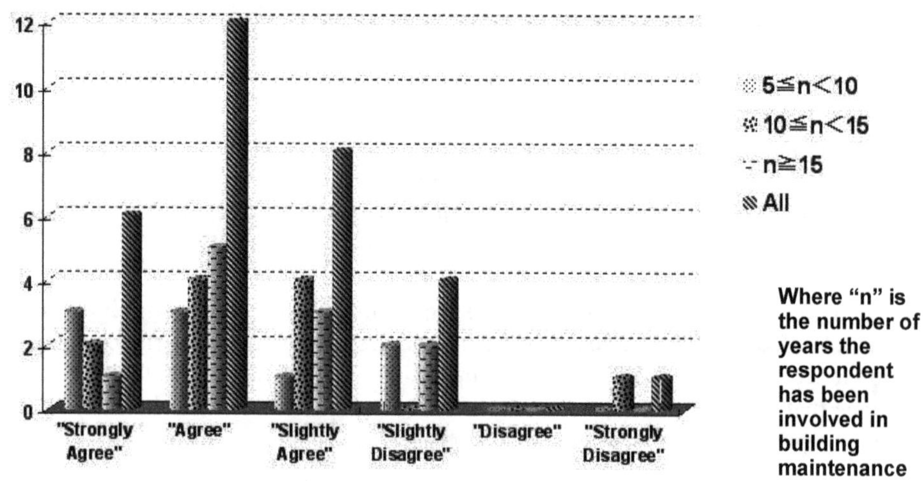

personnel to other similar projects, the mean score lies between "agree" and "slightly agree," implying that it may not be as common as it should be in project-based organisations. Currently, two principal criteria appear to be used in selecting members for a project. The first criterion relates to past and similar experience. For example, if a building surveyor has previous experience in spalling concrete structures, he or she

Table 8. Building surveyors like to share their knowledge and experience

Building surveyors in practice like to share their knowledge and experience in building maintenance with others		
Group of respondents with years of experience (n)	Mean	Standard deviation (S.D.)
$5 \leq n < 10$	2.22	1.20
$10 \leq n < 15$	2.55	1.37
$n \geq 15$	2.55	0.93
All	2.45	1.15

Key: in 6-point scale - 1: Strongly Agree - 6: Strongly Disagree

Table 9. Current practices in knowledge and experience management in building maintenance

Item	Mean	Standard deviation (S.D.)
• The most common way in the industry to reuse the knowledge and experience of a participant in building maintenance is to reassign that person to other similar projects	2.74	0.77
• There are enough ways in the profession to enhance the sharing of knowledge and experience in building maintenance	4.00	1.06
• At present, the management of knowledge and experience in building maintenance is regarded as an important aspect in the profession/industry	3.26	1.53
• The management of knowledge and experience in building maintenance would be beneficial to the profession	1.48	0.81

Key: in 6-point scale - 1: Strongly Agree - 6: Strongly Disagree

would be a logical choice for appointment to a project where this is a key requirement. However, there is a possibility that candidates suitable for certain projects may be irrevocably engaged elsewhere and unavailable for immediate involvement in any new project. With such scenarios in mind, some companies prefer not to allow too much specialisation, since this can limit staffing flexibility. Moreover, should specialists leave, they take away a wealth of knowledge and experience that may be lost to another, rival consultancy. Some companies prefer to establish wide-ranging diversity in staff members so that certain employees have similar, mutually interchangeable capabilities and skills. The second criterion used to select project team members was that of availability. This can be called a "pool system of managing staff resources"—whoever is available will be allocated to the project.

Respondents slightly disagree on the statement that there are enough ways in the profession to enhance the sharing of knowledge and experience in building maintenance. This may be due to the fact that there are not sufficient channels for professionals to share their knowledge in this specific field. One may argue that the professional society for surveyors can be a channel for disseminating and sharing knowledge. However, building maintenance is only part of the scope of their professional service. In addition, it is not easy to gather all the like-minded people together to share their practices. Secondly, the competitive nature of the construction industry may dissuade people from sharing with professionals from rival firms.

The management of knowledge and experience in building maintenance is regarded as of neutral importance in the profession/industry: There are almost equal percentages of people who support or do not support this statement. In Hong Kong, due to the enormous number of facilities that are constructed, the effort of maintaining and keeping them in good condition is of paramount importance. Since the outbreak of Severe Acute Respiratory Syndrome (SARS), maintaining healthy buildings has become even more important. Perhaps professional societies and universities, in collaboration with practitioners, can act as a driving force in promoting the sharing of knowledge and experience in proper building maintenance.

Generally, respondents range between "strongly agree" and "agree" regarding whether the management of knowledge and experience in building maintenance would be beneficial to the profession. Doing so can enhance one's experience through learning from others' experience or best practice. In addition, it broadens the professional horizon of practitioners to know that there may be better ways of doing things.

Figure 5 shows that most of the respondents agree or slightly agree that reassigning personnel to other similar projects is the most common way of reusing the knowledge and experiences of a practitioner in building maintenance.

Figure 6 reveals that respondents hold varying views on whether there are enough ways in the profession to enhance the sharing of knowledge and experience in building maintenance. The same number of respondents chose "slightly agree," "slightly disagree," and "disagree," and in total they accounted for 87.1% of all respondents.

Figure 7 shows that respondents hold widely varying views on the emphasis placed on the management of knowledge and experience by practitioners in the profession. Of the 31 professional building surveyors who responded to the survey, 54.8% expressed at least slight agreement with the statement, while 45.2% of them expressed at least slight disagreement. The standard deviation also indicated a spread of opinions on the matter (Table 10).

The majority of the respondents agree that the management of knowledge and experience in building maintenance would be beneficial to the profession (Figure 8). In fact,

Figure 5. Reassignment of personnel to similar projects is the most common method of reusing knowledge and experience in building maintenance

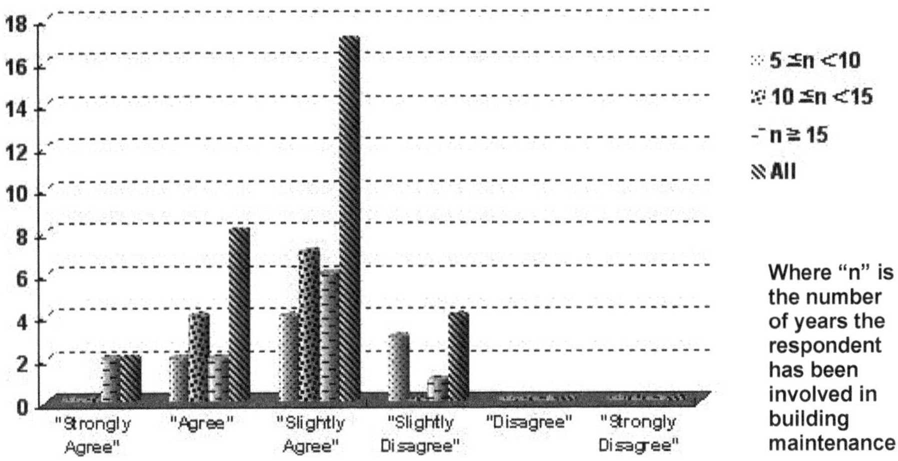

Figure 6. There are enough ways in the profession to enhance the sharing of knowledge and experience in building maintenance

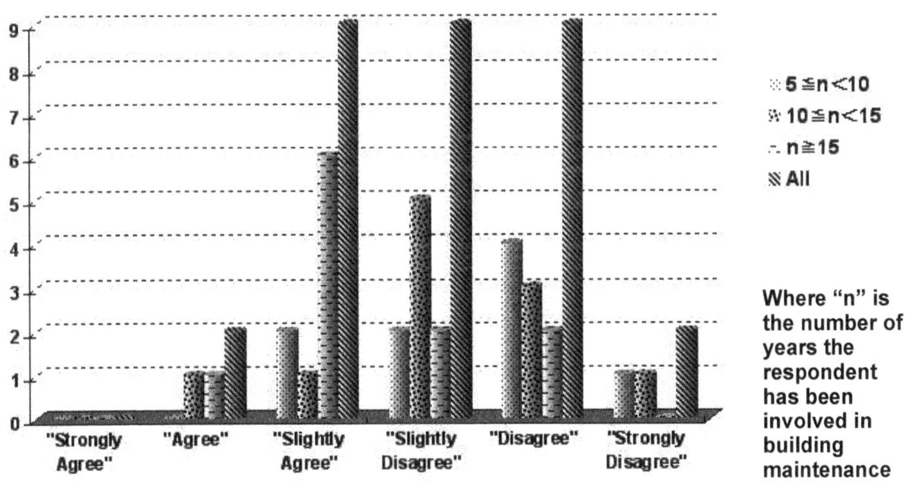

Figure 7. At present, knowledge management in building maintenance is regarded as an important aspect of the profession

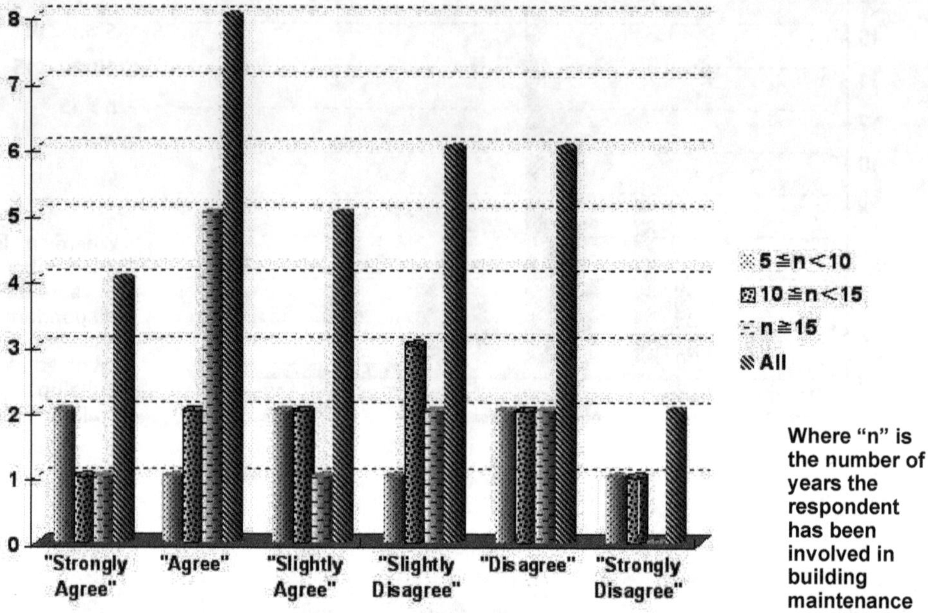

Table 10. Current emphasis on knowledge and experience management

At present, knowledge management in building maintenance is regarded as an important aspect of the profession		
Group of respondents with years of experience (n)	Mean	Standard deviation (S.D.)
5 ≤ n < 10	3.33	1.80
10 ≤ n < 15	3.55	1.51
n ≥ 15	2.91	1.38

Key: in 6-point scale - 1: Strongly Agree - 6: Strongly Disagree

Figure 8. Management of knowledge and experience in building maintenance would be beneficial to the profession

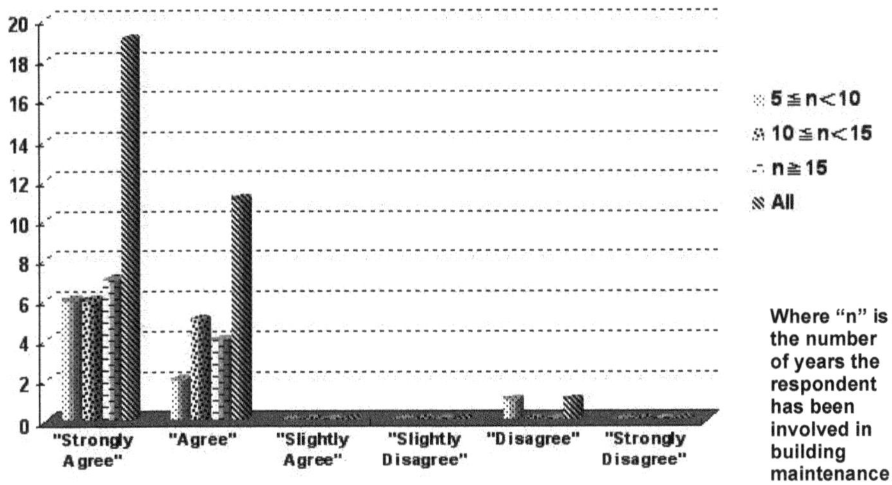

professionals would like to improve on their professional service rather than keep on "reinventing the wheel" or repeating the same mistakes.

Future Trends

This study could act as a cornerstone in research on capturing and reusing knowledge and experience in building maintenance.

The development of a Web-based knowledge management system should aim to facilitate the retrieval of knowledge and experience. This means that the system should be designed with proper functions to enable retrieval as well as sharing. Since knowledge in building maintenance is context specific and cannot be generalised; this implies that the system should not be designed to provide any general rules in building maintenance. Instead, it should contain enough cases on building maintenance to allow users to find relevant ones and retrieve the knowledge and experience of other practitioners in order to facilitate their own decision-making process in building maintenance.

Figure 9 shows the main page of the proposed Web-based building maintenance knowledge management system that has been designed. The primary aim of the prototype system is to facilitate the reuse of knowledge and experience in building maintenance. Through capturing, storing, and transforming knowledge and experience in building maintenance, knowledge and experience can be reused by retrieving from the system. Of

Figure 9. Main page of the prototype system

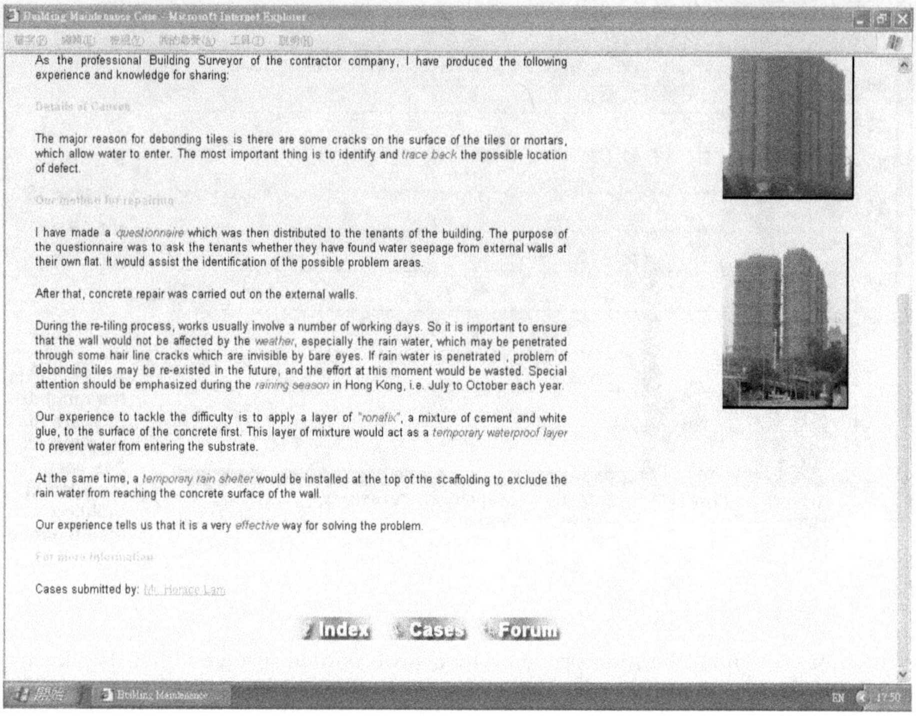

course, the system will only be useful when practitioners start to share successful as well as failed experiences.

In addition, building maintenance involves the participation of many professionals—not only professional building surveyors, but also building services engineers, architects, facilities managers, and so forth. Professionals usually need to make their decisions based on the decisions of the other professionals involved, thus it would be beneficial to all professions if the system could consider the capture and reuse of knowledge and experience from different professionals in building maintenance. Further studies could be carried out on the subject. In addition, the knowledge management system could be used by other professionals during the design stage to prevent the repetition of mistakes that have caused problems in a completed building due to an inappropriate design.

The fragmented way in which the construction industry is organised means that efficiency in project delivery is less than expected, resulting in dissatisfied clients and low profitability for construction firms. The temporary nature of project organisations also prevents the effective transfer of knowledge from one project to the next. Furthermore, the multi-profession nature of project teams makes knowledge management even more problematic and has significant implications for organisational learning (Bresnen, Edelman, Newell, Scarbrough, & Swan, 2003). It is these factors that motivate the adoption of knowledge management through the project life cycle in order to improve quality, business performance, competitiveness, efficiency of project delivery, client satisfaction, decision making, innovation, and project knowledge retention and learning.

Conclusion

Knowledge is regarded as the most important asset for sustainable success in today's knowledge-based economy. Many professions have started to place more emphasis on the management of knowledge in their industry, and many researchers and scholars in different fields have carried out studies on this aspect.

As was found in the results of the survey, building maintenance is by nature a complex process that requires a lot of knowledge. The knowledge required cannot be obtained by simply referring to published materials such as codes of practice, standards, and guidelines. Instead, some knowledge is built up by personal experience, and becomes part of the personal assets of a practitioner in the field. A similar point of view has been expressed by Wetherill et al. (2002) that knowledge in the field usually resides in the minds of the individuals working within the domain.

The implication is that whether other practitioners can learn from the knowledge and experiences of others will depend greatly on the willingness of the practitioner who has acquired the knowledge and experience to share them with others. In general, the survey found that most of the respondents were willing to do so, as they thought that the sharing of knowledge and experience would benefit them and others, and would contribute to the development of the whole profession.

However, it was revealed that there were varying opinions on whether building surveyors in practice like to share their knowledge and experience with others. This uncertainty may be a possible barrier to the sharing of knowledge and experience in building maintenance. Ives, Torrey, and Gordon (2000) state that the sharing of knowledge cannot be fostered without genuine trust. If building surveyors do not have a clear idea of whether or not others are willing to share their expertise, it will be difficult to build up the required level of trust.

In addition, similar to the findings by Karama et al. (2002) on current practices in knowledge management in the architecture, engineering, and construction industries, the survey revealed that most of the respondents agreed that reassigning practitioners with the relevant knowledge and experience to other similar projects is the most common way of reusing knowledge and experience in building maintenance, but that somehow the actual practice is different from the expectation. Also, most of the respondents agreed that there are insufficient ways of enhancing the sharing and reusing of knowledge and experience in current practices.

In view of this, it would be worth considering devising a system to facilitate and enhance the reusing and sharing of knowledge and experience in building maintenance.

Practical Tips and Lessons Learned

- Any knowledge management system should be designed to enable capture and retrieval, as well as sharing.
- Whether one can learn from the knowledge and experiences of others will depend greatly on one's willingness to share one's knowledge with others. In simple words, the more you share, the more you receive.
- Knowledge usually resides in the minds of the individuals, and it is through sharing and socialisation that this knowledge is elicited.
- Sharing of knowledge can be fostered by cultivating social capital.
- Knowledge is context specific and cannot be generalised.
- Failures and mistakes provide valuable lessons to be learned.

References

Allan, D. (1993). What is building maintenance? *Facilities Management, 2*(3), 7-12.

Anand, V., Glick, W.H., & Manz, C.C. (2002). Thriving on the knowledge of outsiders: Tapping organizational social capital. *Academy of Management Executive, 16*(1), 87-101.

Bergmann, R. (2002). *Experience management: Foundations, development methodology and Internet-based applications.* Berlin: Springer.

Bresnen, M., Edelman, L., Newell, S., Scarbrough, H., & Swan, J. (2003). Social practices and the management of knowledge in project environments. *International Journal of Project Management, 21*(3), 157-166.

Brett, P. (1997). *An illustrated dictionary of building: An illustrated reference guide for practitioners and students.* Oxford: Butterworth-Heinemann.

Cohen, D., & Prusak, L. (2001). *In good company: How social capital makes organizations work.* Boston: Harvard Business School Press.

Disterer, G.. (2002). Management of project knowledge and experiences. *Journal of Knowledge Management, 6*(5), 512-520.

Egbu, C.O., & Botterill, K. (2002). Information technologies for knowledge management: Their usage and effectiveness. *ITcon, 7,* 125-136.

Fong, P.S.W. (2003). Knowledge creation in multidisciplinary project teams: An empirical study of the processes and their dynamic interrelationships. *International Journal of Project Management, 21*(7), 479-486.

Gillham, B. (2000). *The research interview.* London: Continuum.

Hicks, B.J., Culley, S.J., Allen, R.D., & Mullineux, G. (2002). A framework for the requirements of capturing, storing and reusing information and knowledge in engineering design. *International Journal of Information Management, 22,* 263-280.

Hong Kong Institute of Surveyors. (2002). *The Hong Kong Institute of Surveyors directory & annual report 2000/2001.* Hong Kong: Hong Kong Institute of Surveyors.

Horner, R.M.W., El-Haram, M.A., & Munns, A.K. (1997). Building maintenance strategy: A new management approach. *Journal of Quality in Maintenance Engineering, 3*(4), 273-280.

Ives, W., Torrey, B., & Gordon, C. (2000). Knowledge sharing is a human behaviour. In D. Morey, M. Maybury, & B. Thuraisingham (Eds.), *Knowledge management: Classic and contemporary works* (pp. 99-138). London: MIT Press.

Kamara, J.M., Augenbro, G., Anumba, C.J., & Carrillo, P.M. (2002). Knowledge management in the architecture, engineering and construction industry. *Construction Innovation, 2,* 53-67.

Kelleher, D., & Levene, S. (2001). *Knowledge management: A guide to good practice.* London: British Standards Institution.

McCaffer, R., & Adham, T. (1992). Education for effective project management. *Proceedings of the World Conference on Engineering Education* (pp. 565-570), Portsmouth Polytechnic, UK, September.

Rowley, J. (1999). What is knowledge management? *Library Management, 20*(8), 416-419.

Thamhain, H.J. (1991). Developing project management skills. *Project Management Journal, 22*(3), 39-44.

Wetherill, M., Rezgui, Y., Lima, C., & Zaril, A. (2002). Knowledge management for the construction industry: The E-Cognos project. *ITcon, 7,* 183-196.

Wong, C.F. (2001). *Knowledge management in construction project management.* Unpublished MPhil Thesis, Hong Kong Polytechnic University.

Yeung, C., & Holden, T. (2000). Knowledge re-use as engineering re-use: Extracting values from knowledge management. *Proceedings of the Third International Conference on Practical Aspects of Knowledge Management* (PAKM 2000) (pp. 24/1-24/10), Basel, Switzerland, 30-31 October.

Zavadaskas, E., Bejder, E., & Kaklauskas, A. (1998). Raising the efficiency of the building lifetime with special emphasis on maintenance. *Facilities, 16*(11), 334-340.

Chapter VI

Diffusion of Information and Communication Technology: A Community of Practice Perspective

Vachara Peansupap
RMIT University, Australia

Derek H.T. Walker
RMIT University, Australia

Abstract

This chapter explains the community of practice (COP) concept from the perspective of three major Australian construction contractors. It also describes and provides an analysis of their COPs from the perspective of the individual, work group, and organisation. We provide maps of these COPs for each of the three contractors studied. The intention of using this approach is to offer readers insights into how these COPs can be developed and supported. The study we base this chapter upon focussed on groupware as an ICT initiative being rolled out using COPs as an integral part of the innovation diffusion strategy. Case study findings relating to these COPs reveal that they are not only supported by ICT, but also that ICT itself supports these COPs in a self-referential and synergistic way. The important contribution that this chapter makes to our understanding of the studied phenomenon was not only the 'what' or descriptive nature of these COPs, but also insights that help us understand the 'how' of the process, so that lessons learned may be absorbed and diffused more widely in the construction industry.

Introduction

Construction organisations comprise geographically dispersed, virtually linked sub-organisations that work together to realise projects. They increasingly do so using information and communication technology (ICT) to communicate, coordinate their activities, and solve complex problems. One salient problem they face is how to effectively use requisite ICT tools. Communities of practice (COPs) generate knowledge networks that enhance and sustain competitive advantage, and they are also used to help COP members actually use ICT tools. Etienne Wenger defines communities of practice as "groups of people informally bound together by shared expertise and passion for a joint enterprise" (Wenger & Snyder 2000, p. 139). This 'chicken-or-egg' issue about needing a COP to use the tools that are needed to effective broaden COPs (beyond co-located these groups) led us to explore how best to improve the process of ICT diffusion through construction organisations—primarily using people supported by technology that improves knowledge sharing.

We present insights gained from recent PhD research results in this area. A semi-structured interview approach was used to collect data from ICT strategists and users in the three large Australian construction organisations that are among the 10 or so first tier companies by annual dollar turnover in Australia. The interviewees were categorised into five organisational levels: IT strategist, implementer, project or engineering manager, site engineer, and foreman. The focus of the study was on the organisation and the way that it implements ICT diffusion of a groupware ICT diffusion initiative.

From the three Australian cases, several types of COP networks emerge: within-organisation COP; institutional, implementer, or technical support; project manager/engineer focussed; and collegial support. Also, there are cross-organisational COPs that organically emerge as a result of people sharing an interest or experience in something significant. Firstly, an institutional network is defined as a strategic group, interested in development of technology innovation within an organisation. This COP principally links business process domain experts with an ICT strategist.

We have structured this chapter as follows. First we provide some background to ICT diffusion in general and how COPs can use groupware ICT applications. We follow this with a discussion of theory underpinning the research project reported upon in this chapter, and we explain the differences between teams and COPs. We then report research findings from three case studies and present our analysis. We conclude the chapter with some helpful tips and hints for readers wishing to make practical use of our findings.

Learning Objectives

1. Readers should gain a sound understanding of how ICT is diffused through construction organisations using a human infrastructure supported by ICT—specifically from the way that COPs influence how their members actually diffuse ICT.

2. While we cannot generalise from our research findings, presentation of our findings resulted in comments that the concepts presented may be more generally applicable to any emerging technology introduced to a business entity. Thus, an additional learning objective is to trigger in readers a measure of reflection of how our findings and arguments may be applied more broadly.

Background

Innovation and its effective diffusion and deployment is an essential competitive driver in the construction industry (Slaughter, 1998, 2000). It provides the capacity for delivering two process improvement or product development benefits. Process innovation is focused on the '*how to*' capacity that leads to an improvement or change in traditional work processes by introducing smarter or more effective ways to do things—this in turn can lead to construction management process productivity improvements. Product innovation is focussed upon developing new products in response to market forces (Meyers, Sivakumar, & Nakata, 1999).

ICT innovation is a major potential innovation improvement that can deliver real benefits to the construction industry and its participants (Duyshart, Mohamed, Hampson, & Walker, 2003). Much of the paper-based information exchange process that takes place during the construction phase that underpins decision making involves duplication, continual translation, and transcription from one medium or form to another, as well as loss of information (Duyshart, 1997). ICT has not only been used to decrease integration problems, but also can be used as an effective way for experts to share knowledge and jointly solve problems. For example, the BP virtual office facilitated complex problems to be solved using the expertise of a global network of virtually linked experts (Prokesch, 1997). A case study of a knowledge-intensive firm of people familiar with e-mail has been shown to be more effective as a tool for low-level knowledge than expected (Robertson, Sørensen, & Swan, 2001). Thus collaborative enabling ICT is a relevant and important tool for improving productivity, and so effective ICT diffusion becomes even more important in this context. A paradox that has been emerging is that groupware and ICT applications that effectively connect geographically dispersed people are self-referential. Benefits from the very technology that is used to allow people to collaborate are best achieved by effective ICT diffusion, yet these very applications require an effective ICT diffusion process.

Our discussion in this chapter focuses upon the roles of COPs that support the diffusion of innovation within three construction organisations. The literature argues that the COP concept has an essential role in knowledge sharing and in turn can develop a more knowledge-productive culture of learning in construction management organisations (Wenger, 1999; von Krough, Ichijo, & Takeuchi, 2000; Wenger & Snyder, 2000).

Theory Underpinning the Research Project

Much has been written on the need for firms to develop core competencies (Prahalad & Hamel, 1990) that provide sustainable competitive advantage either from cost or product/service differentiation (Porter, 1990). ICT innovation can be argued to support both a cost competitive advantage (through management efficiency reducing wasted management and administrative energy), as well as providing sophisticated clients with service level improvements that deliver a distinctive and differentiated qualitative competitive advantage. Many firms develop their own competitive advantage through initiating innovation. This can be achieved and may be optimised by effective ICT diffusion occurring at both the individual and organisational levels.

At the organisational level, diffusion of innovation depends on how well organisations can absorb external sources of innovation, as well as develop their internal capacity through trial and error experimentation and piloting, research and development, and supporting learning systems (Cohen & Levinthal, 1990). Absorptive capacity is one of the essential factors that sustain innovation and its diffusion through building up an experience and knowledge base that can be drawn upon when needed to develop or diffuse innovation. This infrastructure capacity helps provide not only the technical and knowledge enablers of innovation and its diffusion, but also can be used to help build an organic organisation, often unofficial, that utilises external and internal sources of knowledge to enhance the internal innovative process or product. Due to the individual features of an organisation's business processes, the adoption of external innovation needs to be modified to suit its specific business objective.

At the individual level, innovation diffusion depends upon information or knowledge gatekeepers who help transfer innovative knowledge from external and internal sources to the internal unit of organisation. These gatekeepers interpret or transform knowledge into simple language to fit the environmental context of known target groups (Rogers, 1995). Diffusion could not be achieved if individuals within an organisational unit did not accept the innovation and convince others to use it.

At the group level, people naturally tend to form knowledge networks to share and reframe knowledge that they routinely or occasionally use. History provides many such examples of learning communities. Trades and guilds of Europe are one medieval example, with more recent cases of COPs being documented in many organisations. One example from the Daimler Chrysler Corporation reports groups of people being clustered around a particular skill to form 'tech clubs' (Wenger, McDermott, & Snyder, 2002). The power of people forming small groups to learn from each other has triggered a great deal of interest and led to the concept of communities of practice. Lave and Wenger (1991) first introduced this term when studying forms of apprenticeship and social groups, and studied how these communities shared not only knowledge, but also the culture of access to knowledge and its use to diffuse complex tacit knowledge. A COP shares knowledge and skills, and sustains its members through obligation to exchange knowledge, providing access and accessibility to shared insights and knowledge about the practice of work (Wenger et al., 2002, p. 4).

An ICT diffusion study of 117 people from three large Australian construction organisations reveals 11 broad factors affecting ICT diffusion. These 11 factors were grouped into four clusters of influencing characteristics: management, individual, technology, and environment. ICT diffusion is impacted by management, individual, and technology characteristics that affect each other. All these are nurtured or inhibited by organisational culture environmental characteristics (Peansupap, Walker, Goldsmith, & Wilson, 2003b). One of the more interesting findings from this ICT innovation diffusion study was that people-support, in terms of COPs, was an evident element of the four clusters of characteristics. Thus human capital infrastructure appeared to provide a pivotal role supporting ICT innovation diffusion. Therefore, a fruitful way forward in developing improved ICT diffusion is to map, facilitate, build, and support the development and maintenance of an effective mechanism for people to form groups that help each

Table 1. Differences between teams and COPs (adapted from Wenger et al., 2002, p. 42; Bourne, 2003, p. 14)

	Teams	**COPs**
Nature of formation	Selected through resourcing processes	Generally formed through voluntary and informal processes
Term	Temporary and finite: until the project is complete	It depends: the COP will remain as long as its members consider it has a purpose
Duration	Set up a specific time period for team joining; team member will be desegregated after finishing the project	Do not specify time; the member COPs will be finished if the members feel that it is not in their interest
Structure	Each team member will have a specific role and 'place' in the team	Peers with a common purpose
Purpose	Deliver the 'result' (building, system, change); organisation determines a specific goal and time framework to achieve a goal.	Sharing of knowledge; members determine group interest and share their knowledge in natural manner
Operating principle	Command and control	Collaboration and commitment
"Legitimation" (power+authority relations)	Formal hierarchy /leadership	Informal and dynamic/ fluctuating membership; status in COPs must be earned
Essential success ingredients	Trust, shared 'vision', and purpose; commitment	Trust, shared 'vision', and purpose; commitment
Type of knowledge	Explicit knowledge and information (documentation, processes, report)	Tacit knowledge (stories, practical experience, lessons learned, tips)
Communication mode	Formal (formal meeting)	Informal (social activity and peer group, conference); *ad hoc* and informal meeting

other by sharing knowledge (particularly about how to best use ICT applications that support knowledge management).

People helping each other by sharing knowledge can occur within the context of 1-2-1 relationships as either isolated or continued instances, and indeed this is a normal feature of working on joint problem-solving activities. More often people are working in groups so that there is a many-2-many, 1-2-many, or many-2-1 communication patterns. Groups can become teams, but they are distinguishable. A team is defined as "a small number of people with complementary skills who are committed to a common purpose, performance goals, and approach for which they hold themselves mutually accountable" (Katzenbach & Smith, 1993, p. 45). However, as they argue (1993, p. 89), "A working group uses its purpose solely to delineate individual roles, tasks, and responsibilities. Those roles typically match formal organisational positions. To get their assigned tasks done, working group members, especially at senior levels, usually delegate the real work to others beyond the group." Groups exchange information, knowledge perspectives, and insights, as well as act as a point of further reference; however, there is unlikely to be the depth of focus that is evident in teams working toward realising a particular project.

The distinction between teams and COPs is palpable and relevant because there is often a different purpose for the two, though our research indicated that there can be a designed development of integrating COPs and teams in helping realise team goals and objectives. As Wenger et al. (2002) have demonstrated, companies that creatively engage with COPs realise a significant set of knowledge assets and competencies that would otherwise be unobtainable.

Table 1 assists us in understanding the differences between teams and COPs. It has been developed from ideas offered by Wenger et al. (2002, p. 42) and Bourne (2003, p. 14). It highlights the difficulty in assuming that teams naturally are COPs or that they can naturally become a COP where sharing knowledge becomes the cultural foundation that distinguishes them from other forms of associations of people solving complex problems. The informality and self-referential nature of COPs versus teams is an important distinguisher, as is the term of the association and duration of their activities. Motivation also differs between these groups, as does operating principals, legitimacy, and ingre-

Figure 1. Formal/informal organisation structure for teams and COPs

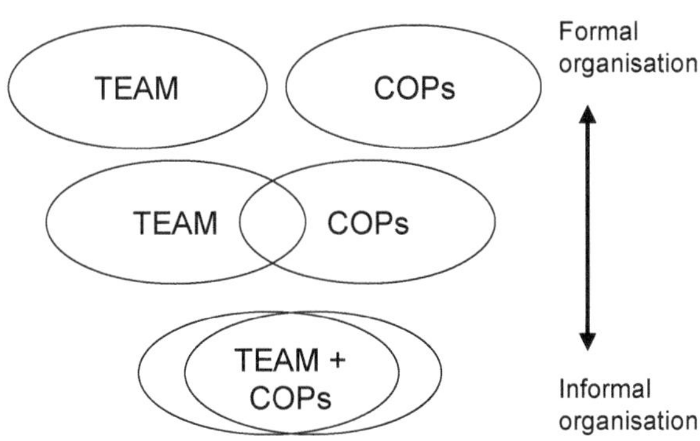

dients of success. The type of knowledge exchanged is generally different and the communication mode also different between teams and COPs.

This begs the question, how can organisations harness the power of COPs in engaging with teams to solve complex problems? Part of the answer lies in the levels of formality as indicated by the study results of ICT innovation diffusion, cited earlier by Peansupap et al. (2003b). These stressed the importance of cultural factors and a management style that facilitates learning and professional development, as well as actively supports interaction, questioning the status quo in order that a full repertoire of potential problem solutions can be explored and examined in an atmosphere of open and honest discussion. This can be further explained with reference to Figure 1.

This idea may help to explain the relationship between team and COP differences by level of formal and informal structure/style of management of organisations. The two groups (teams and COPs) in a formal organisation may be more separated than those in informal organisations. If an organisation can inject a combination of informality as well as its formal team governance arrangements, then it has an opportunity to integrate the potential knowledge assets of COPs with that of any given team. The formal-informal requirement also has another interesting dimension—trust. In one sense, formal team governance can engender trust because it establishes a set of rules that everyone knows should be adhered to; thus, there is an element of predictability and near-certainly about the way that activities take place. Also Wenger et al. (2002, p. 51) argue that a set of guiding principles needs to be addressed when developing or nurturing COPs, and they promote seven principles for cultivating COPs that is akin to a formal team working on a project. Their principles do appear to engender trust from rules and principles that apply, namely:

1. designing the community of practice for evolution;
2. having an open dialogue between inside and outside perspectives;
3. inviting different levels and intensity of participation;
4. developing both public and private community spaces;
5. having a focus on value, and combining familiarity and excitement; and
6. creating a rhythm for the community—where activities fall into a natural flow that supports relationship building and knowledge exchange, and where people's contribution ebbs and flows. People enter various levels of a COP, from core participants who actively participate and share knowledge through to 'lurkers' whose contribution ebbs and flows with their engagement.

An interesting case study of a project spawning a community of practice is the Xerox Transition Alliance reported by John Storck and Patricia Hill (2000). This was an example of a strategic community of practice of 50 IT specialists responsible for maintaining 70,000 workstations and 1,200 servers in which Xerox top management deliberately established the Alliance to serve as a vehicle for diffusing new ICTs. The interesting aspect of this case that is relevant to our discussion in this chapter is that this community of practice was designed and developed for ICT diffusion, and that it was judged a success that endured as a more recognisable community of practice after the project aim of ICT diffusion was achieved. It was facilitated and formally enacted and recognised many of the informal imperatives that characterise COPs; however, it has been termed a *strategic* COP. Six principles were highlighted:

1. design an interaction format that promotes openness and allows for serendipity;
2. build upon a common organisational culture;
3. demonstrate the existence of mutual interests after initial success at resolving issues and achieving corporate goals;
4. leverage those aspects of the organisational culture that respect the value of collective learning;
5. embed knowledge-sharing practices into the work processes of the group; and
6. establish an environment in which knowledge sharing is based upon processes and cultural norms defined by the community, rather than on other parts of the organisation (Storck & Hill, 2000, pp. 73-74).

The implications of the theory proposed by Wenger et al. (2002) and Storck and Hill (2000) for the construction industry is that a COP approach to encouraging ICT diffusion in particular, and knowledge sharing in general, is both valid and to be encouraged.

Australian Case Studies: Methodology

A semi-structured interview approach was used to collect qualitative data from ICT strategists and professional users in the three large Australian constructors at several organisational levels using a case study approach (Yin, 1994). Key contact people within the IT department who understood the research aims nominated interviewees who were asked to identify ICT users already using ICT in their work so that a better understanding of how these organisations approached ICT diffusion can be found. Thus the sample is not a random sample but a purposeful one drawn from ICT professional users in major construction companies that principally operate in Australia, but also do so as global construction contractors.

The focus of the study was on the organisation and the way that it implements ICT diffusion of a groupware ICT diffusion initiative. The research question is directed at understanding how and why observed behaviours took place in diffusing an ICT groupware initiative. It concentrated upon the 'latest wave' of ICT innovation facing major construction contractors, and the aim was to gain a better insight into how several of the major global players in this industry sector approached ICT innovation. It was anticipated that the study would allow—through comparing and contrasting the organisations applied to ICT diffusion—identification of better business practices and deeper mechanisms underpinning these, so that they can be unearthed and better understood. Lessons learned may be offered for general acceptance or adaptation; results are not intended to be seen as a general factual status (audit) either within the organisations concerned or as being representative of all the top tier contractors under study. The sample used was too small to generalise from, but the study does enable findings to be used to shed light on ICT diffusion best business practice elsewhere.

Australian Case Studies: Research Findings

Interviewees presented in Table 2 can be grouped into five levels: IT strategists (senior-level management champion and initiative driver), implementers (given the task of encouraging diffusion of the ICT groupware initiative), project managers (responsible for construction teams on projects using this technology), and site engineers and site foremen (both direct users of the technology in coordinating the physical and administrative work being undertaken on-site). The reason for this approach is to gain understanding of the factors influencing ICT diffusion from multiple perspectives.

Case Study A (CSA)

CSA is a large construction contractor with well over AUD$1 billion in annual global turnover. The chief executive officer of CSA established an IT quality assurance strategy vision in 1996. He envisioned IT assisting integration of construction information within CSA. From this vision, the group of regional managers and quality assurance managers

Table 2. Categories of interviewee in the three case studies

Interviewee	Case study		
	CSA	CSB	CSC
IT strategist	1	1	1
Implementer (L1)	1	1	1
Project/Engineering Manager (L2)	4	1	1
Site Engineer (L3)	1	3	2
Foreman (L4)	1	1	1
Total	8	7	6

Figure 2. Case Study A's communities of practice

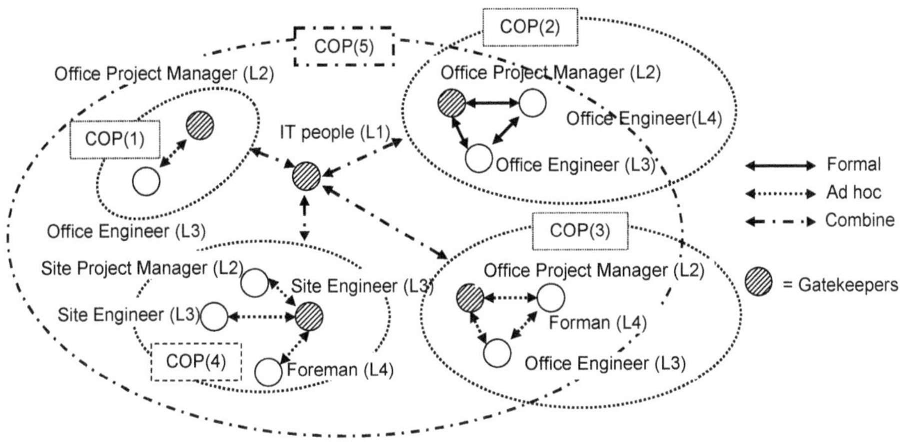

had meetings to discuss and explore ICT that could be used for effective communication and coordination between project members within CSA. During the development period, IT staff in CSA worked closely with managers, key end users, and champions who have relevant construction work processes experience.

The development of the ICT application was based on a software package that provides the basic communication functions for general business needs. Traditionally, most construction information transfer relies on paper-based systems, so the software package required design and customisation of user interfaces to suit the traditional construction approach. The ICT application had been customised to be compatible with organisational forms and work processes, and this encouraged users to familiarise themselves with entering information using ICT instead of paper. The modules of ICT used by CSA consisted of main processes such as tendering, project communication, and construction database applications. This study focused on project communication.

Figure 2 illustrates five examples of COP that facilitated the ICT diffusion (Peansupap et al., 2003a). One of the senior engineers had a role in validating the ICT initiative and informally created the development of COP(1) with staff from the firm's IT department. When he experienced problems with using ICT applications from a practical and/or technical perspective, he would resolve them with the IT people.

One office project manager helped establish COP(2) by providing time in the morning to talk and exchange ICT knowledge with his colleagues. He spent his morning time providing specific training and discussion about ICT problems with his subordinates and encouraged their feedback and participation. COP(2) assisted the diffusion of ICT through this team because it shared problems and new ideas on how to apply ICT to assist traditional work processes. In addition, COP(2) facilitated additional feedback to be channelled between users and ICT tool developers. COP(3) is a different group of individuals that often communally solved ad hoc problems through the gatekeeper, usually by phone, if not through IT people. COP(4) also solved ad hoc problems, but on a one-to-one basis through the gatekeeper linked to IT people.

COP(5) used an e-mail discussion group to facilitate ICT use, to communicate with each other to get help. COP(5) helped users who had problems with relation to the ICT use. Members posted their questions and the IT staff for other COPs to respond to. Users shared their experiences and problems and also suggested solutions. This reduced repeated questions on the use of ICTs and reduces IT staff workload in repeatedly responding to the same problems.

Case Study B (CSB)

CSB is one of the largest construction companies in Australia. The company consisted of many core business units, including design, construction, and project management. In 1997, the managers from the construction e-business unit set up the IT strategy based on a perceived benefit of gaining competitive advantage from Internet technology use. The company evaluated many available groupware integration applications; however, there was no suitable ICT application available at that time that suited the company's needs. Therefore, CSB decided to design and develop its own ICT groupware interface tool linked to standard e-mail, Web, and so forth.

The development of ICT was initiated by a CSB group that consisted of experts from many construction fields such as architecture and design, project management, and construction. These experts worked together with developers from an IT consultant company that has the key responsibility in programming the ICT, whereas experts provided the knowledge of technical aspects of construction processes. Additionally, the group was strongly supported by an ICT project-initiating sponsor.

Although the objective of the ICT application was to improve, communicate, and

Figure 3. Case Study B's communities of practice

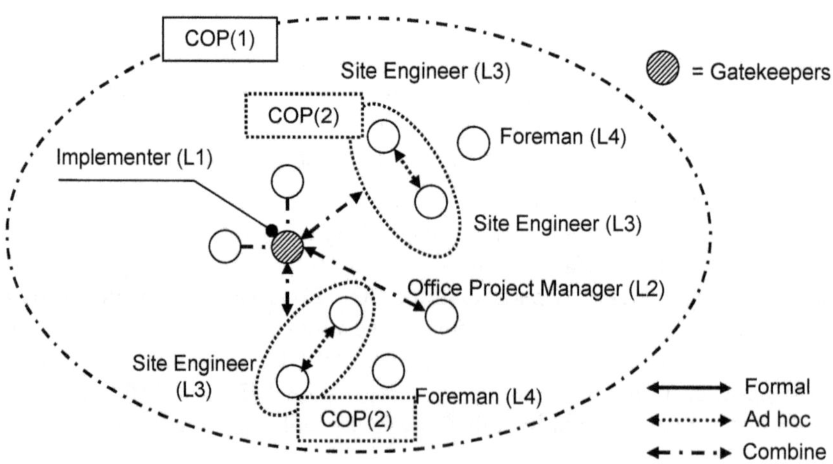

coordinate construction project information, participants provided mixed opinions about ICT usefulness. The reason for this is that the ICT was designed to facilitate managing construction projects. Therefore, it may not be perceived as useful for design managers who were not involved with construction activities. In addition it was rarely used by all project parties, thus inevitable duplication of both electronic and manual document transfer effort occurred.

Figure 3 illustrates two examples of COPs that facilitated the ICT diffusion (Peansupap et al., 2003a). At the organisational level, ICT use depended upon the implementer who had most technical knowledge. The implementer was at the centre of and developed COP(1) associated with use of the all-embracing ICT Web application. He transferred his knowledge to users, received feedback from COP(1), and used e-mail and phone extensively to communicate and help solve problems of ICT users. Thus, a virtual COP to improve effective ICT use was created. In general, CSB presented a strong culture of helping each other (not limited to IT issues, but also people can ask help from many experts in CSB). At the project level, ICT discussion and help by COP(2) started with colleagues who worked on the same construction project. Most participants confirmed that they sometimes share and exchange knowledge of ICT use. However, they had time and ICT knowledge limitations. Technical ICT assistance was mainly received from implementers. In addition, construction foremen have more responsibility for construction technique aspects, while site engineers, project administrators, and project managers handled most administrative tasks. COP(2) indicates the frequent situation where very small groups that helped each other formed and disbanded mainly for ICT issues to be solved at a low functionality level, because they tended to get help from the ICT implementer when trying to use the ICT application at higher functionality levels.

Case Study C (CSC)

CSC is a highly innovative Australian construction contractor. The company has an annual global workload of over AUS$1 billion. Their projects include building and civil engineering infrastructure. This company received several awards relating to construction innovation and has a sound absorptive capacity. The company has strong policy support for improvement of construction productivity and safety. At the time of this study, CSC's commitment to using ICT was project-by-project based. The company briefed project participants to provide them with an understanding of the benefits of using ICT. After obtaining commitment for ICT use from CSC project managers and other main project participants (such as client, design consultants), the implementer would provide training for project participants and expected CSC users.

An external project Web service developed the ICT application. The implementer also had a role in facilitating customisation of the ICT application to suit the company's work processes and provided strong support for diffusion of ICT within CSC and other project participants. The strength of COP(1) in Figure 4 (Peansupap et al., 2003a) is mainly dependent upon the implementer who had the knowledge and background of both the construction and computer context. He started his own communities by providing training on how to use the application for his teams and main project participants.

Figure 4. Case Study C's community of practice

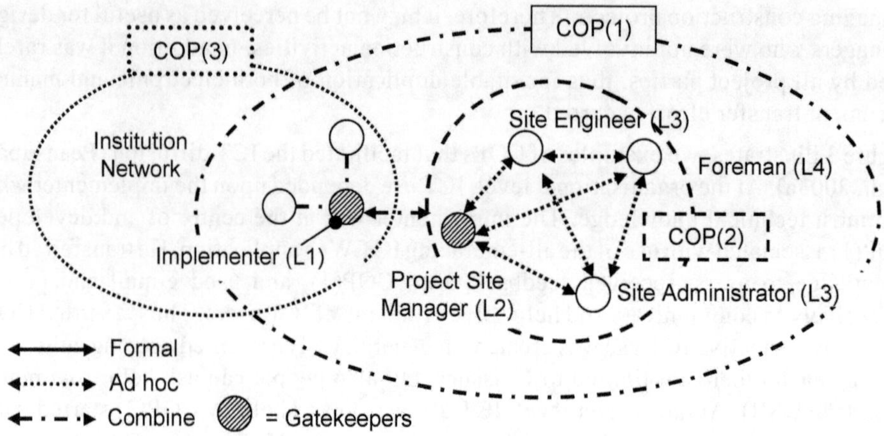

In addition to COP(1)—the implementer—a COP of colleagues interacts in COP(2). Several COP(2)s existed for collegial help; they have a positive influence on novice engineers who did not receive any training sessions to develop skills and be able to use the ICT application. A COP(2) member mentioned that his senior engineer helped him use the ICT application. The project manager also encouraged and helped his subordinates by providing advice on ICT use. Although he has limited knowledge, he tried to resolve problems regarding its use through the implementer.

COP(3) is the organisation-wide network for each business group. CSC started the technology centre with its key functions being to promote and expand a range of ICT technological innovations into business units (BUs). The centre aimed to improve work performance, safety, and quality in construction work processes. It consisted of people who were 'the experts' from different BUs across the organisation. They dedicated time to meet every three to four months. The technology centre provided support and advice to BUs on adopting ICT innovation initiated and was supported by all BUs.

Australian Case Studies: Discussion

In all three cases it was clear that IT groups held a significant and pivotal position in the ICT diffusion process through a combination of help-desk, 1-2-1, and 1-2-many communication channels. The use of groupware for e-mail and intranet access for resolving diffusion issues was also substantial in all three cases. Rich communication channels were also used—the COP provided reflection and feedback, facilitating interrogation and ideas clarification/testing. From the three cases, several types of COP networks emerge: institutional, implementer, or technical support; project manager/engineer focussed; and collegial support. Each has its individual focus, resources, and behaviour/attitude implications.

It is important to realise that the literature on COPs stresses that COPs are already 'out there', they exist as social networks of friends with shared interests and experiences that want and need to communicate. COPs can often evolve in an ad hoc manner from an association of 'mates' past co-workers who have gained a bond of trust and commitment through joint effort of working on projects or specific tasks together. They can also develop from membership of a special interest group such as a professional association where members may well have never worked together before, but share a passion and commitment to sharing and re-conceptualising knowledge about specific issues. Thus there is a relationship within organisations, as highlighted in the case study reported above (Peansupap et al., 2003a) and between organisations. The study was restricted to within-organisation COPs; this will be discussed first, before moving on to discuss COPs across organisations.

Institutional Network

An institutional network is defined as a strategic group interested in development of technology innovation within an organisation. Development of ICT required staff that had construction processes experience in all three cases—the aim is for business process needs to drive ICT development; thus, customisation, testing and piloting, feedback, and fine-tuning are required to be delivered by construction management experts and ICT experts. This COP principally links business process domain experts with an ICT strategist. Resources required include high-level domain expert input and substantial face-to-face time with reflective management behaviour.

Introduction of CSA's ICT was developed from a meeting of regional managers responsible for quality assurance-responsible IT experts and for health and safety and the environment. This group dedicated time to testing ICT and designing simple user-interfaces. This type of COP was a temporary one, formed until the objectives of ICT development had been defined and rolled out. After the development of the ICT initiative, the IT department took over responsibility for implementation. Similar to CSA, ICT in CSB was developed by senior mangers with expertise in construction from the e-business group and IT consultants who identified relevant ICT opportunities. This group was formed to design and develop the ICT tool for developing and managing construction projects. Therefore, the initiative group of CSA and CSB may be classified as *a task-oriented team* (Storck & Hill, 2000).

CSC initiated a technology centre to promote and explore technological innovation relevant to its business units (BUs). Groups of people with backgrounds from various BUs across the organisation dedicated time to formally explore and discuss the opportunities of using the technological innovation in construction processes. Thus the initiative group of CSC may be classified as *a strategic community alliance* (Storck & Hill, 2000).

Implementer or Technical Support Network

CSB and CSC had a key champion with sufficient drive and enthusiasm to be the ICT initiative implementer who envisioned the ICT strategy. A COP would then be built or emerge, nurturing the ICT implementer whose role is to transfer ICT knowledge within project teams to expected users. The implementer plays a significant role in training and being a mentor to users who have a background of construction processes sufficient to understand any potential problems and/or implications of using ICT in construction processes. This person would also be involved with the software company that provided the ICT service and would participate in the development of the ICT initiative. CSA was dependent on the ICT team for training and development. Typically, implementers reside at the hub of the COP; they are the organisation's experts and main resource in making sense of the ICT and its development. They require resources to sustain the COP that links and permits ICT initiative knowledge to be cross-levelled and diffused widely across the organisation, as suggested by Nonaka and Takeuchi (1995). Their behaviour is supportive, inclusive, and that of an enthusiastic knowledge activist (von Krough et al., 2000).

Project Manager/Engineer and Collegial Support Network

It became clear from the interviews that project managers play a significant role in ICT diffusion by developing their own community of practice. One of the office managers in CSA mentioned that his team uses part of the morning time to discuss how to best use ICT. He found it very helpful for new engineers who received ICT training, but may take time to understand how to apply it in their work. He also attempted to support resources for delivery and feedback from end users to IT staff to improve the application to meet end-user needs.

Furthermore, the collegial network being considered as the first source of help to users (who have limited ICT knowledge) is consistent with a study of 2,000 aerospace engineers who found that well-informed technological gatekeepers with an intimate knowledge of the technical tasks being undertaken were the preferred first choice for finding salient information or knowledge (Anderson, Glassman, McAfee, & Pinelli, 2001, p. 151). Personal communication with a peer who knows the context of the problem provides a rich and clearer communication channel for assistance. It often allows users to observe and learn from real examples by real or virtual demonstration (Wenger et al., 2002). It is easier and quicker to get help from colleagues relating to ICT use.

The existence of the within-workgroup COPs such as CSA COP(2, 3, 4) and CSC COP(2) link colleagues together, as well as provide gatekeepers to the ICT support COP members. In CSA COP(4) the pattern was individuals interacting on a one-to-one basis with the gatekeeper through to the ICT support group. In this kind of COP, from a collegial perspective, there is a dyad relationship in which the gatekeeper supports the user and learns, filters, and consolidates typical difficulties colleagues experience with ICT application. This is then fed back to the ICT developers, and the gatekeeper becomes a valuable focal point in that COP. However, the value of colleague interaction is minimised, compared to the more connected CSA COP(2 or 3) in which there is more cross-group interaction. A more isolating COP model is evident in CSB COP(1), where most of the help is gained from a dyad relationship between the ICT implementer and people in that COP. In this example it is necessary for temporary or small-scale COP to emerge as illustrated in COP(2), but often their skill level for answering urgent questions is limited.

The resource implications for maintaining collegial support is a need for committed gatekeepers who are provided with the means and motivation to support the COP. Additionally, any COP that is linked via groupware needs that ICT application to be effectively diffused for it to be of any use. There needs to be an attitude and behaviour consistent with openness, knowledge-sharing, and also motivation and rewards for participation, even if these rewards are intrinsic (Nahapiet & Ghoshal, 1998).

Relationships of COP within an Organisation

The three case studies show that COPs can help construction firms diffuse ICT knowledge by COP members sharing experiences and insights, and helping each other to solve problems related to their ICT use. This discussion underlines the concept of a COP by focusing on relationships between the identified within-firm COPs. These relationships play a significant role in linking several communities together. From our case studies, the institution of COPs has the complementary role of sharing the message of potential and realised ICT benefits through construction business managers. For example, one senior construction manager may need to find a solution to improve his company's work business processes. Members who have experience with ICT capability may suggest a solution to the manager. At this stage, the solution may be a preliminary concept to improve the business process. Within institutional COPs, this intra-relation-

Figure 5. Three types of COP within an organisation

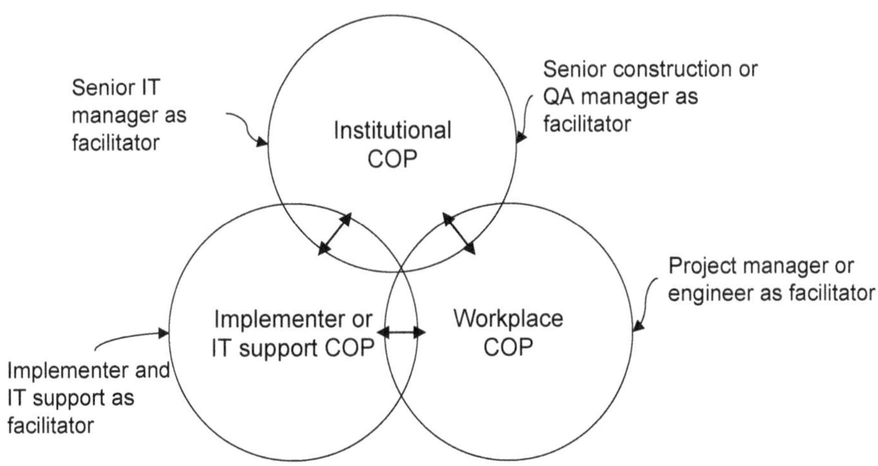

ship can help members be aware of the availability of current ICT tools or to develop interest in the use of an ICT application. Facilitators of an institutional community of practice (who provide leadership and support) should comprise key representatives, such as top management, to reinforce ICT diffusion throughout the organisation. Such people provide policy drivers. Perhaps a senior quality manager who is involved in improving construction processes could be appropriate; likewise, senior project managers who have practical experience in the field or perhaps senior IT managers who have a strong background of IT knowledge contribution could be appropriate. These representatives may have significant roles in connecting and linking COPs, and facilitate knowledge sharing between COPs.

Figure 5 illustrates that the first relationship in attempting to integrate or link COP within an organisation is the connection between the institutional COP and any COP facilitated by the ICT implementer or IT support groups. A senior IT manager can play an important role in linking these two COPs. If the senior IT manager is a member of these two groups, he/she can work with IT members to seek more technical information on how best to apply ICT applications to improve construction work processes. Groups of implementers or IT support personnel may share their technical experience with senior IT managers, and this in turn provides feedback to COP members within the institutional COP. On the other hand, the decision adoption of ICT application to support the business need may depend on several business constraints in which groups of implementers or IT support staff may not have sufficient experience. Therefore, the relationship between these two groups may help IT people to understand business operation realities, which in turn provides a suitable ICT application for business needs to address the issue of improving such processes through the use of ICT-supporting infrastructure. Meanwhile, people in business units can improve and update their understanding of ICT application knowledge.

The second relationship is a connection between the institutional COP for any project manager or a collegial COP that may be based around workplaces. It could be suggested that a senior quality manager or senior construction manager can play an essential role in linking these two COPs. Experienced construction managers[1] may help share their experience of how to encourage colleagues to use ICT applications. The experience may be useful to business managers in adapting strategic management plans or to improve ICT implementation strategy. Some experience may be useful in highlighting best-practice knowledge, where BUs receive help on decision making for future projects. Senior construction managers may also learn from business planning people, who could have a positive impact on their future projects. For example, if a construction firm successfully used Web-based project management tools, then this message may be fed through a COP in which business managers and construction managers can learn and share options about how to best diffuse these tools.

Finally, the connection between 'implementer or IT support COP' and the 'workplace COP' (see Figure 5) can help people share experiences of combining both construction and IT technical knowledge. Most members in workplace COPs have high levels of experience of construction procedures and processes. Thus, sharing this construction experience with members of implementer or IT support COPs should be helpful for improving ICT applications to suit users' requirements. On the other hand, knowledge about ICT applications also helps members of workplace COPs to understand limitations associated with ICT applications. Linking COP types can improve mutual understanding of ICT innovation support staff and operational users to help each party improve their application of ICT, rather than blaming each other for poor implementation of any ICT initiative.

Relationships of COPs between Organisations

The above discussion centred on within-organisation COPs. This section focuses on inter-organisational COP relationships. These groups also provide benefits for each member organisation because they facilitate sharing and learning from each other. Figure 6 illustrates these relationships.

It is important to recognise that these COPs already exist, cannot be controlled by the organisation in any heavy-handed way, and if suppressed will merely go 'underground'. Therefore, it is wise for organisations to embrace them, and to actively and proactively help shape them to align organisational objectives where appropriate with the COP, rather than vice versa. For example, members of professional institutions may share information about standards of using ICT in the construction industry, and these may develop into a useful collaborative network such as the Construction Industry Trade Exchange (CITE) in Australia and similar cross-industry initiatives elsewhere. In some cases, knowledge alliances are formed around these COPs. In the oil and gas industry for example, the Transfield Worley Best Practice Alliance developed in Australia, where a group of competing companies and supply chain specialists formed a community of

Figure 6. Relationships of COP between organisations

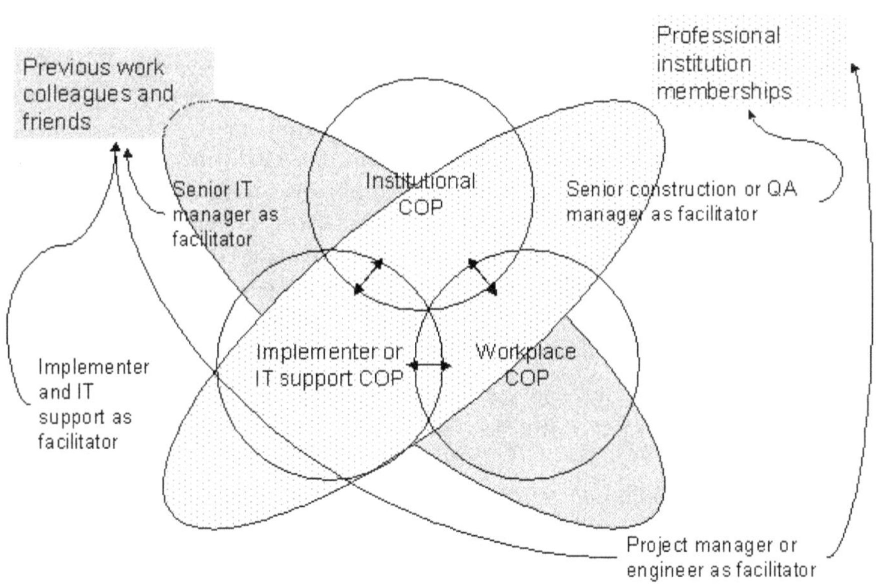

practice supported by Web tools to link geographically dispersed people and to organise forums, discussion groups, conferences, and COP activities within that knowledge alliance.[2]

Additionally, previous co-workers and friends may influence sharing the knowledge of ICT applications through membership of formal or informal social networks and COPs. People feel comfortable asking those close to them that they trust, therefore these COPs could sometimes help staff learn how to use ICT applications even though the individuals concerned are affiliated with competing organisations. Teigland (2000) observes from a study of an IT-intensive firm that people in organisations generally have conflicting loyalties between their organisation and their COP. Professional knowledge flows across organisations through COPs that span many members from many different organisations. The strategy of organisations managing this knowledge resource that has ready access to external competitors should not be to pretend that such knowledge leakage does not occur, but to openly recognise that they stand to gain as much (if not more) than they stand to lose. When an individual needs specialised help on a complex problem and he/she has to go to an external COP and get help, a valuable transaction has taken place. The value obtained could have been expensive if purchased (and probably was unobtainable on the 'open market') and represents barter or exchange tokens of favours through a process of trust building—favours can potentially be retrieved at some later date. By supporting, monitoring, and engaging in such COPs, organisations help ensure that they get their fair share of any rewards to be gained from this natural and organic form of collaboration. As cross-organisational COPs have the power to effectively diffuse ICT knowledge between organisations, these organisations should be aware of this reality and reinforce the COPs by encouraging staff to join them.

Conclusion

In this chapter we argued that there are three types of within-organisation COPs that can be developed to positively contribute to ICT diffusion. *Institutional COPs* help set the strategic direction for ICT development and validation. *Implementer or technical support COPs* link users with ICT support staff through gatekeepers that can help with the process of interpretation and re-framing problems and difficulties. Supporting workgroup COPs provide technical support COP gatekeepers in the project manager/engineer network and collegial support COPs that provide much of the necessary one-to-one or small-group support. We also indicated how the knowledge from *organisation-external* COPs may be effectively harnessed. Two of these were previous work colleagues and friends, and professional institution members.

Effective ICT and knowledge sharing is also emerging as a competitive advantage in project alliancing. For example, Bovis Lend Lease won the right to construct the National Museum of Australia as a result of its information management strategy (Duyshart et al., 2003; Walker & Hampson, 2003).

Practical Tips and Lessons Learned

- COPs are already 'out there', organisations should recognise their usefulness and power for knowledge creation and sharing.
- COPs can be guided, supported, and nurtured by organisations, though a heavy-handed approach should be discouraged.
- COPs can be an effective support mechanism for ICT innovation diffusion.
- The three Australian case studies demonstrate that COP types need to be established within organisations that service differently the needs of people at various levels within organisations.

Acknowledgments

We acknowledge the Cooperative Research Centre in Construction Innovation for their financial and moral support. The research study work reported upon here was part of a broader study into ICT diffusion and knowledge management under CRC Project 2001-004 (2B)—Delivering Improved Knowledge Management and Innovation Diffusion.

Further, we acknowledge that the section relating to the Australian research case study contains results first presented and published in the proceedings of the 19th ARCOM Conference, 3-5 September 2003, in Brighton UK, volume 2, pages 673-683. The paper authored by V. Peansupap, D.H.T. Walker, P.W. Goldsmith, and A. Wilson is entitled, "Developing Within-Company Information and Communication Technologies (ICT) Innovation Diffusion Networks: A Study of Three Australian Major Contractors."

Endnotes

1 It should be noted that experience relates to relevance—it is probable that younger organisational staff may have considerable ICT use experience to share with older staff, and older staff may have deep insights from their industry experience on how best to apply ICT tools in a practical manner.

2 Roger Brennan, Manager of the Transfield Worley Best Practice Alliance, IQPC www.iqpc.com.au Seminar in Brisbane on 25 August 2003.

References

Anderson, C.J., Glassman, M., McAfee, R.B., & Pinelli, T. (2001). An investigation of factors affecting how engineers and scientists seek information. *Journal of Engineering and Technology Management, 18*(2), 131-155.

Bourne, L. (2003). *Knowledge management reflective learning.* Assignment Paper, Doctor of Project Management Program (25), RMIT University, Melbourne, Australia.

Cohen, W.M., & Levinthal, D. (1990). Absorptive capacity: A new perspective on learning and innovation. *Administrative Science Quarterly, 35*(1), 128-152.

Duyshart, B., Mohamed, S., Hampson, K.D., & Walker, D.H.T. (2003). Enabling improved business relationships – how information technology makes a difference. In K.D. Hampson (Ed.), *Procurement strategies: A relationship-based approach* (pp. 123-166). Oxford: Blackwell Publishing.

Duyshart, B.H. (1997). *The digital document.* Oxford: Butterworth-Heinemann.

Katzenbach, J.R., & Smith, D.K. (1993). *The wisdom of teams – creating the high-performance organization.* Boston: Harvard Business School Press.

Lave, J., & Wenger, E.C. (1991). *Situated learning – legitimate peripheral participation.* Cambridge: Cambridge University Press.

Meyers, P.W., Sivakumar, K., & Nakata, C. (1999). Implementation of industrial process innovations: Factors, effects, and marketing implications. *Journal of Product Innovation Management, 16,* 295-311.

Nahapiet, J., & Ghoshal, S. (1998). Social capital, intellectual capital, and the organizational advantage. *Academy of Management Review, 23*(2), 242-266.

Nonaka, I., & Takeuchi, H. (1995). *The knowledge-creating company.* Oxford: Oxford University Press.

Peansupap, V., Walker, D.H.T., Goldsmith, P.W., & Wilson, A. (2003a). Developing within-company information and communication technologies (ICT) innovation diffusion networks: A study of three Australian major contractors. *Proceedings of the 19th ARCOM* (pp. 673-683), Brighton, UK, 3-5 September.

Peansupap, V., Walker, D.H.T., Goldsmith, P.W., & Wilson, A. (2003b). Factors influencing information communication technology diffusion: An Australian study. *Knowledge Construction: Joint International Symposium of CIB Working Commissions W055, W065, W107* (pp. 415-426), National University of Singapore, 22-24 October.

Porter, M.E. (1990). *The competitive advantage of nations.* New York: The Free Press.

Prahalad, C.K., & Hamel, G. (1990). The core competence of the corporation. *Harvard Business Review, 68*(3), 79-91.

Prokesch, S.E. (1997). Unleashing the power of learning: An interview with British Petroleum's John Browne. *Harvard Business Review, 75*(5), 147-168.

Robertson, M., Sørensen, C., & Swan, J. (2001). Survival of the leanest: Intensive knowledge work and groupware adaptation. *Information Technology & People, 14*(4), 334-352.

Rogers, E.M. (1995). *Diffusion of innovation*. New York: The Free Press.

Slaughter, E.S. (1998). Models of construction innovation. *Journal of Construction Engineering and Management, 124*(3), 226-231.

Slaughter, E.S. (2000). Implementation of construction innovations. *Building Research & Information, 28*(1), 2-17.

Storck, J., & Hill, P.A. (2000). Knowledge diffusion through "strategic communities." *Sloan Management Review, 41*(2), 63-74.

Teigland, R. (2000). Communities of practice at an Internet firm: Netovation vs. on-time performance. In E. Lesser, M.A. Fontaine, & J.A. Slusher (Eds.), *Knowledge and communities* (pp. 151-178). Boston: Butterworth-Heinemann.

von Krough, G., Ichijo, K., & Takeuchi, H. (2000). *Enabling knowledge creation*. Oxford: Oxford University Press.

Walker, D.H.T., & Hampson, K.D. (2003). Project alliance member organisation selection. In D.H.T. Walker & K.D. Hampson (Eds.), *Procurement strategies: A relationship-based approach* (pp. 74-102). Oxford: Blackwell Publishing.

Wenger, E.C. (1999). Communities of practice: The key to knowledge strategy. *The Journal of the Institute for Knowledge Management, 1*(Fall), 48-63.

Wenger, E.C., & Snyder, W.M. (2000). Communities of practice: The organizational frontier. *Harvard Business Review, 78*(1), 139-145.

Wenger, E.C., McDermott, R., & Snyder, W.M. (2002). *Cultivating communities of practice*. Boston: Harvard Business School Press.

Yin, R. (1994). *Case study research*. Thousand Oaks, CA: Sage Publications.

Chapter VII

Community of Practice Software Management Tools: A UK Construction Company Case Study

Mark Jewell
Addept Solutions, Australia

Derek H.T. Walker
RMIT University, Australia

Abstract

This chapter provides insights from a large UK construction organisation case study where communities of practice have been supported through use of a software tool and management approach that encourages their spread across the organisation. We provide a descriptive characterisation of what the community of practice (COP) software tool does, how it evolved, and anecdotal evidence from interviews with its users of its value to the UK case study organisation. We recognise the need to investigate COP value generation more formally, and we have developed a research proposal to undertake further work in a collaborative study with industry to provide useful COP performance measures to be undertaken. This chapter provides valuable insights from several years' reflection upon the tool's use and application, and we highlight both drivers and barriers to its deployment. The objective was to provide a practical example of what COP management tools could and should address.

Introduction

In the previous chapter of this book, an argument was made that a *community of practice* (COP) generates knowledge networks that enhance and sustain competitive advantage, and that it is also used to help COP members actually use ICT tools. In that chapter a number of lessons learned from the way in which COPs operated in three Australian construction contracting organisation case studies were discussed.

In this chapter, we provide further practical insights into COPs using a large UK construction organisation as a case study. COPs in this example were supported through a software tool and management approach that encourages the spread of COP beyond the institutional, technical support, and collegial support COPs highlighted in the three Australian case studies reported upon in Chapter 6.

We provide a descriptive characterisation of what the COP ICT software tool does, how it evolved, and anecdotal evidence from interviews with its users of its value to the UK case study organisation. The research required to investigate this dimension more formally is recognised and is subject to further work in developing an argument for a collaborative industry-academic study that is anticipated and currently undergoing refinement. Thus, for this chapter, the approach is anecdotal, but nevertheless based upon several years' reflection upon the tool's use and application and drivers and barriers to its deployment.

We have structured the chapter as follows. First we provide some background on COPs and follow this with discussion of several important theoretical knowledge management concepts. This provides a useful framework for us to describe the COP management tool SIGMA CONNECT, followed by how this tool was used by Carillion PLC in the UK. We then offer our conclusions on COPs and how the software tool was used to affectively encourage, enable, and support these COPs. Finally we offer some practical tips and lessons learned.

Learning Objectives

1. Readers should advance their understanding of how construction organisations using COPs can enhance their knowledge creation and sharing, including understanding the meaning of terms such as *tacit knowledge, sticky knowledge,* and *absorptive capacity.*

2. This chapter also details how one of a number of ICT tools may be used to more effectively create and share knowledge within organisations—this will deepen the reader's understanding of how to realise the potential power of COPs. It provides details of what such software tools can offer in managing and facilitating COPs. This is important because readers will be better positioned to make an informed decision on what features a COP ICT tool should provide to meet their needs.
3. This is a fresh and new area of knowledge for the construction industry in particular, thus this chapter provides a rare opportunity for readers to become better acquainted with a powerful and useful tool available to them.

Background

In the previous chapter of this book, the evolution and development of the concept of COPs were outlined. Etienne Wenger defines a community of practice (COP) as "groups of people informally bound together by shared expertise and passion for a joint enterprise" (Wenger & Snyder, 2000, p. 139). Wenger and others have also described how organisations can encourage and support a COP. Several theoretical aspects of the management of knowledge and knowledge transfer are appropriate topics to introduce in this chapter because they relate to the impact of developing COPs within organisations. COPs are effective tools that beneficially impact upon the capacity of engaged individuals in supporting an organisation's increased learning capacity and innovation deployment. COPs are highly intensive incubators of knowledge creation and transfer. Their ethos is one of sharing knowledge, of voluntary joint problem-solving activities, and building a capacity for learning (Wenger, 1999; Wenger, McDermott, & Snyder, 2002). Understanding underpinning mechanisms of how they do this is vital to their effective deployment.

An important feature of COPs is their capacity for people to share *tacit* knowledge as well as *explicit* knowledge, because this forms a channel for the socialisation of knowledge exchange and its creation. People share not only knowledge, but also perceptions of the context that surrounds the knowledge being shared. Tacit knowledge delivered by people provides an added richness to the nature of the knowledge compared to recorded explicit knowledge (shared for example via manuals, help screens, online tutorials, etc.) because tacit knowledge is embedded with deep personal insights and potentially helpful advice. COPs therefore constitute a very rich and bountiful channel of knowledge exchange. Any tools or mechanisms that can assist and aid a COP's effectiveness facilitate a productive and valuable competitive advantage. Because tacit knowledge is notoriously sticky and difficult to replicate, it also adds further potential for competitive advantage.

We have introduced and briefly explain below knowledge management terms that may be unfamiliar to readers in the context of COPs and knowledge transfer.

Knowledge Management Terms Explained

Tacit Knowledge

Polanyi explored the concept of tacit knowledge, how we can know more than we can tell, through citing an example—face recognition. We can recognise a particular person's face, even someone from the past or someone who we have never met, from the thousands and indeed millions of faces we could have presented to us, yet we cannot explain how we know that particular face (Polanyi, 1997, p. 136).

Tacit, according to the dictionary, means silent, not openly expressed but implied, understood or inferred—from the Latin *taceo,* I am silent (Macquarie, 1987, p. 1727). Tacit knowledge is more difficult than explicit knowledge to create, capture, codify, communicate, and transfer, because these processes are intellectually very highly energy intensive. It requires deep consideration and thought about how to manage this knowledge to benefit individuals and organisations. This is because it involves dealing with people and their motivational drives and inhibitors. It can be argued that people have more complex and unpredictable (and hence unmanageable) characteristics than programmable machines that characterise technology. COPs are ideal for facilitating the exchange of tacit knowledge between individuals. COPs encourage people to talk, exchange insights, and jointly work on problem solving so that the rich context of their knowledge is explicated and tested during their social intercourse.

Sticky Knowledge

Burton-Jones (1999) describes some kinds of tacit knowledge as 'sticky', that is, difficult to codify or explain so that it tends to stick to the person with that knowledge and is only transferred with a fair bit of consideration and effort. Knowledge is sticky and both expensive (in terms of transaction costs) and difficult to transfer because it is more than just facts and information. Knowledge is about context, the history and hidden myriad of inferences and cause-and-effect loops that explain why something did or did not happen in a particular way. Documented manuals and procedures fail to cover all eventualities and are time consuming to access and absorb.

Gabriel Szulanski conducted a series of studies into the transfer (and often failure to fully transfer) of best practice within organisations. He concluded that the three major sources of knowledge stickiness (barriers to transfer of knowledge) are *absorptive capacity, causal ambiguity,* and *the quality of the relationship between source and recipient of knowledge* (Szulanski, 2003). These are briefly explained below.

The most significant source of stickiness is *absorptive capacity.* Cohen and Levinthal (1990, p. 128) define absorptive capacity as the ability of a firm to recognise the value of new external information, assimilate it, and use it for commercial ends. It is a measure of an ability to absorb ideas, information, and knowledge, and applies to both external and internal sources of information and knowledge. Their paper is one of the most influential on this aspect of knowledge management. They detail how firms gain this absorptive capacity; much of it entails both hard work and thoughtful management support of growing its development of a knowledge culture. Building absorptive capacity requires long exposure to experimentation, trial and error, and reflecting deeply on lessons learned through this process. It also requires its people to seek out information and knowledge, both from within the organisation as well as outside it. This research activity need not be 'academic' in a bookish sense, but is more often a practical outcome of people trying their best to make sense out of complex situations when solving problems. The more practice they have in tackling problems as learning exercises and taking the effort and time to reflect upon what they have learned—and transferring this knowledge to others—the greater is their absorptive capacity. When this does not happen, it makes it harder for knowledge to be effectively transferred because these particular strengths are poorly developed. Thus, the wheel gets constantly reinvented and best practice seems impossible to be transferred. Absorptive capacity is a crucial factor in knowledge being transferred either from tacit to tacit or tacit to explicit—the recipient is bounded by his/her absorptive capacity to understand the shared knowledge content and context.

A consequence of poor *absorptive capacity* is often an inability to understand cause-and-effect loops that envelop any opportunity to learn from experience. *Causal ambiguity* is the condition where people have an inability to be able to confidently make a cause-and-effect link. Naturally, if you cannot make this connection, then mistakes are repeated, an inability to replicate best practice is evident, and the management of valuable knowledge becomes extremely difficult. To be able to effectively diagnose situations and be able to confidently read the cause-and-effect linkages requires not only deep knowledge about the context of the situation under study, but also an ability to capitalise upon a strong absorptive capacity. Access to ICT tools such as knowledge repositories have potentially great value, but the skills to fully use this valuable asset are essential to be able to make best use of such knowledge. Unfortunately, electronic knowledge repositories have a limited capacity to store contextual knowledge that can be quickly and easily accessed and understood.

The third major influence on knowledge stickiness is the *relationship between the source and recipient of knowledge*. In terms of electronic sources, they are notoriously cumbersome to engage with—not user friendly. Search engines often either provide too few 'hits' or provide an overwhelming number of them that swamp the user's capacity to deal with the information provided. In terms of people-2-people interactions, the issue of culture and communication plays a major and often subliminal role. An organisational culture can encourage or inhibit knowledge sharing. Personal traits also can influence relationships. Further, organisational leadership style and structure all influence relationships between colleagues and their motivational drivers.

Finally, it should also be understood that when transferring knowledge through a process such as best-practice dissemination, there are four recognised stages of this process (Szulanski, 2003): *initiation* (when the idea/innovation or best practice is being recognised); *implementation* (when planning the dissemination and introduction of the innovation takes place); *ramp-up* (when the innovation is rolled out or a cut-over of the new and existing situations takes place); and *integration* (when the innovation becomes routine and embedded). The three major factors indicated, plus other minor causes of knowledge stickiness, impact upon knowledge transfer in varying intensity and impact at different stages of the knowledge transfer stages.

At the initiation stage, for example, absorptive capacity is highly important because it helps people recognise a best practice and how it may be applied. At the implementation stage, a best practice is planned and introduced, and this poses communication and relevance challenges. The ramp-up stage can be highly affected by causal ambiguity if any cause-and-effect loops for the way in which the best practice is being 'tweaked' in its new setting are not well understood. Finally at the integration stage, backsliding needs to be deterred and any tendency for this lack of integration to be understood so that it can be avoided to allow best practice to be embedded and routinised.

Stickiness of knowledge poses considerable problems for organisations wishing to maximise the conversion of tacit knowledge in people's heads into explicit knowledge that has been codified. However, sustaining competitive advantage relies upon an organisation's competencies being difficult to copy or replicate, so having a knowledge advantage relies upon both codifying knowledge, as well as embedding it in difficult-to-copy repositories such as people's heads and organisational routines, procedures, and culture.

COPs have several roles in relieving the stickiness of knowledge. They increase the absorptive capacity of individuals and groups through joint exploration of issues, generating and sharing ideas and insights on issues/problems, and other project-related topics of joint interest. They may also encompass involvement in seminars, training, and development, as well as research of varying topics. COPs also help link cause-and-effect loops through discussion and sharing insights. Finally, they generally involve improved relationships between participants, especially when membership is voluntary (Wenger, 1999; Wenger & Snyder, 2000; Wenger et al., 2002).

The following section presents a discussion of a UK construction company's experience in using a particular software tool to manage COPs. It provides a relevant and tangible example of what can be achieved, and provides some guidance on how to experiment and pilot the development of COP. The tool is Sigma Connect and it provides a rare example of the construction industry using such tools. Numerous first tier (major construction) contractors use ICT groupware for team communication, but we have found little evidence of the use of COP tools that can manage the nature of the interaction that characterises COPs in the construction industry. Thus, this case study provides a unique or near-unique opportunity for an individual case study, and justifies our focus on one company and its experience with one tool (Yin, 1994).

Characteristics of the *Sigma Connect*™ Tool

In this section we briefly discuss how the tool evolved and describe some of the more important features of the tool. Many of the features discussed share a common heritage with other Web-based applications, intranets, and extranets. This tool helps to administer and manage COPs concerned with ICT applications, and can also embed and integrate this with a wide array of company-specific or general software applications. Such tools evolve and develop over time through feedback and experience gained by users. Sigma Connect has an interesting pedigree.

Origins of the Sigma Connect COP Management Tool

It is interesting to reflect upon its origins because this will help readers understand how such tools can mature and develop with constant improvement and experimentation. The original features are relatively easy to embrace, however as the tool evolved and matured, more tacit knowledge became embedded in its development and this is why we believe it is a good example to use in this chapter—it represents the 'state of the art' or 'bleeding edge' of COP management tools.

Sigma Connect was conceived in late 1996 within the Shared Petro-Technical Resource of British Petroleum (BP), where it was known as 'Connect'. The original desire was to create an expertise-based directory to promote closer working relationships between some 500 exploration consultants and their clients in the business units. The initiative sought to move away from the classical texturally oriented approach, and build a more informal solution in the form of an intranet-based Web site. Upon completion of an initial pilot phase, 'Connect' was taken under the stewardship of the knowledge management community within BP, enabling its use to be extended throughout BP with Addept Computer Services, responsible for the ongoing development of the system which evolved throughout 1997 and 1998 as participation increased. In mid-1998 'Connect' was adopted as a strategic tool for the merger between BP and the American Oil Company (Amoco). It proved invaluable in assisting the integration of people from the two corporate backgrounds. In addition to the proven support as an expertise system, functionality was added to support the management of COPs within the organisation. By early 1999 'Connect' had become a mature application within the BP Amoco group, and in June 1999 the intellectual property rights were transferred to Addept Computer Services, enabling the application to be offered on a commercial basis as Sigma Connect. Currently Sigma Connect has been used with COP interests ranging from construction to banking and from retail to climate change.

Sigma Connect began life as an expertise-finding or "yellow pages" system, enabling people to volunteer information about their knowledge and experience, which they were comfortable sharing. Such systems have been gaining acceptance in major global organisations such as HP's use of 'Connex' (Rollo & Clarke, 2001, p. 49), Xerox's 'Eureka' (Kikawada & Holtshouse, 2001, p. 305), and IBM's 'Community Knowledge Portal' (Rollo & Clarke, 2001, p. 56). The voluntary nature of the system was deemed to be very important, with the emphasis on the quality of the information being captured rather than the quantity. The experience of these other systems, where a mandatory approach has been taken to the gathering and inclusion of data, shows that people are far less likely to respond positively to a request for help originating from a search of such a system. While Sigma Connect provides a sophisticated array of search mechanisms for retrieving information from the system, the real challenge has been on how to encourage and reward users for contributing information. Providing support for COP within Sigma Connect is just one way in which usage of the system can be encouraged and benefits delivered directly to the users. Thus this experience accords with general motivation of COP interactions from the three Australian case studies reported in Chapter 6.

Maturity Level Development and Features of Useful COP Management Tools

Managing COPs with software provides some challenges that need to be overcome for the tool to be effective. Sigma Connect provides a useful model because its development trajectory has triggered a number of useful features to be incorporated. These are briefly described so that readers can discern between this tool's maturity level and other similar tools that are at their nascent stage of evolution.

- *User Authentication*: Sigma Connect supports the Windows NT challenge and response mechanism and/or portal referral authentications models in addition to its own username and password access. This minimises the number of user logins to gain access to the information contained in or linked to Sigma Connect.
- *Integration*: Sigma Connect can be integrated with other systems within the organisation. Upstream integration enables new members to be automatically added when they join the organisation and enables contact information to be kept in-sync with core data repositories. Downstream integration enables other intranet Web sites and applications to make direct use of the information captured by Sigma Connect.
- *Taxonomy*: A taxonomy enables structured searches of the system. This taxonomy is allowed to grow and change to reflect the ongoing business goals. The provision of statistical analysis for the taxonomy usage provides a better understanding of its use.
- *Usability*: The overall look and feel of the portal is one of informal encouragement, rather than dictatorial enforcement. This helps to make people comfortable with using the system and leads people to contribute information and promotes the on-going usage of the system. The terminology that the system uses is very flexible, ensuring it is in line with that already in common use within the organisation, rather than forcing people to adopt unfamiliar phraseology.
- *Structured/Unstructured Fields*: A mix of both structured and unstructured fields provides users with the most productive environment for capturing and searching for expertise. A moderated interface for expanding and managing the content of structured fields enables the systems to grow with the business.
- *Information Delivery*: Using a browser-based interface enables the widest possible access to the system from practically anywhere. There is no reliance on client-side software, and the system can be accessed both from within an intranet or across the public Internet using a low-bandwidth connection, a pre-requisite for many remotely operating construction workers.

- *Encourage Usage*: Any system that fails to remain up to date will rapidly become discredited and rapidly fall out of use. For this reason it is fundamentally important to encourage individuals to use the system and provide incentives for keeping the information up to date. This is achieved in Sigma Connect by providing more than just an expertise finder as part of the system. Providing support for communities of practice, with document sharing and discussion forums, or providing personalised contact management functionality are both good ways of supplementing the basic functionality to encourage usage of the overall system.
- *Dynamic Profiling*: Rather than just being a reactive system, which is driven by user searches, Sigma Connect provides proactive support through the use of dynamic profiling to highlight other people in the system who have similar interests and areas of expertise. This dynamic profiling is also applied to COPs, highlighting COPs that have a high level of membership overlap.
- *Physical Resources*: In addition to addressing the issues of capturing and searching for peoples' skills and experience and managing COPs, Sigma Connect also address the management of the attributes of the organisation's physical resources—meeting rooms, training facilities, application systems, and so forth.

The way that the software tool is used also provides interesting insights. Sigma Connect actively nurtures the growth and development of communities with the objectives of:

1. helping the organisation understand what communities exist and how they operate;
2. stimulating communities to operate globally throughout the organisation;
3. facilitating the re-alignment of communities with the needs of the organisation; and
4. more fully encouraging engagement of people in a proactive manner through the dynamic profiling features—it helps to direct and link people to the range of COP interest groups, thus it potentially expands and enhances COP activities.

Using much the same philosophy as for individuals, Sigma Connect also provides a global platform for communities. It builds on the existing momentum, structures, and facilities, rather than seeking to replace them, stimulating existing initiatives and encouraging new. Through Sigma Connect, an organisation can become aware of all its communities, which in turn benefit from that awareness. With Sigma Connect communities you can:

- distinguish communities by type;
- create sub-communities (to any level);
- search by name or topic;
- distinguish between different members and visitor roles;
- add links, documents, and images; and
- manage community membership.

These features help communities flourish and provide a supportive environment for controlled innovation. The visibility of these communities means that management can ensure that they are aligned with the needs of the business and that task outcomes are more readily available. These features and the experience of its development allow Sigma Connect to be a useful model for aligning support for and development of COPs within organisations.

Use of *Sigma Connect™* by Carillion PLC: The UK Study

We now discuss how a well-established COP management software tool has been used at a major UK construction organisation. We also provide background detail about the UK construction organisation Carillion's experience of using Sigma Connect provided through a semi-structured interview with Gerwyn Jones of Web Development Innovations at Carillion PLC undertaken by an Addept director who was intimately involved with and engaged in supporting Carillion's use of the tool.

Sigma Connect was introduced to the UK construction contractor Carillion PLC as part of its knowledge management strategy. Carillion was formed in 1999 through the demerger of the Tarmac group of companies. Tarmac was a household name in the UK with more than 100 years of experience in the construction industry. Carillion is now focussed on the business and construction services industry with a turnover of 2 billion Euros and over 20,000 employees. The construction industry is rapidly changing, driven by both commercial and public demands for sustainability, safety, and quality. In line with these changes, Carillion's business is also changing, moving away from straight development projects lasting two to three years, to longer service-based projects that not only include the initial construction, but the ongoing facilities management, maintenance, and servicing, which can last for 20 to 30 years or longer.

Sigma Connect was first piloted within Schal, a subsidiary of Carillion, which provides fee-based construction management and consultancy services. A key factor in this business was getting the right people with the right experience involved in each project, process innovation often being the differentiator between competing project bid proposals. The pilot was used to help understand the potential barriers and develop buy-in from the stakeholders of each subsidiary for promoting Sigma Connect's use across Carillion. Cultural barriers were remarkably rare and isolated to a very limited number of individuals: Most employees wanted to promote what they knew and find others that could help them deliver better outcomes for their projects. Eighty percent of Schal's employees were using Sigma Connect within weeks of its launch.

Although Schal could be considered as being experienced construction information technology users, a technical barrier that would affect the sharing culture was identified. Schal's knowledge management strategy noted 'equity' as a key success factor: Every employee had to feel that their knowledge and experience was equally important to the organisation. Sigma Connect suited the equity premise perfectly, but of Schal's 200 employees, not everyone had easy access to a computer that was connected to Schal's intranet, particularly Schal's hands-on site supervisors. The solution employed was to install an additional desktop computer at each project site for hands-on use by the employees. The computer was branded a "Knowledge Centre" and looked physically different to all other desktops (the monitors and desktop cases were black, not beige). This solution was embraced by most of the hands-on site supervisors. Shortly after this successful pilot, the Web Development Innovations team was formed in Carillion with the remit of investigating knowledge management strategies and tools. Gerwyn Jones was impressed by the success of Sigma Connect within Schal, and it was decided it should be rolled out across the whole group.

Carillion views Sigma Connect as an essential part of the knowledge management jigsaw puzzle. It is not a solution in its own right, but plays an important role along with what they term their 'Knowledge Forum', the 'Knowledge and Innovation network', and 'Best Practice Programmes', many of which are substantiated by a COP within Sigma Connect. Carillion has created a number of different COPs within Sigma Connect. These range from the purely social groupings like a five-a-side football tournament, through the typical expertise-based COPs focussed on the use of other ICT products, to a highly structured COP to support the implementation of their internal Fast Track Management Programme.

The main business driver for implementing Sigma Connect and introducing knowledge management into Carillion is the very nature of the modern construction industry—being highly competitive, high risk, with low margins. To succeed in this environment, a business has to be sharper, more efficient, and consistently using its knowledge assets to 'get it right first time' and avoid repeat mistakes. The Business Services side of the business is highly customer focussed in an equally competitive, but growth market.

Carillion's knowledge management initiatives are aiming to avoid people making costly mistakes, but if they do, to learn from it so that no one in the organisation makes the same mistake again. In the UK PFI (Private Finance Initiative) and PPP (Public Private Partnership) tenders are very hotly contested (Akintoye & Hardcastle, 2003). A significant amount of money and resources can be spent tendering for one of these bids, so successful knowledge management is relevant not just to the end project, but in tendering for the business in the first place. Gerwyn Jones stated, "*It is essential that we have the right people, the right lessons learnt, the right knowledge, and all the information at our disposal so we can tailor a bid quicker and better than ever before.*" Before Sigma Connect was implemented, people relied on talking to people they knew, which was limited by the size of their own informal networks. Sigma Connect opened up those networks, breaking down the natural boundaries of location, functional structure, and cultural background to create an environment that welcomes sharing and fosters innovation. This was particularly important to Carillion as it is split into eight major functional divisions, but where any particular project may involve people from three or more of these divisions. Jones also confirmed that Sigma Connect was instrumental in breaking down barriers to knowledge flow within Carillion.

Awareness of the importance of knowledge management in the construction industry has now reached the level where it is not unusual for prospective suppliers of major PFI/PPP projects to be asked what their knowledge management strategy is, within an Invitation to Tender document. As such Sigma Connect is regularly demonstrated to potential clients in bid presentations. This has not only impressed clients, but also generated an additional incentive for staff to ensure their information is current within the system.

Providing a quantitative figure for the return on investment in Sigma Connect is very difficult, particularly as the central support team does not always hear the success stories. But Jones believes, "*Sigma Connect has contributed to the overall success of our knowledge management strategy—expanding the informal networks people used to rely on and improving the quality of our bids. It helps to promulgate a sharing culture within the organisation, one where we capitalise on the lessons we learn.*"

Three examples cited by Jones of how Sigma Connect has helped are:

- With a reduction in new infrastructure in the UK, Carillion Rail decided it should explore the business potential of expanding into Norway and Scandinavia. Sigma Connect was used to find people within the organisation who spoke Norwegian and had the appropriate experience in the railways business. One of the people found subsequently gained a "Value Award" for her performance in achieving business objectives that were above and beyond expectations.
- In the IT department, two engineers were faced with fixing a component in a printer without having any instructions in English. A quick search on Sigma Connect promptly found someone locally who could understand the instructions, and the printer was soon back in action.
- A project was initiated to investigate the creation of a call centre. It was only through Sigma Connect that one of the managers discovered that one of his colleagues, who he had known for over a year, had five years experience in managing a call centre prior to joining Carillion.

The taxonomy structure in Sigma Connect reflects the areas of knowledge and expertise that are relevant to Carillion's business. One enterprising business development manager, responsible for consultancy to property owning and occupying clients, has added areas of expertise to the taxonomies, so that the expertise relevant to his consultancy packages are available. This allows him to immediately identify any new people joining Sigma Connect who have expertise in his areas, and at the same time promotes his part of the business.

Carillion has defined a number of different types and styles of COPs within Sigma Connect. Some examples include:

- Fast Track Management Programme
- Data Protection Group
- Project Eagle - Gaining a better understanding of the business processes
- Hyperion Pillar Users - A software users' COP
- Movement for Innovation - Best-practice groups
- Electronic Data Management - Primarily interested in collaboration tools
- A five-a-side football tournament

The entire active user population of Sigma Connect is in many respects a COP that facilitates the diffusion of ICT, and more importantly of industry knowledge and experience throughout the organisation.

The main benefits of defining a COP within Sigma Connect are:

- Clearly identifies the existence of the COP and its members
- Promotes membership of the COP to people with appropriate expertise profiles
- Contact details and automatic e-mail links are provided for every member
- Group e-mailing of all the members can be accomplished in a single click
- Allows for the management of the membership of each COP in line with organisational objectives
- Provides a central point for storing or linking to documents, files, and Web pages
- Ability to provide an Internet forum linked to the COP
- Dynamic membership profiling highlights potential overlaps between communities
- Information displayed can be different for members and non-members
- Membership can be used to control access to documents and files
- Provides search engine for retrieval of information across multiple COPs

All of these benefits are in addition to the implicit benefits provided by a COP as highlighted elsewhere in this chapter. In rolling out Sigma Connect within Carillion, Jones has utilised the implementer style of COP described earlier in this chapter. Following the initial launch of Sigma Connect, which was supported by a poster campaign, articles in the company newsletter, and e-marketing campaigns, Jones and his team are still championing the system as part of their more general knowledge management initiatives. As such, Sigma Connect is part of a larger culture change programme in Carillion.

Jones believes there are four critical factors for the successful implementation of Sigma Connect:

1. Senior management support
2. Having a dedicated team of people who believe in it to implement and support it
3. Ongoing promotion and coaching
4. Introduction to new employees at induction stage

These four aspects combined constantly revitalise the system and ensure its widespread uptake and ongoing use. To date, approximately one-third of the employees with access to the internal network are now active users of Sigma Connect. Initially Sigma Connect was seen as a stand-alone solution; now Carillion is integrating Sigma Connect with other systems. For example, on the Web site promoting “sustainability within construction,” a link to Sigma Connect could highlight the people within the organisation who are experts in sustainability.

Typically, the COPs in Sigma Connect direct people to other sections of the intranet where information relevant to the COP is stored. Carillion is planning to implement access control security for these areas of the intranet via Sigma Connect. This means that some documentation will only be available to members of a COP, but more importantly, the list of people that have access is visible not only to the members of the group, but also to the visitors who do not have access - further encouraging transparency and an open culture.

Conclusion

We started this chapter with a discussion of three important knowledge management concepts—COP, tacit knowledge, and sticky knowledge. These are linked because they relate to the need for tacit knowledge in particular to be shared by people through the channel of human interaction. We stressed that tacit knowledge is generally very sticky and context bound, and therefore rich channels of communication are necessary to adequately convey meaning—hence knowledge. Further, we stressed that COPs play an important role in facilitating an increase in the absorptive capacity of individuals and groups, of helping to provide the contextual tacit knowledge that helps explain cause-and-effect links, and so reduce causal ambiguity in understanding how innovation or change impacts upon performance. We also indicated that COPs can facilitate the enhancement of the quality of relationships between COP participants. All these factors help to break down barriers that make tacit knowledge sticky. Thus, COPs can be valuable knowledge management aids. With the advent of tools that help build, manage, and maintain electronic interaction of COPs, the scope and scale of COPs can be enhanced and expanded.

We saw in Chapter 6 that COPs are already 'out there' and already impinge upon organisations; thus, it behoves construction organisation leaders to grasp a potentially fruitful opportunity to leverage this highly valuable knowledge asset. The tool illustrated in this chapter provides more concrete guidance for the reader. It shows how COPs that may be currently restricted to user access by phone, e-mail, fax, and person-2-person contact can be facilitated through a Web portal approach that is effectively managed. Features and characteristics illustrated by the Sigma Connect system provided a useful benchmark of system maturity in terms of features offered and issues of security, access, and linkage to other organisational ICT systems.

We also provided a concrete example of how Sigma Connect can be applied in practice; thus, this chapter provides both theoretical results from a case study, as well as practical information on how COPs can be facilitated in the construction industry. Sigma Connect provides an example of a proven, Web-based environment for people to collaborate, sharing their knowledge and experience—connecting people to people and people to information, thereby reinforcing and expanding the scope of their communities of practice and unlocking the implicit knowledge held within the organisation. COP management tools like Sigma Connect also recognise that dialogue is one of the most effective ways of transferring knowledge. Only through dialogue can one ensure that the context of the knowledge flow is in line with the recipients' needs. For this reason, the most effective knowledge channels in an organisation tend to be the personal networks, which form the basis of many COPs. One key to successful knowledge management can therefore be defined in terms of understanding, supporting, expanding, and influencing these networks through the definition and management of COPs and the implementation of appropriate technology to support such COPs.

It is also interesting to reflect that interest in effective ICT and knowledge sharing is also emerging as a competitive advantage in project alliancing. For example, Bovis Lend Lease was part of a winning consortium in Australia which won the right to construct the National Museum of Australia as a result of its information management strategy response to the client requirement for ICT innovation that better facilitates knowledge sharing as well as electronic document transfer (Duyshart, Mohamed, Hampson, & Walker, 2003; Walker & Hampson, 2003).

This chapter is limited by its scope, and a number of interesting issues remain to be explored, such as the degree of ease or difficulty with which tools can be used for these COPs to be developed, the nature and degree of COP connectivity using ICT tools, access to knowledge through use of ICT COP tools, and recruiting the right people with required knowledge and the learning culture environment. However, the chapter's relevance and contribution lies in its connection to the concepts of a COP and the exploration of the nature of the social networks, as well as the case study of Carillion's use of Sigma Connect as an effective tool for managing COP.

Practical Tips and Lessons Learned

- Sticky knowledge is a feature of tacit knowledge, but it can be made less sticky through effective COP actions.
- COPs are already 'out there'; organisations should recognise their usefulness and power for knowledge creation and sharing, and capitalise on their capacity to expand their scope and scale through the ubiquity of the Web.
- The Carrillion PLC case study indicated how a range of COP types can be effectively managed using Web portal technologies—the software needs to be well established to service different needs of people at various levels within organisations, and so considerable 'back-of-house' features do need to be incorporated. The Sigma Connect tool provides a useful benchmark of functionality.
- The dynamic profiling and other smart features of these tools enable people to be more effectively connected through their expressed interests that are tracked by the COP's management tool.

Acknowledgments

We wish to acknowledge the Cooperative Research Centre in Construction Innovation for their financial and moral support. The research study work reported upon here was part of a broader study into ICT diffusion and knowledge management under CRC Project 2001-004 (2B)—Delivering Improved Knowledge Management and Innovation Diffusion. Further, we acknowledge Carillion PLC (www.carillionplc.com) for their cooperation with sharing their knowledge on their use of the Addept Solutions (www.addept-solutions.com) software for facilitating and managing COP.

References

Akintoye, A.M.B., & Hardcastle, C. (2003). *Public-private partnerships: Managing risks and opportunities*. Oxford: Blackwell Science.

Burton-Jones, A. (1999). *Knowledge capitalism*. Oxford: Oxford University Press.

Cohen, W.M., & Levinthal, D. (1990). Absorptive capacity: A new perspective on learning and innovation. *Administrative Science Quarterly, 35*(1), 128-152.

Duyshart, B., Mohamed, S., Hampson, K.D., & Walker, D.H.T. (2003). Enabling improved business relationships – how information technology makes a difference. In D.H.T. Walker & K.D. Hampson (Eds.), *Procurement strategies: A relationship-based approach* (pp.123-166). Oxford: Blackwell Publishing.

Kikawada, K., & Holtshouse, D. (2001). The knowledge perspective in the Xerox Group. In I. Nonaka & D. Teece (Eds.), *Managing industrial knowledge – creation, transfer and utilization* (pp. 283-314). London: Sage Publications.

Macquarie. (1987). *The Macquarie dictionary*. Sydney, NSW: Macquarie University.

Polanyi, M. (1997). Tacit knowledge. In L. Prusak (Ed.), *Knowledge in organizations – resources for the knowledge-based economy* (pp. 135-146). Oxford: Butterworth-Heinemann.

Rollo, C., & Clarke, T. (2001). International best practice – case studies in knowledge management, standard/guideline. *Sydney, Standards Australia, HB 275*(Supplement 1), 207.

Szulanski, G. (2003). *Sticky knowledge barriers to knowing in the firm*. Thousand Oaks, CA: Sage Publications.

Walker, D.H.T., & Hampson, K.D. (2003). Project alliance member organisation selection. In D.H.T. Walker & K.D. Hampson (Eds.), *Procurement strategies: A relationship-based approach* (pp. 74-102). Oxford: Blackwell Publishing.

Wenger, E.C. (1999). Communities of practice: The key to knowledge strategy. *The Journal of the Institute for Knowledge Management, 1*(Fall), 48-63.

Wenger, E.C., & Snyder, W.M. (2000). Communities of practice: The organizational frontier. *Harvard Business Review, 78*(1), 139-145.

Wenger, E.C., McDermott, R., & Snyder, W.M. (2002). *Cultivating communities of practice*. Boston: Harvard Business School Press.

Yin, R. (1994). *Case study research*. Thousand Oaks, CA: Sage Publications.

Section II:

Learning from Experience

Chapter VIII

Knowledge Management: Facilitating Organisational Learning within the Construction Industry

Graham Orange
Leeds Metropolitan University, UK

Patrick Onions
Knowledge Studio, South Africa

Alan Burke
Leeds Metropolitan University, UK

Barbara Colledge
Leeds Metropolitan University, UK

Abstract

This chapter describes an action research project, Building a High Value Construction Environment (B-Hive), involving: two major construction clients, a privatised utility company (Thames Water), and a major leisure services provider (Whitbread Hotels); a large construction company (Taylor Woodrow); two leading construction consultancy firms; and Leeds Metropolitan University and London School of Economics. The research method, action research, was particularly appropriate for this project because it promoted learning and understanding through action and reflection, culminating in the production of a workshop-based approach that facilitates externalisation of knowledge and organisational learning. The approach facilitated the sharing of knowledge amongst construction project partners that would traditionally be very protective, and even adversarial, towards each other.

Introduction

The chapter seeks to address the problem of fragmentation within the construction industry through the processes of knowledge management and organisational learning. Background to the construction industry is provided with references to respected industry commentators such as Latham (1994) and Egan (1998). This is followed by a review of appropriate contemporary research literature on knowledge management and organisational learning leading to the introduction of the Cross Organisational Learning Approach (COLA).

The COLA process was developed as the output of a funded research project designed to facilitate the sharing of knowledge by all partners of construction projects. The chapter describes COLA and identifies outputs from the process. A diagrammatic representation of the learning model is given in the Appendix. The chapter ends by identifying future issues and directions for research.

Background

The issue of fragmentation is well documented (Latham, 1994; Egan, 1998) as being a critical barrier to change since it is seen as a major factor in the poor communications between parties working together on construction projects. The construction industry is organisationally complex and highly fragmented with more than 95% of companies being small to medium sized enterprises (CICA, 1997). In addition, the construction industry suffers from supply chains and relationships that are both dynamic and transient as a direct result of the temporary nature of construction projects, resulting in a poor communication structure. In 1994, the Latham Report (Latham, 1994) recorded that in the UK there were 163,000 construction companies listed in the DETR's statistical register, most employing less than eight people.

This fragmentation means that the ownership and control of separate functions and their associated processes in a construction project life cycle resides in the hands of separate organisations with their own distinctive cultures and working practices. There is ubiquitous recognition of the need to restructure the construction industry to overcome this problem by engendering a spirit of compromise and collaboration. Latham (1994) generally recommended a rationalisation of inter-organisation agreements and methods of communication, and clearer definitions of roles and contract stages. Specifically, Paragraph 6.43 of the report recommends that the client and contractor enter into a partnering agreement. Under this arrangement, the parties agree to work together on a project "in a relationship of trust, to achieve specific primary objectives by maximising the effectiveness of each participant's resources and expertise." Latham wished to create a team spirit, where cooperation replaced conflict, in order to reduce costs, although how this could work was not made clear.

The meaning of a pioneering partnership is still being explored and negotiated within the industry; it represents an opportunity to change the *status quo* and bring about a paradigmatic shift in thinking, but much will depend on the strength of further initiatives if this shift is to be realised.

Associated to this fragmentation at the level of organisational ownership is the associated fragmentation at the level of project team membership which manifests itself in two ways. Firstly, the transient nature of team membership during the course of a particular construction project, in that teams are dynamic in that membership is constantly in a state of flux as individuals move between teams as and when their knowledge and skills are required. Secondly, the fragmentation that exists between projects means that a 'team' is rarely left intact to further build relationships on subsequent projects. Such continuity is essential if learning benefits are to be realised (Barlow & Jashapara, 1998).

Egan (1998) points to the crucial importance of providing this kind of continuity in team composition:

> *"The repeated selection of new teams in our view inhibits learning, innovation, and the development of skilled and experienced teams [and] a team that does not stay together has no learning capability and no chance of making the incremental improvements that improve efficiency over the long term."*

It is thus recognised that this fragmentation and instability has led to chronic knowledge loss when compared to other industries.

Similar recommendations to address this problem have come from two different sources. Egan (1998) suggests that an objective, impartial "Knowledge Centre" (Paragraph 85) be set up to provide access to information regarding good practices, innovations, and experiences for any organisation or individual connected with the construction industry.

The identification of a need for concepts such as the Knowledge Centre appears to be a manifestation of the need for better communication: between individuals, between organisations, and between the present and the future. It is recognition of the need to manage the organisations' knowledge or, as it is now routinely referred to, intellectual capital.

The B-Hive project addressed this need by engaging in action research to design and provide a process, with appropriate information systems and technology support, which could be used to create and capture organisational knowledge. The process seeks to promote organisational learning and informed action through a process of analysis and reflection at various stages throughout the construction process.

This process—Cross Organisational Learning Approach (COLA)—is referred to throughout and, after discussion of its academic underpinning, is described in more detail.

Learning Objectives

The chapter will:

1. Review approaches to facilitating organisational learning within the construction industry.
2. Consider how information systems and technology support might be used to create and capture organisational knowledge.
3. Discuss how organisational learning and informed action might be promoted through a process of analysis and reflection throughout the construction process.
4. Describe an action research project (COLA) that pilots knowledge and organisational learning through reflection in a construction context.

Knowledge and the COLA Process

We suggest that two predominant problems afflict the study and utilisation of knowledge today. The first problem, which we seek to address here, is consensus on a workable and embracing definition for knowledge. The second is a suitable method for the manipulation of knowledge.

The prevailing taxonomy of knowledge models depicts a hierarchical prescriptive structure emphasising categorisation of knowledge into discrete classifications. This facilitates implementation of knowledge management initiatives in situations where a catalogue is beneficial. Such models recognise that knowledge may be implicit, imply that it may be codifiable, but rarely describe how to codify it. There is however no suitable provision for handling the complex interrelationships of tacit and explicit knowledge. Since much of what we do practically relies on knowledge, not information or data, there are practical implications.

Technology today requires that we share our knowledge more openly and using less intrinsic methods than our born senses. Writers of an e-mail for example will use capitals or emoticons such as little smiley faces to communicate their intentions, where previously that meaning would be clarified with body language.

Technology often does not possess a sufficiently compatible interface to biological systems. During externalising and transfer, the storage medium for knowledge is changing and so is the format and substance of knowledge. For example, sales call knowledge moves from neurons to silicon and from an unstructured set of ideas and feelings to tangible and compartmentalised database fields.

While hierarchical views may be valid, one person's knowledge can be another person's information or data. An organisation using an intranet as a knowledge transfer and productivity tool would need to consider the context of the knowledge, relevance to the organisation, creators or codifiers, and the recipient.

Knowledge management, as the contextualisation and integration of information so that useful activity may be informed and performed, therefore requires more clarity as to what knowledge 'is'.

On one level 'knowledge' is often used as a synonym for 'information', so when people talk about knowledge management, they are sometimes referring to information management in a different guise. In this context the term knowledge management refers to technology used to store softer information which previously did not lend itself to formal structuring and storage in computer-based information systems. The advent of newer storage and manipulation systems such as document management systems, hypertext, and Web-based technologies has meant that these 'softer' types of information/knowledge can be tackled. There is of course nothing wrong in using the term 'knowledge' in this synonymous sense, but it is important to be aware of wrongly assuming common understanding of meanings when tackling knowledge management issues.

Developing an understanding of knowledge may best be achieved by considering various contemporary models. Polanyi (1967) arranged knowledge on a continuum from tacit to explicit; from tangible and that which could be expressed, to intangible and retained in the mind of the knower. Choo (1998) added cultural knowledge, the basis for what we deem fair and trustworthy. Vasconcelos, Kimble, and Gouveia (2000) classify knowledge by replacing Polanyi's continuum with a dichotomy, and adding another level to further categorise tacit and explicit knowledge.

Considering knowledge practically, Demarest (1997) identifies scientific, commercial, and philosophical knowledge. Sveiby (2000), in his model of knowledge management practitioners, divides knowledge into objects and processes. The transfer of knowledge has also been treated ontologically. Boisot (1997) regarded knowledge as either codified or uncodified in his C-D Theory. Nonaka and Takeuchi (1995) described the flow of knowledge as a transfer process that involves either tacit or implicit knowledge.

Philosopher Karl Popper (1959, 1963), after deliberating on scientific knowledge, developed his theory of falsification whereby knowledge cannot intrinsically be proven true, only false by means of a refutation, knowledge a belief until replaced with a concrete rebuttal. This implies if a belief is knowledge, then knowledge must be able to exist solely in the imagination of the knower and have no prior corporeal relationship.

Medical opinion has considered whether knowing knowledge and brain function are linked. Bateson (1972, 1979) deduced that information is based on perceived difference in biological systems. Greenfield (1997) separated short- and long-term memory, where the latter has an explicit component and an implicit component that doesn't need to actively and consciously remember something.

Perception however is not necessarily just a faithful replica of what we sense. In discussing data analysis in social systems, Dey (1993) said: "The trouble is, of course, that we tend to see what we want to see and hear what we want to hear... We tend to make more of the evidence that confirms our beliefs and pay less attention to any evidence that contradicts them," suggesting that knowledge may be strongly affected by context and past or other knowledge.

Definitions vary. Samuel Johnson succinctly defined knowledge as: "Knowledge is of two kinds: we know a subject ourselves, or we know where we can find information upon it."

A little more detailed is Wiig's (1993) definition: "Knowledge consists of truths, beliefs, perspectives and concerns, judgements and expectations, methodologies, and know-how. Knowledge is accumulated, organised and integrated, and held over longer periods to be available to be applied to handle specific situations and problems."

Although not a formal enough basis on which to design a digital repository, this definition introduces time and process. Davenport and Prusak's (1998) definition includes the concepts of transience, context, frameworks, metaphors, processes, and the mind of the knower: "...a fluid mix of framed experience, values, contextual information, and expert insight that provides a framework for evaluating and incorporating new experiences and information. It originates and is applied in the minds of the knowers...."

These differences in perception about what constitutes knowledge is an apposite issue within the context of knowledge management. The challenge within knowledge management is to decide when it is appropriate to change knowledge types into forms that can be captured, manipulated, stored, and communicated by a computer. This may involve converting tacit knowledge, which is not formalised and is therefore difficult to communicate to others, into explicit knowledge, which can be fully recorded, communicated, and shared (Willard, 1999), whilst at the same time being aware of that which may be lost in translation. More recently, approaches such as the 3K model (Onions & Orange, 2003) have focussed on better accommodating the dualistic, multi-faceted nature of knowledge, simplifying complexities, and reconciling contradictions between tacit and explicit knowledge.

In order for organisations to learn, it may be necessary to release the tacit knowledge that is stored in organisation 'experts' minds. The passing on of knowledge is an essential part of organisational life and economy. It may be necessary for an organisation to attempt to manage some categories of tacit knowledge, capturing and converting it into an explicit format that supports communication to others. It can then be passed on as information (enhanced by experience) to those who may find it useful, fed back as consequences of others' actions, and used to identify and exploit new opportunities. One objective of knowledge management is to provide a means to externalise and codify appropriate tacit knowledge for exchange and to support the reflection and discussion facilitated by the COLA review process. An information system has been developed to support this process and to record and disseminate appropriate forms of the learning which result from the review.

Individual, Collective and Organisational Learning

The knowledge and experience of an individual shapes the individual's actions through a process known as learning, enabling the individual to change and so deal more efficiently with similar situations and cope with, or invent, different approaches to new situations. Learning occurs when new information is compared to previously received information. The comparison is then reflected upon in order to evaluate a suitable behavioural route to the goals required.

Thus, there are internal and external aspects to learning. The internal aspects address how new information is processed by the individual. External aspects cover the means by which new information is arrived at.

Information can be acquired by an individual using methods that are a combination of the two extreme learning methodologies: taught and discovery. Buckler (1998) finds advantages and disadvantages in both. The taught method is useful for passing on solutions that may be copied and where conformance to specifications is required. The advantage of teaching is that it is a low-risk method of passing on information in a consistent way. The disadvantage of prescriptive teaching—or rote methods of learning—is the increased potential it has for stifling motivation and reinforcing automatic responses to events that might otherwise have presented opportunities for learning.

In contrast, the Discovery method allows theories to be developed and followed through. It provides an opportunity to develop solutions individually, thus allowing creativity and innovation. The drawbacks of discovering new information are that it can consume vast amounts of resources, is subject to a higher risk of failure, and may allow the learning objectives to stray.

This raises the question of organisational learning. What is it? How can it be promoted? How do these theories relate to organisational learning? In order for an organisation to survive and grow, it must therefore be able to learn, yet as a conceptual entity, as Senge explains, "Organizations [can] learn only through individuals who learn." When an organisation's members consciously learn and change their behaviour accordingly, the organisation itself can not be said to have learned. It is only when the effect of an individual member's behaviour has a permanent effect on the behaviour of others within the organisation that it can be said that there has been a "mutual behavioural change," which therefore indicates "mutual learning." It is then that this organisational learning and the capacity to take organisational action can be seen as an emergent property of that collective individual learning.

Senge identifies five 'disciplines' (Senge, 1990) as ingredients that are essential for a learning organisation. The first four disciplines are based around personal or individual qualities, such as a manner of thinking, a commitment to continuous discovery, a way of perceiving issues, encouragement and motivation, and the use of shared views. The fifth discipline, however, deals with the culture of the organisation to provide the conditions that facilitate the creation of interpretative knowledge. It requires that an environment be established where the previous four disciplines are encouraged and can be practised, where there is a desire to operate collectively and no personal aspirations for gain, nor any need for defensiveness. Senge believes that it is in such an open and learning environment where new and different concepts can be raised and built on that groups, and therefore organisations, can learn. That learning thus becomes part of the organisational memory: This memory consists of sets of shared previous experiences that are important shapers of future action.

Epple, Argote, and Murphy (1996), in their discussion of knowledge acquisition and transfer, talk of this concept of an organisational memory and promote Levitt and March's argument (in Epple, Argote, & Murphy, 1996) that organisations not only learn from their own experience, but from the experiences of other organisations (where individuals can pass on that information). As a result of such transfer of knowledge, the learning organisation is said to benefit by increases in efficiency.

In terms of the learning theories discussed previously, a learning organisation must allow both the Taught and Discovery methodologies to operate simultaneously. There must be opportunity for research, innovation, and divergence from the 'normal' course and a provision for an amount of resources to be expended in this way. There must be an acceptance that there is no 'right' answer and an attitude that improvement is always possible and achievable. Nevertheless, there must also be a facility whereby information regarding the best current practice (which might be decades old or have been updated earlier the same day) can be accessed and copied or adapted. A mix of the two methodologies is therefore necessary, though the formula of that mix will depend on the organisation and its environment.

Stata in Senge (1990) claims that, because of the Tayloristic approach of labour division that has been adopted by most industries in the western world, the workers and the managers have become uncommunicative doers and thinkers. Sata believes that an organisation must bring the two together to enable a learning process to develop. This confirms both Egan's recognition that those engaged in the organisation's activities are vital to that organisation's (or industry sector's) improvement and Drucker's (1998) view that feedback is the most important learning principle.

A further requirement for learning seems to be the ability to reflect on past experience to modify future thinking and behaviour. The role of reflection in individual and organisational learning has been considered in detail by Argyris (1989) and Argyris and Schon (1976). Argyris in particular has studied the role of reflective practice in organisational learning.

Perhaps one of the most important principles supporting the COLA review process is that of reflective practice. Reflection manifests itself through enquiry (Day, 1993) and evaluation involving problem solving and reconstructing meaning. It is an iterative process, enabling tacit knowledge to become explicit and allowing for the externalisation of interpretative knowledge at each iteration, progressively refining and enhancing the individual's knowledge and expertise through reflection, which takes place by capturing recurrent issues and problems.

Thus, while it is important that individuals actively learn and modify their behaviour accordingly, it is vital for an organisation that the process of collective learning is fostered by the organisation (and its environment). Reflective practice should not be seen just within the context of the individual since it is a social process (Harvey, 1998). It is: "....a form of self-reflective enquiry undertaken by participants in social situations in order to improve the rationality and justice of their own practices, their understanding of those practices, and the situations in which those practices are carried out" (Schratz, 1993).

A primary function of COLA is to provide a review process forum for the social construction of knowledge. This kind of knowledge may not already exist since it is concerned with the alignment of norms and values within the group; the group needs to establish what can be reasonably expected from each of its members. This knowledge is socially constructed because it is established through discussion and reflection upon past experience and future expectations. Such knowledge, when it is generated, does not belong to an individual, but can be seen to exist collectively within the group and is subject to constant renegotiation.

Reflective practice can have benefits for the organisation as a whole since it contributes to individual learning, and when seen as a social process, it contributes to organisational learning. Individual learning on its own is not sufficient (Jones & Saad, 1998) for the organisation to maximise the benefits to be gained from reflection. Individuals move around the organisation from team to team. They do not necessarily share their knowledge and experience with colleagues because the mechanisms do not exist to support sharing, or perhaps they just do not know how, or culture does not facilitate sharing.

Organisations must therefore be aware of the various methods, views, and attitudes, and how they affect collective and organisational learning.

Various views have been put forward on knowledge management strategies, and many of them can be seen as complementary rather than mutually exclusive (Stewart, 1997).

The National Computing Centre Guidelines suggest two approaches to knowledge management (Vowler, 1999): manage the people or manage the knowledge The first approach can be seen as suitable for dealing with interpretative knowledge, whilst the second is more suitable for dealing with soft knowledge.

Managing the people requires social, cultural, and philosophical skills in order to be able to encourage individuals into the mindset of openness and sharing. This approach is primarily concerned with interpretative knowledge, whilst technologists are primarily concerned with explicit knowledge, which requires the use of databases, networks, and so on.

In reality both of these approaches are often undertaken simultaneously, as judgments are made about knowledge which can be beneficially codified and knowledge which is best held within individuals and the social system. Hansen, Noria, and Tiernay (1999) advise against this, recommending a strong application of either codification or personalisation. The forced introduction of a computer system that stores information previously held informally within the social system is an easy way to alienate those who are intended to use it!

COLA and the COLA Environment: Knowledge and Organisational Learning through Reflections

The product of the research is the Cross Organisational Learning Approach (COLA—further details may be found at is.lse.ac.uk/b-hive), which has theoretical perspectives in the arena of organisational learning, underpinned by the principles of reflective practice and knowledge management, and supported by the development of an information system that facilitates the capture and dissemination of organisational knowledge.

The B-Hive project identified the problems confronting the construction industry as complex, with different participants holding both different understandings of the issues and having often competing goals, and where there are no certain and agreed measures of inputs, processes, and outputs. The set of approaches and methodologies constituting *problem structuring methods* (Rosenhead, 1989) were adopted both as tools for the academic/industry team to agree on their approach, and later as part of the interventions themselves. An exercise using Strategic Options Development and Analysis (SODA) (Eden, Jones, & Sims, 1979) led to the identification of the issues of post-completion review in the leisure company and the Management of Project Changes in the utility company as key aspects of the process, where unexploited opportunities for learning were being created in a partnering and value-sharing environment. B-Hive has run one-day post-completion review workshops with the leisure company. The first two exploratory, post-completion workshops were run on the basis of value engineering techniques (Connaughton & Green, 1996), drawing on the experience that the leisure company's representative had in applying this technique. A major constraint on designing a workshop process for project review is the availability of staff and the length of time that they are willing to spend at a workshop. The experience of the exploratory workshops indicated that attention had to be paid to pre-workshop activities, and some activities that are normally carried out within a workshop had to be moved to the pre-workshop phase.

These considerations led to the development of a process model (a diagram of the COLA process and description of its elements may be found in the Appendix). This was an attempt to identify processes to assist in project review in order to: promote the sharing of knowledge across organisational boundaries, and identify the role that information systems can play in supporting the process. It also played a major role as a communication tool to build consensus amongst the B-Hive project team and to inform other workshop members about the process. The model enabled a dialogue between project members about the status of organisational knowledge and learning and, through debates about the concept of a bank of acquired knowledge, the acceptance of the notion of tacit knowledge (Polanyi, 1967). This also enabled an approach that placed emphasis on managers' understanding and interpretations of the world in which they work as a basis for moving towards action (Introna, 1997).

The COLA review process records and monitors key issues, decisions, and actions surrounding these key issues, ensuring that implicit learning is not lost throughout or on completion of the construction project. The term 'review process' is used which allows a great deal of flexibility in that many different procedures may be performed for review (e.g., workshop, meeting, teleconference, videoconference, or any combination of these). The B-Hive project mainly used workshops.

There are many situations that may necessitate the review process being triggered, for example:

- Programmed Review
- Post Completion, leading to a review where the scope encompasses the construction project as a whole

- Stage Completion, scoped for a particular stage of the construction process
- Time Based (e.g., period end, monthly)
- Non-Programmed Review
- Issue Resolution, necessitating a review to address a particular problem of high priority, for example running late or over budget, or perhaps a technical difficulty
- Innovation, where a team has been innovative either in process or use of materials; this experience should not be lost

However all reviews facilitate critical reflection on past activities, a focus on individual and organisation learning, and allow change to future actions to increase value to all participants.

Information Systems and Technology

Within B-Hive we have developed the information system required to support the COLA workshops. Technology can assist organisations in making effective use of their assets (Solvet Ltd., 1997). Such claims, while implying a more efficient utilisation of the artefacts an organisation has purchased, such as machinery, computers, and so on, can also apply to the concept of intellectual capital (Edvinsson & Malone, 1997; Stewart, 1997), that is, the availability of useful knowledge within an organisation.

This is particularly true with organisational learning, since no single application or technology holds the knowledge from individuals within an organisation. Rather, it requires a collection of people and many application types, for example, databases, spreadsheets, word processors, CAD systems, multimedia, workflow, document management, and e-mail. The challenge therefore is twofold—to convert relevant categories of soft knowledge into information that can be stored, and to make the stored information available to those who require it irrespective of which part of an organisation they work for or their physical location. An intranet is a mechanism that can be used to provide this access. As Peddler, Burgoyne, and Boydell (1997) explain, the networking of computers allows individuals' experiences to be fed back into the 'collective organisation'. Peddler and colleagues claim this makes information available that had previously been passed on informally, for example "by the coffee machine." However, it would also be true to say that this is necessary since computers, re-organisation, and 'progress' in general have reduced the opportunities for such informal communications.

An extension of this combination of technology and philosophy is the extranet, which provides access to information across organisational boundaries. This is especially applicable for organisations that are cooperating within a group. The use of extranet technology can provide a collaborative virtual environment where individuals can access (where permission is granted) any partner organisation's intranet. In this way individuals, organisations, and the partnership can make use of previously gathered information and past experiences, and learn from them more easily, regardless of their location, either physically or within the partnership.

The COLA information system supports pre-review event, review event, and post-review event activities. Pre-review event information is requested (via questionnaire) to identify major problems and issues throughout the life cycle of the construction project. Currently the questionnaire is circulated to the construction project team members representing, where possible, all organisations of the partnership.

As a result of this data-collection exercise, problems and issues, arising from the construction project, are collated, categorised, and prioritised, ensuring that only those perceived to be of major importance by the participants are presented for consideration in the review event. During the review event, actions, decisions, and responsibilities are assigned and recorded against each problem or issue. The review process is designed such that soft knowledge is elicited, yet interpretative knowledge is simultaneously created and externalised. The information system will monitor the performance and value of decisions made and actions taken. This information is currently held on a Microsoft Access database.

The database holds the descriptive information on the problems and issues. Each will have supporting historical data dispersed throughout the partner organisations held in many formats such as word-processed documents, spreadsheets, databases, and drawings. Where feasible, there will be links between the data held on the COLA project review database and these supporting files, and access will be via an extranet.

The COLA information system is not perceived as consisting solely of the Access database system. The primary function of the database system is to capture and process data centred on the review process, for example the following data may be captured and held:

- Pre-review event
  - Data on construction projects and partnerships enabling project team members to access construction project information required to set the context of issue(s) under review.
  - Data on events that may trigger a reflective review, for example the end of a construction stage or post completion or a significant issue.
  - Issue—this may be as a result of a major problem or perhaps where a course of action has resulted in a positive impact and the partnership does not want to lose the associated knowledge. The system facilitates the classification and prioritisation of the issues prior to the review so that a focus is maintained on key business issues during the review event.
- Review event
  - Major decisions, actions, and responsibilities from both the historical perspectives in terms of the construction project and those defined in the review event are captured online during the review event.
- Post-review event

- The effectiveness of decisions and actions taken is monitored to assess any value improvements resulting from the review process.

This information is made available to all partner organisations regardless of physical location and may be accessed through an extranet utilising World Wide Web-based technologies.

Research Outcomes

- *Skills and knowledge transfer:* The COLA review process adopted useful, practical techniques that were not dependent on technology. The use of these techniques and the workshop structure made a substantial contribution towards convincing people of the benefits of adopting a knowledge management approach.
- *Awareness of corporate knowledge assets:* The research project raised the awareness of knowledge as a corporate resource. All organisations now understand what their valuable knowledge assets are, and have a mechanism to identify them in future. More importantly, the same mechanism also shows how that knowledge may then be used.
- *Increase utilisation and management of corporate knowledge assets for maximum return:* The research targeted the knowledge chain, identifying and capturing knowledge, facilitating use and reuse. The unambiguous way in which this was done ensured that the knowledge needed for specific tasks can be identified, defined, and located throughout its life cycle.
- *A central knowledge repository showing links with information and data:* The large number of disparate and often tacit sources of knowledge hampered centralised collection of knowledge across the organisations. However, the research clearly demonstrated the links between data, information, and knowledge to the individual and eventually to the work processes. This showed that certain knowledge may be explicitly mapped and codified, and linked to other internal and external sources of knowledge, information, and data.
- *Improve accessibility and information sharing between employees:* The knowledge mapping described by the COLA model allows accuracy in locating and transferring tacit and explicit knowledge, speeding access and simplifying retrieval and transfer.

Conclusion

To date COLA has been successful within the context of the research project in that many of the key issues surrounding organisational learning, as discussed in this chapter, have been addressed. One of those key issues is that during the life of a construction project, problems may arise or events occur that result in a non-standard approach being adopted. Normally this would lead to actions being taken that give rise to the benefit of the project (e.g., cost reductions, shorter construction times, new working practices). This leads to the generation of knowledge of how to deal with future occurrences amongst the project team. However, this knowledge is socially constructed and may lose context and meaning as the project team is split on completion of the building. Thus, in reality this knowledge is very rarely made available to other projects. B-hive and the COLA process have provided the means by which some of this knowledge may be generated, externalised, and stored, readily accessible to all partner organisations.

The COLA approach has most definitely fostered a spirit of trust and cooperation amongst, at the project team level, individuals from the different companies and, at a higher level, between partner organisations, thus addressing one of the major concerns of Latham. This was found to be particularly true when COLA reviews were conducted on Whitbread Hotel construction projects.

However, as with many projects, a number of issues have been raised by the research that could not be addressed at the time.

Practical Tips and Lessons Learned

- Individual learning on its own is not sufficient for an organisation to maximise the benefits from reflection.
- Reflective practice benefits the organisation and, when seen as a social process, contributes to organisational learning.
- Mechanisms for sharing individual knowledge and experience need to be addressed explicitly by organisations.
- Technology can assist organisations in making efficient use of their assets, including "intellectual capital."
- The pilot project (COLA) resulted in improved accessibility and information sharing between employees, enabling knowledge and learning captured to be shared for the benefit of future projects.
- The process fostered a spirit of trust and cooperation amongst project individuals, organisations, and stakeholders.

Future Research

It is intended that the COLA approach will be incorporated into future research initiatives so that the approach can be further refined and that such issues may be addressed. In particular there is work on-going that aims to integrate the 3K model, the Integrated Knowledge Architecture Model—IKAM (Onions, 2002, 2003), and the COLA approach.

IKAM suggests a holistic and systemic development of a knowledge strategy that focuses on all aspects of an organisation. IKAM is a programmatic investigation and analytic tool and approach that commences with strategic and operational objectives; factors in the organisation, culture and communication, people, and processes; then examines the knowledge itself and how it is known; introduces the technical infrastructure; and concludes with a selection of appropriate tools, technology, and implementation plan. Some preliminary work has already been undertaken as part of a project at the Department for Public Enterprise in South Africa.

Future research will centre on the following major issues:

- What approach and techniques? There are many approaches and techniques that are put forward as solutions to knowledge management and organisational learning issues. 3K and IKAM help organisations to identify what is best for the organisations and help develop an appropriate LM and OL strategy. COLA will facilitate the development and implementation of that strategy.
- Ownership—who owns the captured 'knowledge'? The research project was concerned with the capture, storage, and dissemination of knowledge from different organisations. Whilst organisations operate as separate entities, the concept of ownership of knowledge is relatively simple and becomes a matter of concern only within the individual organisation's boundaries. Sharing of knowledge with other organisations has always occurred, but has tended to be through informal processes. Furthermore the nature and content of that knowledge has been highly selective, that is, only letting others know as much as is needed and no more. The B-Hive project throws away these old conventions and encourages openness and sharing, promoting the creation of a rich repository of knowledge contributed by and accessible to all participating organisations. So who has control over this resource? Where does it reside? How is it communicated and who controls its communication? What happens when one of the organisations leaves the partnership? Do they still have rights of access to the repository of knowledge that they have contributed to?
- What is lost through elicitation and externalisation of knowledge? While COLA has demonstrated that it is effective, it is still a relatively young, and perhaps crude method, and cannot lay claim to being able to fully store and disseminate all appropriate organisational knowledge. Furthermore, when tacit knowledge is externalised, is meaning and context lost? Can explicit knowledge still be classed as knowledge? Once it has been codified, does any loss of meaning and context render it information? If so what is the value to the organisation of that which is lost?
- The technical infrastructure required to support knowledge sharing will need to be defined. This will involve identifying the processing and communications hardware and appropriate software and applications to facilitate access to knowledge

by all participating organisations. This architecture will, of course, require effective management to deliver the level and quality of network service required to support such knowledge sharing in a dynamic, evolving environment (Lupu & Sloman, 1999; Puliafilto & Tomarchio, 2000). It is also the case that the technical infrastructure of the early 21st century will involve both static and mobile networked systems (Sahai & Morin, 1998).

References

Ackoff, R.L. (1989). From data to wisdom. Presidential address to ISGSR June 1988. *Journal of Applied Systems Analysis, 16.*

Argyris, C. (1976). Single loop and double loop models in research on decision making. *Administration Science Quarterly.*

Argyris, C., & Schön, D. (1987). What is an organization that it may learn? In R. Paton, S. Brown, R. Spear, J. Chapman, M. Floyd, & J. Hamwee (Eds.), *Organizations: Cases, issues, concepts.*

Barlow, J., & Jashapara, A. (1998). Organisational learning & inter-firm "partnering" in the UK construction business, *The Learning Organisation, 5*(2), 86-98.

Bateson, G. (1972). *Steps to an ecology of mind.* New York: Chandler.

Bateson, G. (1979). *Mind and nature.* New York: Bentam Books.

Boisot, M. (1987). *Information and organizations.* London: Fontana Collins.

Buckler, B. (1998). Practical steps towards a learning organisation: Applying academic knowledge to improvement & innovation in business process. *The Learning Organisation, 5*(1), 15-23.

Choo, C.W. (1998). *The knowing organization.* Oxford: Oxford University Press.

Connaughton, J.N., & Green, S.D. (1996). *A client's guide to value management in construction.* London: CIRIA.

Davenport, T.H., & Prusak, L. (1998). *Working knowledge.* Boston: Harvard Business School Press.

Day, C. (1993). Reflection: A necessary but not sufficient condition for professional development. *British Educational Research Journal, 19*(1).

Demarest, M. (1997). Retrieved from www.hevanet.com/demarest.

Dey, I. (1993). *Qualitative data analysis: A user friendly guide for social scientists.* London: Routledge.

Drucker, P.F. (1992). *Managing for the future: The 1990s and beyond.* Oxford: Butterworth–Heinemann.

Eden, C., Jones, S., & Sims, D. (1979). *Thinking in organizations.* London: Macmillan.

Edvinsson, L., & Malone, M.S. (1997). *Intellectual capital: The proven way to establish your company's real value by measuring its hidden brainpower.* England: Judy Piatkus.

Egan, Sir J. (1998). *Rethinking construction.* Department of the Environment, Transport and the Regions.

Epple, D., Argote, L., & Murphy, K. (1996). An empirical investigation of the microstructure of knowledge acquisition and transfer through learning by doing. *Operations Research, 44*(1), 77-82.

European Conference for Industrial Use of Product Data Technology. (1997). CICA, Sophia, Antipolis, April 1997. *Taking control of the building process.* Presented on behalf of the CONCUR project Partners.

Greenfield, S. (1997). *The human brain* (p. 125). London: Weidenfeld and Nicholson.

Hansen, M.T., Nohria, N., & Tiernay, T. (1999). What's your strategy for managing knowledge? *Harvard Business Review*, (March-April).

Harvey, M., Palmer, J., & Speier, C. (1998). Implementing intra-organizational learning: A phased-model approach supported by intranet technology. *European Management Journal, 16*(3), 341-354.

Introna, L. (1997). *Management, information and power.* Macmillan, Basingstoke.

Jones, M., & Saad, M. (1998). Loop the loop to get ahead—organisational learning as a tool to manage change. *Project,* (June).

Latham, S.M. (1994). *Constructing the team: Final report of the government/industry review of procurement and contractual arrangements in the UK construction industry.* London: HMSO.

Lupu, E.C., & Sloman, M. (1999). Conflicts in policy-based distributed systems management. *IEEE Transactions on Software Engineering, 25*(6), 852-869.

Nonaka, I., & Takeuchi, H. (1995). *The knowledge creating company.*

Onions, P. (2002). *Knowledge, efficiency and effectiveness in the organisation.* MBA Dissertation, Oxford Brookes University.

Onions, P. (2003). *Knowledge strategy blueprint for the Department of Public Enterprises, South African government.* Johannesburg: The Knowledge Studio.

Onions, P., & Orange, G. (2003). The 3Ks: A model of knowledge that supports ontology and epistemology. *Proceedings of the 6th World Conference on Systemics, Cybernetics and Informatics,* Orlando, Florida, July.

Peddler, M., Burgoyne, J., & Boydell, T. (1997). *The learning company: A strategy for sustainable development.* McGraw-Hill.

Polanyi, M. (1967). *The tacit dimension.* London: Routledge and Kegan Paul.

Popper, K. (1959). *The logic of scientific discovery.* London: Hutchinson.

Popper, K. (1963). *Conjectures and refutation: The growth of scientific knowledge* (p. 42).

Puliafilto, A., & Tomarchio, O. (2000). Using mobile agents to implement flexible network management strategies. *Computer Communications, 23,* 708-719.

Rosenhead, J. (1989). *Rational analysis for a problematic world: Problem structuring methods for complexity, uncertainty and conflict.* Chichester: John Wiley & Sons.

Sahai, A., & Morin, C. (1998). Mobile agents for enabling mobile user aware applications. *Proceedings of the 2nd International Conference ACM Autonomous Agents.*

Schratz, M. (1993). Researching while teaching: Promoting reflective professionality in higher education. *Educational Action Research, 1*(1).

Senge, P.M. (1990). *The fifth discipline, the art & practice of the learning organization.* Century Business.

Solvet Ltd. (1997). *Information and communication technologies—implications for small and medium sized enterprises and their workforce—a strategy model.* England: Calderdale and Kirklees Training and Enterprise Council.

Stewart, T.A. (1997). *Intellectual capital: The new wealth of organizations.* Nicholas Brealey.

Sveiby, K. (2000). Retrieved from www.sveiby.com.au/.

Vasconcelos, J., Kimble, C., & Gouveia, F.R. (2000). A design for a group memory system using ontologies. *Proceedings of the 5th UKAIS Conference* (pp. 246-255). McGraw-Hill.

Vowler, J. (1999). If it quacks and it waddles... it's a duck. *Computer Weekly*, (10 June), 62.

Wiig, K. (1993). *Knowledge management foundations* (Volume 1). Arlington, TX: Schema Press.

Willard, N. (1999). Knowledge management: foundations for a secure structure. *Managing Information, 6*(5), 45-49.

Appendix: The COLA Process

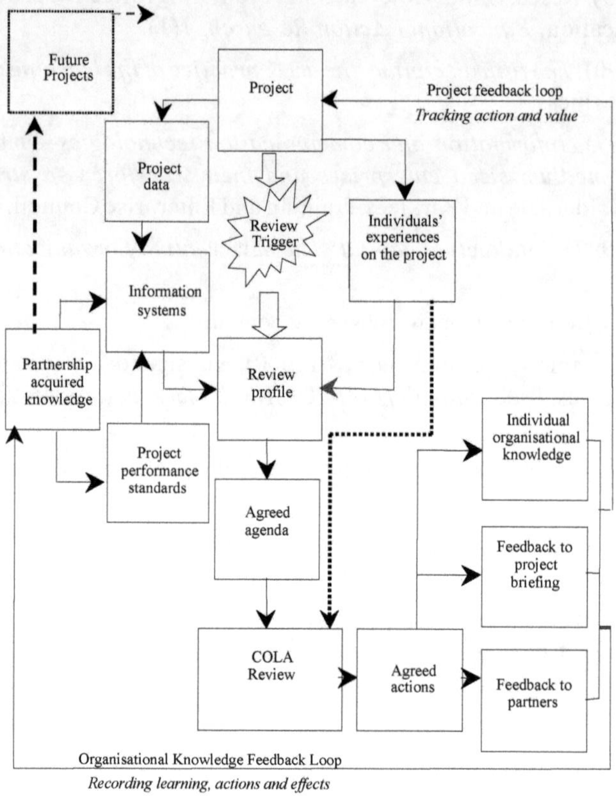

Model Descriptors

- **Partnership Acquired Knowledge.** The knowledge built up within a partnership. This is in part written and formal, and in part the accepted ways of working developed within a partnership.
- **The Project.** The construction project under review.
- **Project Data.** Briefs, plans, meeting records, conditions, incidents, performance information, people and organisations, business results, and so forth.
- **Project Performance Standards.** Criteria by which the project is measured: requirements, benchmarks, model costs, building regulations, and so forth.
- **Information Systems.** These may be paper based or computer based. B-Hive has developed, as part of COLA, a model for a system to collate, analyse, and report on the comparison between the current project and past projects. The specification

includes analysing questionnaire returns and prioritising issues, as well as monitoring the effectiveness and value of actions.

- **Review Trigger.** The event that sparks the review—this may be the end of a stage of the project or a critical event in the life of the project.
- **Individuals' Experiences on the Project.** Each individual will have a different view of the project depending on their role and experience. The **COLA** process values and uses this knowledge of the project.
- **Review Profile.** The individual views are collected in the pre-review stage—through questionnaires for more formal reviews. These are set alongside the data and performance standards held in the information systems.
- **Agreed Agenda.** Prior to the review there will be individual discussion on the profile to determine the key areas for discussion and decision to meet the participants' objectives.
- **COLA Review.** The review may take different forms. Depending upon the risk and value attached to the issues to be considered, it may take the form of a day workshop or it may be part of a regular project meeting. A workshop for a high-risk/value review would use an appropriate Problem Structuring Method or problem resolution technique, and require an external facilitator. B-Hive has developed a workshop methodology for such reviews. A COLA review leads to a set of owned and agreed actions.
- **Agreed Actions.** It is crucial to the development of organisations and partnerships that actions are placed upon individuals who have the power and authority to see them through to completion. Commitment to an action also includes agreement of criteria for monitoring its implementation and measuring its effectiveness.
- **Individual Organisational Knowledge.** Each member organisation of the partnership will take the learning gained during the review process, and may use it in other projects and to develop their competitive advantage.
- **Feedback to Project Briefing.** The agreed actions are fed back to be used to improve current and future project briefs.
- **Feedback to Partners.** Within the partnership it is critical that all organisations share the knowledge of the agreed actions. This adds to the Partnership Acquired Knowledge—allowing the benefits of partnering to be realised and shared between all members to keep the partnership ahead of the competition.
- **Feedback Loops.** The **COLA** process is based upon feedback that supports the partnership by tracking the agreed actions, their value impact in the current and future projects, and feedback that records learning and makes it available for future collaborative ventures.

Chapter IX

Knowledge Management in Higher Education and Professional Development in the Construction Industry

Fung Fai Ng
The University of Hong Kong, Hong Kong

Abstract

Development of a 'knowledge society' affects not only enterprises and organisations, but also individuals. Lifelong learning and continuing professional development are essential for graduates and workers to remain competent and competitive. Hence, knowledge management is not only important in business processes, but also in education. As the importance of knowledge management in the construction industry is growing, competence in knowledge management is essential for graduates of curricula related to the construction industry. Design and implementation of such curricula should take into account methods for enriching the knowledge management competences of students in higher education. This chapter reports on lessons learned from the design and implementation of a particular module where principles of knowledge management are integrated into the learning activities.

Introduction

Since the early eighties, knowledge management has become a 'hot' issue. Modern organisations consider knowledge as an important resource and a source of competitive advantage. Our society is evolving from an 'information society' to a 'knowledge society'. Advances in information technology and the accessibility to huge amounts of information on the Internet have made everyone aware of the potential for using and creating information and knowledge. The impact is not only on enterprises and organisations, but also on individuals. Lifelong learning and continuing professional development are essential for graduates and other workers to remain competent and competitive. Hence, knowledge management is not only important in business activities, but also in education.

The impact of knowledge management in higher education is multi-levelled. Educational administrators and teachers have begun to look at how they might use information systems to assist in creating effective learning environments. Knowledge management can also be used to support both educational administration and the teaching and learning environment. As the importance of knowledge management in the construction industry is growing, competence in knowledge management is essential for graduates of curricula related to the construction industry. Design and implementation of such curricula should take into account methods for enriching the knowledge management competencies of students in higher education. This chapter reports on lessons learned from the design and implementation of a particular module where principles of knowledge management are integrated into the learning activities. The module is Surveying Studio in the BSc (Surveying) curriculum of the Department of Real Estate and Construction at The University of Hong Kong.

Processes in knowledge management are not new. However, the knowledge management approach provides a more systematic way in which to design and facilitate learning activities. Students working in a team environment to solve problems are exposed to issues related to knowledge management such as communication, knowledge creation, knowledge acquisition, knowledge sharing, and collaborative work. Students are introduced to various tools to support knowledge management, including dialog mapping software, collaborative authoring, and concept mapping. The goal is to integrate principles of knowledge management into a constructivist learning environment so as to enhance students' competence in knowledge management.

Learning Objectives

1. Appreciate the relationship between learning, professional development, and knowledge management.
2. Integrate the principles of knowledge management in designing a course.

3. Design a course with a constructivist learning environment to promote self-learning and to develop skills of knowledge management.

4. Develop and implement learning activities that make use of knowledge sharing tools.

Background

In an 'information society', competency in information literacy is an important requirement in business and in education. According to the Information Literacy Competency Standards for Higher Education (American Library Association, 2000), an information-literate individual is able to:

- Determine the extent of information needed
- Access the needed information effectively and efficiently
- Evaluate information and its sources critically
- Incorporate selected information into one's knowledge base
- Use information effectively to accomplish a specific purpose
- Understand the economic, legal, and social issues surrounding the use of information, and access and use information ethically and legally

The emergence of the 'knowledge society' and the knowledge-based economy signify a new era for knowledge workers. Globalisation has led to ever-increasing competition and shortening of the life cycles of products and services. New information and communication technologies continue to transform workplaces and work practices. There is a requirement from industry for continuous professional development and lifelong learning. Based on the report of the Careers Education and Placement Centre (2001), employers today expect undergraduates to have core competencies in:

- Communication skills
- Analytical reasoning
- Lateral thinking
- Interpersonal sensitivity
- Practical orientation
- Information technology proficiency
- Work attitude
- Emotional stability

The growing complexity and dynamics of professional work increasingly require teamwork. The above requirements call for a new paradigm for education and training which goes beyond information literacy. It encompasses creative and critical thinking, problem solving and communication skills, as well as the ability to interact with others in a collaborative manner. The requirements are evolving from competency in information literacy to competency in knowledge management.

The trend of lifelong learning and learner-centred approaches to education actually reinforce the need for knowledge management competency. In the curriculum design of the BSc (Surveying) course, in addition to core subjects, is a module called Surveying Studio. The aim of Surveying Studio is to allow students to undertake work under their own initiative, and to test their abilities and understanding of professional skills and techniques. Surveying Studio originally started with a task-oriented approach. Feedback from both staff and students was collected and the design was revised. Other successful ideas from courses related to problem-based learning and peer learning were incorporated. The common feature of all these approaches is constructivism. Theories of constructivism suggest that for learning to be useful, the learner needs to be actively involved in constructing new knowledge within meaningful contexts, not merely absorbing it (Duffy, Lowyck, & Jonassen, 1993). Jonassen (1999) lists a number of design principles that can be used to develop what he calls the "constructivist learning environment." These design principles are:

- Create real-world environments that employ the context in which learning is relevant.
- Focus on realistic approaches to solving real-world problems.
- The instructor is a coach and analyser of the strategies used to solve these problems.
- Stress conceptual interrelatedness, providing multiple representations or perspectives on the content.
- Instructional goals and objectives should be negotiated and not imposed.
- Evaluation should serve as a self-analysis tool.
- Provide tools and environments that help learners interpret the multiple perspectives of the world.
- Learning should be internally controlled and mediated by the learner.

The outcome is a change from a transmission model of learning to a constructivist learning paradigm. In a student-centred approach to education, students are responsible for their own self-directed learning. Based on a student-centred framework, a conceptual model of Surveying Studio is shown in Figure 1.

There are four major components in the conceptual model: case study, studio team, student group, and the problem scenario. The case study is a real-life project in the field of real estate and construction. Information and documentation for the case study are collected from various organisations involved in the project. The case study is intended to set Surveying Studio in an authentic context or theme. The studio team is composed

Figure 1. Conceptual model of Surveying Studio

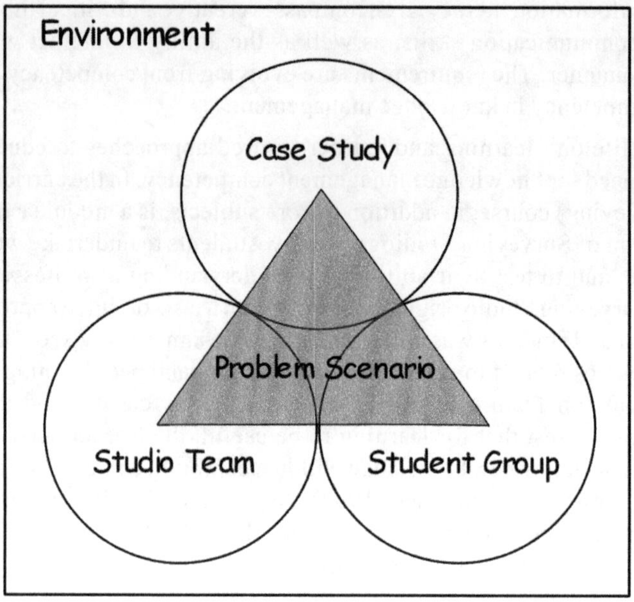

of four or five members of staff including the studio coordinator. The studio coordinator is responsible for planning, scheduling, and coordinating all studio activities. The studio team works together to develop problem scenarios with respect to the selected case study. A problem scenario is an open-ended and ill-structured problem in an authentic context. It allows students to explore the problem from different viewpoints, different pathways, with different learning styles, and to arrive at different solutions. The pathways are complex enough for a variety of investigation approaches by different ability levels. Students are divided into groups with four to six members. Group members remain the same for the whole academic year. Student learning groups work through the problem scenarios to achieve desired learning outcomes.

Studio space and furniture are provided within the department. Students have their own desks, which are usually grouped together according to their learning groups. Network connection points and a wireless network are available to enable the students to connect laptop computers to the Internet from their desks. The studio space is an environment in which students are able to work on their own and conduct informal group meetings. A weekly meeting between the facilitator and student learning group is conducted in a small meeting room. The problem scenario is delivered to the students. The student learning group discusses and defines the problem, distributes workload among group members, continues to explore the issues, and shares the findings and learning experiences with each other. At the conclusion of each problem scenario, the group produces the final learning outcome and deliverable(s). Also, students keep a learning journal to document and reflect on their learning experience and progress. As continuous assessment is employed, there is no written examination for Surveying Studio.

The above description is from the point of view of the module designer. As it is a student-centred learning framework, looking from the student's perspective would show the learning environment as a knowledge management system.

Constructivist Learning Environment as a Knowledge Management System

Marshall and Rossett (2000) noted that knowledge management systems "are composed of two complimentary parts: one technical, the other social" (p. 26). The technical component seeks to "capture, package, and distribute tangible, documented products," while the social side "enables collaboration, connection, and reflection among system users." To adopt the student-centred learning paradigm, a constructivist learning environment should be created as a knowledge management system for the students. The technical components of the system include the infrastructure and resources to support the knowledge management processes carried out by students. The social components include the time, space, and culture to support collaborative work and peer learning.

Davenport (1998) indicates that knowledge is created invisibly in the human brain, and only the right organisational climate can persuade people to create, reveal, share, and use it. Learning is a process of creating knowledge. Learning how to learn is the spirit of knowledge management. Thinking in processes stresses the dynamic nature of knowledge management. Basically, there are two main categories of knowledge management processes:

- *Acquire and create knowledge*: When students search, read, organize, and understand information, they acquire and create knowledge in their minds and integrate the new knowledge with their existing knowledge. Hence individual learning is about individual knowledge development and enhancing competences.
- *Share and use knowledge*: Facing a problem, students apply and use their knowledge to solve the problem. Working in a team environment, students share their knowledge to develop new knowledge, and apply knowledge for problem solving. Sharing knowledge is a way to use knowledge. Knowledge sharing is also the basis of organisational learning.

The relationship between individual learning and collaborative learning is shown in Figure 2.

In both education and business, it is important to create a culture and an environment to maximize individual knowledge development, as well as knowledge sharing, through collaborative work.

Figure 2. IDEF01 diagram of knowledge sharing

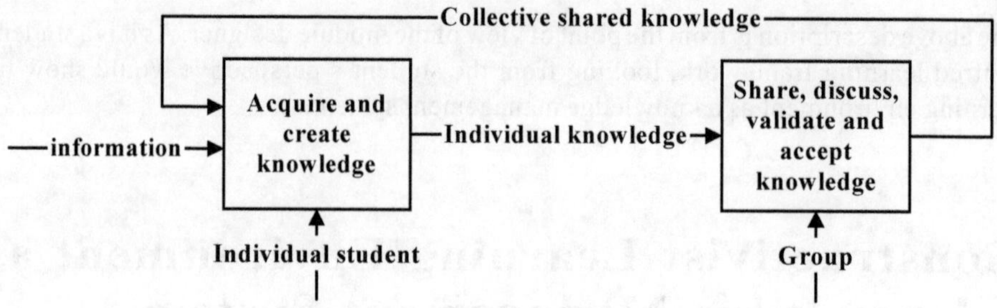

Learning Environment to Support Knowledge Sharing

Individual students are responsible for their own self-directed learning to acquire and create knowledge. The learning environment should provide access to various information sources, including libraries and resources on the Internet. It should also facilitate the articulation and exchange of knowledge. Effective knowledge creation depends on an enabling context, which can foster ideas and facilitate the articulation, creation, and evaluation of experiences and knowledge. Creating a 'right context' is crucial to student-centred learning.

In Surveying Studio, a problem scenario is designed to stimulate the students' need for more knowledge to solve the problem. When the problem scenario is presented to a student group, the students brainstorm and clarify the issues during their first meeting with their facilitator. Individual students contribute ideas based on their existing knowledge. They need to identify the nature of the problem and the knowledge required for solving the problem. At the same time, they identify gaps in their existing knowledge, and the measures and actions required to fill the gaps.

Knowledge Sharing: A Technical Perspective

Although students discuss the problem scenario in a face-to-face meeting, communication and knowledge sharing are not necessarily effective. A learning environment should enable the group to share their thinking and reformulate their assumptions as they expand their understanding. A dialog mapping software tool[2] is used to facilitate discussion in a group meeting. The tool provides a visual presentation for tracking and visualizing the argument process. A laptop computer running the dialog mapping software is connected to a data projector. The projected dialog map consists of questions, ideas, and arguments generated from the discussion process (Figure 3). During the meeting, students raise issues, questions, ideas, and arguments. One student, acting as the group scribe, records the issues, ideas, and so forth, using the dialog mapping software on the laptop computer.

Figure 3. Using dialog mapping to augment discussion

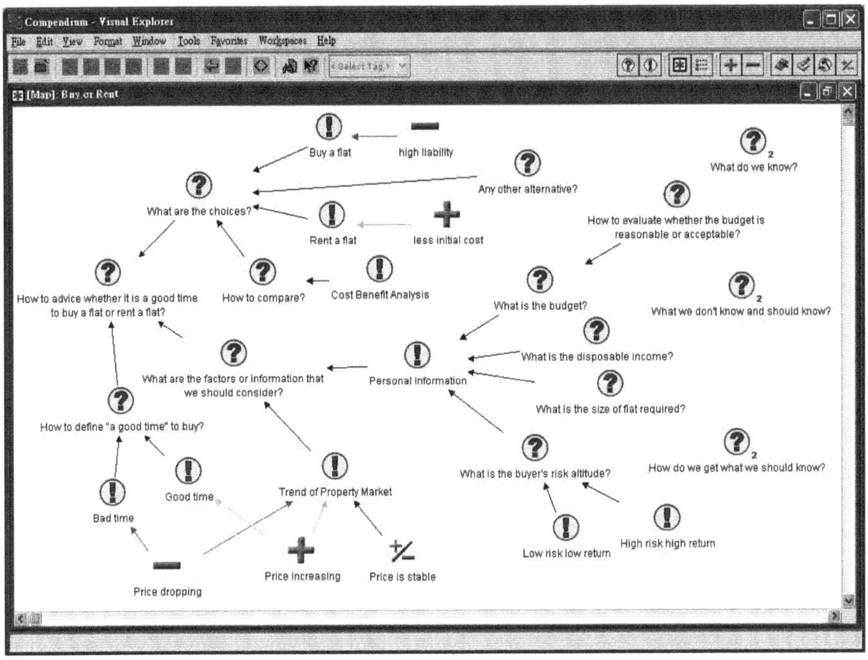

The projected dialog map serves as a group memory, encouraging students to make their assumptions visible, while helping to clarify their thinking and improve on other students' ideas. The combination of information on the dialog map and visual thinking leads to a rich discussion. The following verbatim passage is part of a group discussion after the facilitator has delivered the problem scenario to the group. Student A is the chairperson and Student B the scribe.

Student A: Today, our task is to advise Janis whether she should buy or rent a flat. In my opinion, I would suggest her to rent a flat.

Student C: If Janis buys her own flat, she'll feel more secure. I think she should buy instead of rent a flat.

Student B: Me too. How about we simply vote to determine our recommendation?

Student D: I don't think so. We should not only tell Janis about our recommendation. We need to advise and explain our suggestion to her.

Student E: In that case, we should compare the pros and cons of the two choices to see which one is better. I think timing is one of Janis's concerns also.

Facilitator: Good. How about establishing a focus question first to guide our discussion?

Student A: So, our question is whether it is a good time to buy or rent a flat.

Student D: Besides buy or rent, are there any other options?

Student B: I don't know. Anyway we'll put this down as a question.

Student C: How are we going to compare the two options?

Student A: We should find out the factors affecting the choice.

Student D: I think the decision may be a personal preference.

Student C: That means we have to consider some personal factors such as monthly income, family size, etc....

Student C: Don't forget that we have to judge whether it is a good time to buy a flat or not.

Student D: It depends on the current status of the property market. I think buying a flat is a big investment.

Student A: Do you know what are the costs involved in buying a property?

Student D: I don't know the details, but I know that there is legal fee, agent fee, and the buyer needs to arrange a mortgage loan with the bank....

Student A: So, in the coming week, (Student C) and I would find out more information about the status and trend of the property market.

Student B: I'll work with (Student D) to find out all the costs involved in renting a flat and buying a flat.

Student E: May be we should ask our parents, as they had made that choice already?...

At the end of the meeting, the group has a road map to solve the issue and an action plan to guide the work of individual members. Each student then acquires further knowledge from various sources and shares the knowledge through other channels before the next meeting. The initial design intended to create a discussion forum on the Internet for students to share their findings and ideas in an asynchronous mode after the meeting. The result was not satisfactory. The outcome was a long list of threaded discussion postings. Some messages were subjective, without supporting arguments or follow up. Others were similar or simply repeating other member's ideas. The problem was that the collection of postings in a discussion forum was not consolidated to represent the shared-group knowledge. In fact, students preferred to use their usual channels, for example, ICQ for chatting and discussion rather, than using the prescribed discussion forum.

From the initial experience, we found that students are smart enough to find their own way to communicate and share ideas. What was missing was a repository to capture the collective shared group knowledge generated from the discussion process. Therefore, a knowledge repository was established on the Internet using a Wiki server.[3] Instead of posting individual messages, students create and edit Wiki pages in a collaborative authoring environment. Students continue to use their preferred channels for discus-

Figure 4. Concept map integrated with Wiki page

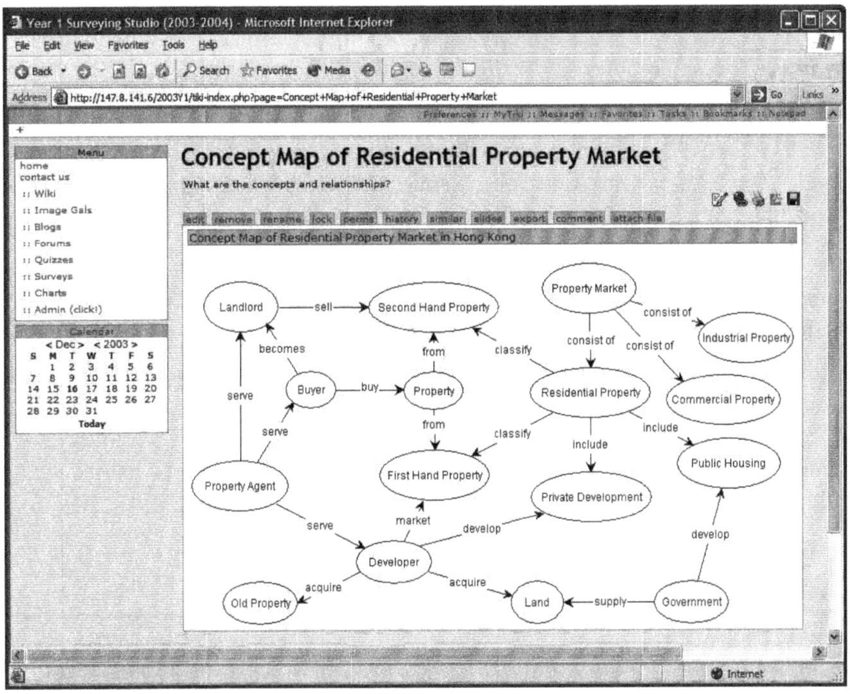

sion, but they are also required to create Wiki pages to consolidate their findings and ideas generated from discussions. Every member can edit the Wiki page to modify exiting information or add more information using a standard Web browser. The collective shared-group knowledge is represented as a set of linked Wiki pages. Students read the Wiki pages to acquire and create their own individual knowledge. However, students are not just passive receptacles of information, they are also creators. The system allows students to post and share documents. It also allows them to edit documents, to view the history of changes to a document, and to create and contribute documents.

To support meaningful learning from the shared-group knowledge, the technique of concept mapping is also introduced. Meaningful learning is a process in which new information is related to an existing relevant aspect of an individual's knowledge structure (Novak, 1998). A simple graphic editing tool[4] is used for the students to create their own concept map. The tool also allows real-time collaborative editing of concept maps, as well as asynchronous editing integrated with the Wiki server (Figure 4).

To reflect and learn from one's own experience is an essential element in knowledge management. Students are required to write and post their learning journal on the Wiki server. A learning journal promotes self-assessment and evaluation which is critical in acquiring and creating individual knowledge. Both positive and negative learning experiences are found in the students' learning journal, for example:

Student V: I enjoy searching and exploring new knowledge.

Student W: I feel extremely stressful. I always don't know what to do at the beginning.

Student X: I learn how to work effectively within limited time and how to cooperate with other group members.

Student Y: Group members cannot compromise on issues. Lack of communication between group members is obvious.

Student Z: We learn how to think, how to organize working schedule, and how to solve problem as a group.

As students reflect on their learning experiences, they identify the way to improve or to solve the problem.

To summarize, the tools and technologies that support the constructivist learning environment include laptop computers, data projector, Internet connections, dialog mapping software, Wiki server, and a concept mapping tool. Technology helps to connect students so they can think and work together. However, to achieve effective knowledge sharing, the social and cultural context is also important.

Knowledge Sharing: A Social Perspective

Students learn on their own and together as a group. They see each other's thinking as they solve problems together, in peer reviews, or in arguing with each other. A critical factor to support effective knowledge creation and sharing is the necessity for students to be physically close to each other in space and time. The sense of "nearness" is crucial in stimulating the exchange of ideas, tacit knowledge, and experiences. Therefore, a studio space where students have their permanent working place is desirable. It establishes a stimulating social environment, and generates social pressure and a sense of mutual responsibility. In their studio, students usually group their desks together according to their group memberships (Figure 5).

The design of a constructivist learning environment faces challenges from various aspects. The social challenge is to develop communities that share knowledge and still maintain enough diversity of thought to encourage thinking rather than copying. The management challenge is to create an environment that truly values sharing of knowledge with trust and reward. The personal challenge is to be open to the ideas of others, to be willing to share ideas, and maintain a thirst for new knowledge.

Before the commencement of Surveying Studio, an orientation and explanation of the student-centred learning approach is delivered to all students. It is important to make sure that all students understand that assessment of their performance is criterion based and not norm referenced. Assessment is based on group performance and individual contribution to the group. Hence, there will be less competition between students. Competition between groups to achieve higher grades may exist, but group members are willing to share and help each other so as to perform effectively as a team.

Figure 5. Studio space to augment knowledge sharing

In addition to the formal meeting to discuss the problem scenario in the presence of the facilitator, the main portion of learning takes place as peer learning in informal meetings among students. Students continually learn from each other on a daily basis. They learn a great deal by explaining their ideas to others and by participating in activities in which they can learn from their peers. They develop skills in organizing and planning learning activities, working collaboratively with others, giving and receiving feedback, and evaluating their own performance. Much peer learning occurs informally without staff involvement.

The design of a problem scenario focuses on the setting of open-ended issues for students, working as a group, to solve the issues. Students engage in exploration of ideas and knowledge, and learn to learn. A member of staff, acting as a facilitator, hands over more control to the students. The student group determines individual roles within the group and it is the personal sense of learning that signifies collaborative learning. Critical thinking, problem solving, making sense of complex issues and personal transformation, the social construction of knowledge—exploration, discussion, debate, criticism of ideas—are the essence of collaborative learning. Bruffee (1999) names this approach 'constructive conversation'—an educative experience in which students learn by constructing knowledge as they talk together and reach consensus or dissent. Dissent, questioning each other's views within a group, is a necessary part of learning.

Current Results

The design and development of Surveying Studio is an ongoing process. There have been both success and failure scenarios. Feedback and lessons learned are used to improve ongoing design and development. Preliminary observations indicate positive results in enhancing students' knowledge management competence in the following aspects:

- *Working with others*: A student group is a learning community. Students develop a sense of responsibility for their own and others' learning. They also develop confidence and self-esteem. They learn to acknowledge the backgrounds and contributions of other students they are working with to develop collaborative skills.
- *Critical enquiry and reflection*: A challenge to existing ways of thinking arises from more detailed interchanges between students in which points of view are argued and positions justified. Opportunities are present for formulating questions rather than simply responding to those posed by others. Students are able to articulate what they understand and to be more open to critique by peers, as well as learning from listening to and criticizing others.
- *Articulation of knowledge*: Concept development often occurs through the testing of ideas on others and the rehearsing of positions that enable students to express their understanding of ideas and concepts. It is often only when they are expressed and challenged that students appreciate whether they have a good grasp of what they are studying.
- *Learning how to learn*: The learning activities in Surveying Studio require students to develop self-management skills and management of others. Students take collective responsibility for identifying their own learning needs and planning how these might be addressed. Learning to cooperate with others to reach mutual goals is a prerequisite for working in a complex society.
- *Self- and peer-assessment*: The learning activities require students to identify criteria to assess their own learning and to practice self- and peer assessment. This is a key element of sustainable assessment needed for lifelong learning.

Surveying Studio enhances students' experience of learning, and enables them to engage in the learning processes that reflect the working practices in which they will be expected to participate later, when they are in the workplace.

Future Trends

Information and communications technologies are changing the nature of instructional design. Technology and the use of the Internet and World Wide Web have allowed major changes both in teaching practice and learning, and consequently, in the management of information and knowledge. Since the inception of the World Wide Web in the early 1990s, increasing numbers of educators have used this medium to make information, relevant to their courses, available to students. The ability to use the Internet to place information online for others to access, as well as to collaborate with others, creates a new way of thinking about the management of knowledge, information, and documents.

The challenge faced by both business and education is not only how to best manage the volume of information so that it is easily accessible to search and retrieve materials, but also how to provide people with the ability to post information and to enhance collaboration. It is important to create a culture where instructors and students are willing to share their information and knowledge, as well as contributing to the generation of new knowledge. The application of knowledge management principles and the desire to deliver documents to learners has opened the door for creating a more effective learning environment in higher education and professional development.

The rapidly changing knowledge economy has led to an increased interest in lifelong learning. Workplaces are in need of employees who can learn new skills and adapt quickly to social and technological changes. Educational institutions are faced with the challenge of providing students with the knowledge and skills to enable them to adapt successfully to job-related changes after completing their formal education.

The growing complexity and dynamics of professional work increasingly requires teamwork. Professionals piece information together, reflect on their experience, generate insights, and use those insights to solve problems. Continuous learning while working is obligatory to meet the performance requirements of the workplace. In the above context, there are many questions regarding new ways of working and learning still to be resolved.

Conclusion

In the developed world, the basis of the economy has changed from manufacturing products to creating and managing knowledge. In addition to the possession of professional knowledge, the ability to continue learning has become more and more important. Education is of increasing importance in its form as a lifelong and continuous learning process, both in business and in the individual context. The significance of lifelong learning and continuous learning raises the need for new teaching and learning paradigms. For example, lifelong learners typically plan their own learning, apply existing knowledge and skills effectively, evaluate their own learning, locate information from different sources, use different learning strategies in different situations, and resolve complex problems with multiple or uncertain solutions. These requirements need to be considered in both curriculum design and course delivery.

Although knowledge management has been developed within a business context, it can be applied to education. Knowledge management is not only essential at an organisational level. It should be a key competence at an individual level. Knowledge management should be integrated into higher education and professional development programs related to construction industry. The design and development of Surveying Studio is a step forward in the trend of integrating knowledge management principles to create a constructivist learning environment in higher education. This learning environment promotes students working as a team and augments knowledge sharing in both a technical and socio context. Various tools to support knowledge management are introduced to students, including dialog mapping, collaborative authoring, and concept mapping tools. Problem scenarios and learning activities are designed to integrate the use of knowledge sharing tools and knowledge management strategies. The primary objective is to enhance students' competence in knowledge management so that, after graduation, they are able to cope with the knowledge-based working environment.

The design and development of Surveying Studio is an ongoing process. We are learning from our experience by acquiring and creating our own knowledge. We are using our knowledge to improve, and also sharing our knowledge through this publication.

Practical Tips and Lessons Learned

- Competence in knowledge management is important for graduates and professionals working in a knowledge-intensive industry.
- Learning how to learn is an essential skill of knowledge management.
- Learning to cooperate with others to reach mutual goals is a prerequisite for working in a knowledge society.
- Encourage the use of all possible types of communication channels (formal and informal, physical and virtual) to support knowledge sharing. The community will find the most effective way to share knowledge.

Endnotes

[1] IDEF0 is a method designed to model the decisions, actions, and activities of an organisation or system. It is a public domain modelling system, and in fact it is an American standard (FIPS 183).

[2] The dialog mapping software used is called Compendium, available at Compendium Institute, www.compendiuminstitute.org.

[3] The Wiki server is established using TikiWiki, freely available at tikiwiki.org.

[4] The graphic editing program is called JGraphPad, freely available at jgraph.sourceforge.net.

References

American Library Association. (2000). *Information literacy competency standards for higher education.* Chicago: The Association of College and Research Libraries.

Boud, B., & Solomon, N. (Eds.). (2001). *Work-based learning: A new higher education?* Buckingham: SRHE and the Open University Press.

Boud, D., Cohen, R., & Sampson, J. (2001). *Peer learning in higher education.* London: Kogan Page.

Bruffee, K. (1999). *Collaborative learning: Higher education, interdependence and the authority of knowledge.* Baltimore: The Johns Hopkins University Press.

Candy, P. (2000). Knowledge navigators and lifelong learners: Producing graduates for the information society. *Higher Education Research & Development, 19*(3), 261-277.

Careers Education and Placement Centre. (2001). *Report of Careers Advisory Board Working Group on Competency Profile.* Hong Kong: The University of Hong Kong.

Davenport, T., DeLong, D., & Beers, M.C. (1998). Successful knowledge management projects. *Sloan Management Review, 39*(Winter), 287-300.

Duffy, T.M., Lowyck, J., & Jonassen, D.H. (1993*). Designing environments for constructive learning.* Berlin: Springer Verlag.

Hartley, D.E. (2000). *On-demand learning: Training in the new millennium.* Amherst, MA: HRD Press.

Homer-Dixon, T. (2000). *The ingenuity gap.* New York: Knopf.

Jonassen, D.H. (1999). Designing constructivist learning environments. In C.M. Reigeluth (Ed.), *Instructional design theories and models: A new paradigm of instructional theory* (Volume II, pp. 215-239). NJ: Lawrence Erlbaum.

Knapper, C.K., & Cropley, A.J. (2000). *Lifelong learning in higher education.* London: Kogan Page.

Marhsall, J., & Rossett, A. (2000). Knowledge management for school-based educators. In J.M. Spector & T.M. Anderson (Eds.), *Integrated and holistic perspectives on learning, instruction and technology: Understanding complexity.* Boston, MA: Kluwer Academic Publishers.

Novak, J.D. (1998). *Learning, creating, and using knowledge: Concept maps as facilitative tools in schools and corporations.* NJ: Lawrence Erlbaum.

Palmer, S. (2001). Engineering flexible teaching and learning in engineering education. *European Journal Engineering Education, 26*(1), 1-13.

Raven, J., & Stephenson, J. (2001). *Competence in the learning society.* New York: Peter Lang.

Chapter X

Empirical Investigation of Organisational Learning Ability as a Performance Driver in Construction

Irem Dikmen
Middle East Technical University, Turkey

M. Talat Birgonul
Middle East Technical University, Turkey

Tunca Ataoglu
GAMA Al Moushegah Arabia Ltd., Kingdom of Saudi Arabia

Abstract

In this chapter, the impact of organisational learning competency on the performance of construction companies is investigated. A conceptual model is proposed for the measurement of organisational learning competency. The main components of the model are learning sources, learning mechanisms, and the organisational setting. Organisational learning competency is assumed to be high only if an appropriate organisational setting exists, as well as the mechanisms used for management of knowledge acquired from various sources. A questionnaire is designed to collect data about organisational learning factors and performance. Findings of the questionnaire answered by 85 Turkish contractors demonstrate that there are statistically significant differences between the performances of contractors grouped according to their learning ability. It is empirically proved that as the organisational learning ability increases, firm performance also increases.

Introduction

Organisational learning (OL) is about management of knowledge and formation of an organisational memory so that the organisation improves itself in light of past experiences and information gathered from the external sources. A learning organisation is defined as an organisation skilled at creating, acquiring, and transferring knowledge, and modifying its behaviour to reflect new knowledge and insights (Garvin, 1993). The success of learning organisations demonstrates the importance of "learning capability" as a major source of competitive advantage.

Learning Objectives

The objectives of this chapter can be summarised as follows:

1. To propose a conceptual framework to model the OL process within construction companies.
2. Based on the proposed conceptual model, to define an organisational learning index for the measurement of OL competency.
3. To investigate the possible correlation between performance and OL ability by referring to the findings of a questionnaire responded by 85 Turkish contractors.
4. To discuss the difficulty of measuring OL ability and shortcomings of the proposed methodology.

Previous Work on Organisational Learning in Construction

When the literature on organisational learning (OL) in construction companies has been investigated, it is clear that the body of empirical research is comparatively small when compared to conceptual work (Garvin, 1993; Kululanga, Price, & McCaffer, 2002). Apparently, this is partly due to the fact that learning ability is hard to quantify, as it is mostly about managing tacit knowledge and developing the cognition of an organisation. Previous work on OL in construction is more related with how construction companies can improve their learning capabilities using specific tools and learning sources. Alwani-Starr (1997) pointed out the importance of training and learning of individuals as enablers of OL. Hakansson, Havila, and Pedersen (1999) stressed the critical role of networking as it provides a major learning opportunity for all parties. Similarly, Orange, Boam, and Burke (1998) identified partnering as a facilitator for learning and suggested a collaborative virtual environment to enhance project learning. Moreover, the importance of organisational culture as an enabler of learning and the strategic perspective necessary to incorporate OL as a part of culture have been stressed by many researchers (Robinson, Carrillo, Anumba, & Al-Ghassani, 2001; Huemer & Ostergen, 2000).

There are also studies on building a "system" for becoming a learning organisation where means and ends of OL have been discussed from a wider perspective. Love, Li, Irani, and Faniran (2000) linked total quality management (TQM) culture with OL and identified TQM as a means for OL. Holt, Love, and Li (2000) have drawn attention to the relation between level of customer satisfaction, alliancing, and learning. Vakola and Rezgui (2000) explored the relations between OL, business process restructuring, and innovation. Knuf (2000) suggested benchmarking as an important OL tool. Furthermore, integrated conceptual frameworks have been proposed by researchers to model the OL process. For example, Kululanga and colleagues (1998, 1999) proposed a framework for learning and identified some critical learning mechanisms. Based on the proposed models, empirical investigations have also been carried out and the OL competency of firms measured (Kululanga, Edum-Fotwe, & McCaffer, 2001; Kululanga et al., 2002).

The current study is based on the idea that a systems approach is necessary to model OL process, and it should be tied to the industrial context, as OL is grounded in a firm and its business environment. The overall OL process—rather than its parts—has been modelled. For example, although knowledge management is a vital part of OL, if the cultural setting does not foster learning, managed knowledge may not turn into organisational knowledge. Thus, the OL process is perceived as a system that may work effectively only in the existence of right strategies, culture, organisational infrastructure, and when the necessary learning mechanisms are utilised to institutionalise the individual knowledge gathered from various learning sources. Industrial characteristics are also important as they indirectly affect the organisational infrastructure and culture. As the construction industry is project based, project learning constitutes the major part of organisational learning. The mechanisms that permit the transfer of project knowledge into organisational knowledge should be incorporated within the learning framework. Moreover, a high number of parties involved within the construction value chain makes "learning from other parties" a critical issue for OL. A framework, specific to the construction industry, has been developed to facilitate the measurement of OL which is further used to collect empirical evidence to test the impact of OL ability on the performance of construction companies.

In many research studies, it is theoretically supported that OL has a positive influence on the competitive advantage of construction companies (Egbu, 2000). However, there are limited efforts to explore this link empirically (Dulaimi & Ling, 2003), mainly due to the fact that OL process has unquantifiable components. The firm capability in construction is usually associated with its performance in a project and measured by the level of conformance with pre-set project success criteria (time, cost, quality, etc.). An improvement at the project level is possible by OL, if knowledge obtained from the previous projects can be transformed into organisational knowledge and it can be revised as a result of new knowledge coming from current projects. In this way, previous mistakes are not repeated and work processes are improved. Firms that learn from competitors, partners, and clients stand a better chance of sensing and acting upon events and trends in the marketplace. DiBella and Nevis (1998) identify organisational learning as a catalyser in overcoming limitations in an organisation's performance. Also, as denoted by Kale and Arditi (2003), most of the benefits of OL are exhausted over time. Thus, OL competence has its major impacts on the performance of a construction company in the long run.

Resource-based view suggests that resources that are valuable, rare, inimitable, and lack substitutes are central to competitive advantage. The OL process is aimed at improving the knowledge sources that have distinctive properties that help to create and sustain competitive advantage. Those knowledge resources, "know-how" and "know-why," are embedded in a firm's abilities. Although, they are very hard to articulate explicitly, such as how to deal with bureaucracy in an international market or how to price risk in a bidding decision, they are critical pieces of knowledge that may directly affect a firm's competitive position. Thus, by increasing the firm's operational and strategic capability through improvement of knowledge resources, OL has a potential to increase performance of construction companies in the long run.

Organisational Learning Process

Learning as an integral part of working occurs naturally in organisations as the individuals adapt new tools to solve current problems, share success and failure stories, and so forth. Unlike individuals who are automatic learners, OL is a conscious activity in the organisational context, which should be enhanced by "supporting mechanisms and firm infrastructure." OL necessitates knowledge handling through communities (Kahkonen & Kazi, 2003). For successful handling of data, both digital and social communities are needed in balance, as well as the supporting tools. Thus, OL competency does not increase by chance; it necessitates a system to promote individual learning and convert individual knowledge into organisational knowledge for modifying its behaviour. Learning is organisational to the extent that: first, it is done to achieve organisational purposes; second, it is shared or distributed among members of the organisation; and third, learning outcomes are embedded in the organisation's systems, structure, and culture (Snyder & Cummings, 1998). Senge (1994) argues that lessons are transformed into knowledge when they are analysed, systematised, disseminated, and internalised within an organisation through an "evaluative process." The keys to this transformation has been denoted by Nonaka (1996) as personal commitment, employees' sense of identity with the enterprise, and its mission.

Learning disabilities of some companies can be attributed to the industry in which they operate. The major risk in construction companies is the knowledge loss at the end of the project and resulting organisational amnesia. Hence, OL competency is strongly related with the efficiency of the mechanisms used to convert project knowledge into organisational knowledge. Project knowledge does not only consist of explicit knowledge that can be documented in quantifiable terms, but also tacit knowledge that is hard to express and store. Some of the experiences gained within a project are about the market, parties worked with (subcontractors, client, partner, etc.), and impact of strategic decisions on the project value chain. Moreover, organisational knowledge in construction companies are more than the summation of knowledge obtained from the projects.

Organisational knowledge may well be revised as a result of "unrealised projects" where the related experiences of head office functions like bidding, marketing, business development, financial management, and so forth are of paramount importance. Measuring the OL competency in project-based industries is a harder task as OL is through project learning and knowledge sources are usually very hard to document and quantify.

The Proposed Organisational Learning Framework

The representation of the conceptual OL framework is given in Figure 1. There are three major components within the proposed framework: "learning sources," "learning mechanisms," and "organisational setting." Learning sources are defined as the internal and external sources used by individuals to acquire knowledge. External sources are created outside the organisation, whereas the internal sources are communicated within the organisation. Learning sources can be individuals/parties within or outside the organisation. Learning mechanisms are those tools used to transform individual knowledge into organisational knowledge. Knowledge management mechanisms are used for acquisition, sharing, storing, and deployment of knowledge. Mechanisms cover the tools (such as IT tools) and activities (such as post-project evaluation) utilised for institutionalisation of knowledge. Furthermore, an appropriate firm infrastructure should exist so that the learning mechanisms can work effectively, which is covered under the heading of "organisational setting." The organisational setting is a part of the "informa-

Figure 1. The OL framework

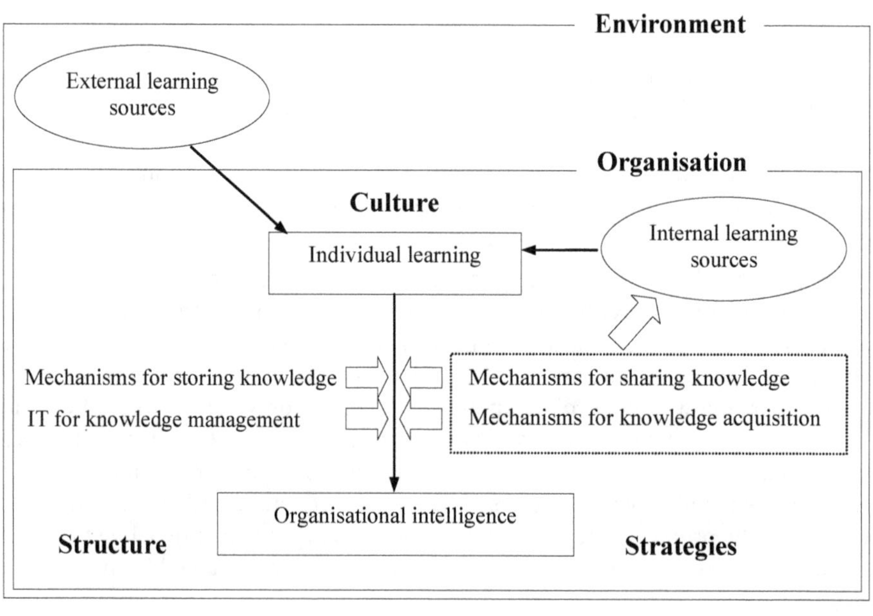

tion ecology" (Davenport, 1997) which covers the totality of the environment and conditions under which an organisation lives. Organisational setting comprises three factors: structure, culture, and strategies. OL competency is strongly related to the organisational structure. Rigid boundaries between different departments, bulky and an unmanageable size of the organisation, and bureaucracy hinder learning in an organisation, whereas flexibility, openness, freedom, and opportunity nurture it (Marquardt, 1996). Also, as argued by many researchers, OL has a cultural foundation. Culture of an organisation is related with its values, beliefs, common approaches, and solutions to the problems. If the culture within the organisation supports learning by encouraging training, motivating employees to suggest new ideas, providing rewards for increased learning, and so forth, OL can be said to be embedded into the culture of the organisation. Culture is both the consequence of the organisation's prior experience and learning, as well as the basis for its continuing capacity to learn (Schein, 1997). OL process is also related to the mission, vision, and strategies of an organisation. Whether the aim of increasing OL competency is incorporated into strategic plans and necessary strategies are systematically utilised to enhance learning are indicators of the OL capability. Senge (1990) denotes that companies should have an OL strategy and continuously try to incorporate OL as a part of their company culture in order to become a "learning organisation." The strategies proposed by Senge (1990) are: developing the discipline of dialogue within the organisation, building team-learning skills, encouraging multi-cultural and global mindsets, and changing the mental model relative to learning.

In the framework, it is assumed that OL competency is high when;

- the learning sources are utilised frequently and effectively,
- necessary learning mechanisms are used effectively to build the organisational knowledge, and
- an appropriate organisational setting exists to support OL within the organisation.

In case all of the above-mentioned conditions are met, OL competency of the company is assumed to be high. As shown in Figure 1, the business environment is also one of the components of the framework. Although it does not have a direct effect on OL, it includes the parties within the business environment with whom the company interacts, and also it indirectly impacts the organisational setting, particularly the culture of an organisation.

Measurement of OL Competency

The conceptual framework can be used to define an OL index that reflects the OL competency of a construction company. To calculate the OL index, the determinants of OL ability have to be identified and quantified. For this purpose, all possible items under each component (sources, mechanisms, organisational setting) are described and a

Table 1. Learning factors listed under learning sources and mechanisms

Component	Listed items
Knowledge acquisition: External learning sources for individual learning	External seminars (given by external bodies), Internet, congresses, exhibitions, tradeshows, publications (books, journals, etc.), academic studies (graduate studies)
Knowledge acquisition: Learning from other parties	Clients (national), foreign clients, domestic JV partners, foreign JV partners, competitors, universities, consultants, governmental bodies (related with construction industry), foreign organisations other than clients (World Bank, etc.), non-governmental organisations (Chamber of Civil Engineers, etc.), associations (Association of international contractors, etc.)
Knowledge acquisition: Organisational learning mechanisms	Internal benchmarking, competitive benchmarking, value chain analysis, environmental scanning, process-based project learning (post-project appraisal, after action review), documentation-based project learning (learning histories, case-based articles)
Knowledge sharing: Knowledge-sharing mechanisms among employees	Teamwork, formal/regular meetings, informal/unplanned meetings, internal seminars/presentations, job rotation, brainstorming sessions, informal communication during social activities
Storing of knowledge: Knowledge repository	Project information (unit costs, productivity, equipment) and strategic information (about subcontractors, performances of employees, competitors, market-country information)
Knowledge deployment : Dissemination of knowledge by learning mechanisms	Intranet, online access to databases, artificial intelligence-based decision support systems (neuronets, expert systems, case-based reasoning, etc.), Web-based project management systems

questionnaire form is prepared in which each identified learning factor corresponds to a question. The learning factors under each learning component, listed within the questionnaire, are presented in Table 1.

Apart from the mechanisms and sources, some hypothesis about organisational setting have also been listed in the questionnaire about the utilised strategies to increase OL, availability of funds to enhance OL, cultural drivers/barriers, communication environment, and management style.

For each factor listed as a learning source or mechanism, the following two values have been questioned:

1. *The importance of each learning factor for organisational learning ability (weight):* The respondents are requested to evaluate the significance of each source and mechanism by considering the contribution of each item to the overall OL ability. For this purpose, a 1-5 Likert scale has been used where 1-5 denote "very low" and "very high" importance levels, respectively. Thus, a subjective reporting approach has been utilised where each respondent denotes his/her own idea about the importance of each learning factor for OL competency and puts his/her perspective into the index.
2. *The effectiveness of a learning mechanism/source, frequency of using a learning source, or availability of stored knowledge (rating):* For each learning mechanism, the respondents are required to evaluate the effectiveness of that mechanism in their company using a five-point Likert scale (where 1 and 5 show very low and very high levels, respectively). Similarly, for each learning source, the respondents evaluate the frequency of learning from a party/source, by using the 1-5 Likert scale where 1, 2, 3, 4, and 5 show "never," "seldom," "sometimes," "usually," and "frequently," respectively. Availability of stored information is rated by assigning a binary scale (available: 5; unavailable: 1).

Table 2. Learning indices and OL index

Abbreviation	Learning Index	Calculation Method
LI_1	Individual learning	Weighted average of frequency ratings with respect to importance scores
LI_2	Learning from other parties	Weighted average of frequency ratings with respect to importance scores
LI_3	Organisational knowledge acquisition	Weighted average of effectiveness ratings with respect to importance score
LI_4	Knowledge sharing	Weighted average of effectiveness ratings with respect to importance score
LI_5	Storing of knowledge	Weighted average of availability ratings with respect to importance score
LI_6	Deployment of knowledge	Weighted average of effectiveness ratings with respect to importance scores
LI_7	Organisational setting	Average of validity ratings
OL	Organisational learning index	Average of LI_1, LI_2, LI_3, LI_4, LI_5, LI_6, and LI_7

In order to evaluate whether an appropriate organisational setting exists within the company, respondents are required to evaluate the validity of the listed hypothesis considering their company and by using a five-point validity scale (1: "strongly disagree"; 5: "strongly agree"). Importance weight of each hypothesis is assumed to be the same.

As shown in Table 2, seven learning indices can be defined for the learning components presented in Table 1. The value of each learning index (LI) is calculated by multiplying the importance weight of each item (average of all values assigned by the respondents) with the performance rating (availability, validity, frequency or effectiveness) of the company for that learning item, summing them up for all items under the same learning component and normalising by dividing that number by the summation of weights. LI is a value between 1 and 5 where the OL ability is higher when the value approaches to 5. The overall OL index can be calculated by taking the average of seven LIs. Thus, OL index reflects the effectiveness of all learning mechanisms, frequency of using learning sources and appropriateness of the organisational learning in a construction company, as well as the contribution of each learning item to OL ability within the construction sector.

Measurement of Performance

In order to carry out an empirical study to test the impact of OL ability on the performance of construction companies, just like the OL index, a performance index should be identified. Performance measurement systems are mainly based on quantification of financial indicators about profitability, revenues, and so forth, by using financial statements and accounting data. However, financial indicators are claimed to be short-term indicators and insufficient for understanding the dynamic business environment in which contractors operate (Robinson et al., 2001). Also, financial information is usually confidential and not easily accessible. In this research, financial performance is proposed to be measured by a subjective reporting approach (Dess & Robinson, 1984). It has been assumed that performance can indirectly be measured by looking at the final outputs of profitability and workload. The subjective reporting approach has been used by many researchers (e.g., Kale & Arditi, 2002) and claimed to be a satisfactory indicator of performance for construction companies. In this research, two performance indicators have been specified: "the growth in contract awards," which indicates the ability to get a job, and "profitability," which shows the ability to do work. Respondents are asked to evaluate their company against major competitors by considering their performance over the last three years by using a five-point Likert scale. This score is denoted as "performance rating." Additionally, they are asked to evaluate the importance of these two criteria as an indicator of performance by using a five-point Likert scale, which is named as the "importance score." The performance of a construction company is calculated by multiplying the performance ratings with corresponding importance scores, summing them up, and then dividing that number into the summation of importance weights. Thus, performance index is a weighted average of performance ratings and it indicates the level of competitive advantage in a company. Like the OL index, performance index can take a value between 1 and 5, where the higher values denote higher performance.

Research Methodology

The application of the proposed framework and corresponding measurement method has been demonstrated by a questionnaire study conducted in the Turkish construction industry (Birgonul, Dikmen, & Ataoglu, 2002; Ataoglu, 2002). The construction industry is one of the major industries within the Turkish economy having an approximate 6-8% share in GNP and an output of $8 billion to $9 billion per year. Just like in other developing countries, infrastructure projects constitute the major percentage within the overall construction output. By using the experience gained within the domestic market, the Turkish contractors achieved a considerable competitive advantage within the international market. Currently, Turkish contractors are operating in 56 countries and completed jobs amounting to $50 billion. Their share in the global construction market is around 2-3%. The major sources of competitive advantage are low cost of labour, rapid construction, and high risk-taking ability.

The target population of the questionnaire has been decided to be the mid-high-scale Turkish contractors, the majority of which are members of the Turkish Contractors Association (TCA). TCA is the organisation that embraces the largest construction companies in Turkey. Before administering the questionnaire, a pilot study with three companies was conducted and relevancy of the framework as well as the questionnaire was validated.

The aforementioned questionnaire has nine parts, seven of which are related to the OL index (as in Tables 1 and 2). The first part includes demographic information (age, size, etc.), whereas the last part is about the calculation of performance rating. A document that summarises the OL framework, calculation of the OL index, and aims of the study accompanied the questionnaire. The majority of the questionnaires have been administered during face-to-face interviews, whereas some of them are sent by airmail. The questionnaires were sent to 150 companies and were responded to by 85 Turkish contractors. Respondents were mostly top-level managers like: general manager, business development manager, member of the executive board, and so forth, so they are assumed to have enough knowledge about the organisational culture and strategies. Seven out of 85 returned questionnaires were incomplete and were discarded from the data set. The data from 78 respondents constitutes the basis of statistical analysis. The companies are medium-large size contractors that have an average age of 20 years, \$100 million of construction works completed till date, and the majority of them operating within the international construction market.

Four sets of variables have been identified. These are the control variables (age, turnover, partnering frequency, and the ratio of overseas construction volume within the total construction volume), performance index, learning indices, and the OL index. Statistical analysis has been carried out by using SPSS 10.0. In the forthcoming parts, some of the results of the aforementioned analysis will be reported.

General Findings about Organisational Learning within Turkish Construction Companies

Table 3 summarises the most and least frequently used sources/mechanisms of OL in the Turkish construction companies. The most and least frequently used items coincide also with the most and least significant tools/mechanisms that may increase OL according to the respondents. The general tendency is that knowledge storage embraces storing of project information rather than strategic information about the market and competitors. Similarly, post-project evaluation is widely used by contractors to convert project information into organisational knowledge; however, business-level mechanisms such as value chain analysis and benchmarking are less frequently used. The Internet comes out to be the most frequently used source of individual learning. Apart from the questionnaire results, face-to-face interviews carried out with the respondents point out the fact that general barriers against learning in construction such as its being project-based, fragmented structure, and so forth also hold true for the Turkish construction industry. Also, the low utilisation rate of some learning mechanisms and sources may partly be attributed to the lack of a professional management approach within the majority of the companies and cultural barriers.

About the organisational setting comprising culture, structure, and strategies, the majority of the respondents agree that:

Table 3. General findings about elements of the OL framework

	Most frequently used	Least frequently used
Learning source for individual learning	Internet	Academic studies by employees
Organisational learning mechanism	Post-project appraisal	Internal benchmarking
Type of stored knowledge	Project information—productivity and unit price rates	Strategic information—information about competitors
Knowledge-sharing mechanism	Teamwork and unplanned/ informal meetings	Job rotation
IT for knowledge deployment	Intranets	Web-based project management systems
Other party as a learning source	Clients	Competitors

- There is mutual trust and collaboration between the employees.
- Knowledge is perceived to be one of the most valuable assets by the company.
- OL competency is seen as a source of competitive advantage.
- The organisational setting supports innovation and creativity.

Results imply that there is high awareness between the construction professionals on the importance of OL and knowledge as an important competency and a strategic resource, respectively. Also, the collaborative environment between the employees fosters learning by increased information sharing within the company. On the other hand, most of the respondents denote that necessary funds and time are not reserved for professional training and personal development of employees. Thus, unavailability of financial resources reserved for learning usually constitutes a major OL barrier. Also, general tendency reflects that an environment for collaboration, information sharing, and mutual learning usually does not exist between the companies and their competitors. This result is also in accordance with the low values of "learning from others index" that will be discussed in detail later in the chapter. Moreover, most of the respondents argue that a rewarding mechanism to encourage OL does not exist within their companies which is also one of the most significant barriers against OL. As a conclusion, one can argue that, although the companies are aware of the importance of learning, the performance measurement systems, reward mechanisms do not support OL within the companies, the learning opportunities within the competitive environment are not utilised effectively, and necessary funds are not usually reserved for training. However, it should be kept in mind that these conclusions are based on general findings and may not hold true for all companies answering the questionnaire.

General Findings about Organisational Learning Indices

Table 4 summarises the general findings on learning index values. Results show that "organisational setting index" and "organisational knowledge acquisition index" have the highest mean values, whereas the "learning from others index" and "knowledge-sharing index" have the lowest values. Companies are best at using certain mechanisms such as post-project reviews, environmental scanning, and so forth to form the organisational knowledge, and they have the necessary organisational setting that may foster OL. However, as mentioned earlier, Turkish contractors usually do not have necessary mechanisms and culture to facilitate learning from other parties which may significantly plug the way of OL. Although, mechanisms are used to create organisational knowledge, usually unavailability of knowledge-sharing mechanisms decreases their OL competency. The overall organisational learning index that includes all learning elements has an average value of 3.40, meaning that in general, the OL competency of Turkish contractors is medium-high, although there are significant differences between individual companies.

There are also significant correlations between the index values. When the Pearson correlations between learning indices are investigated, it has been found that the highest level of association exists between "individual learning sources" and the rest of the learning indices. It has substantial positive association with knowledge-sharing mechanisms (0.662), learning from others (0.605), knowledge-storage mechanism (0.551), organisational setting (0.549), and moderate positive association with knowledge-deployment mechanism (0.444) and organisational learning mechanisms (0.403) at the 95% significance level. This result is expectable, as knowledge acquisition is the initial step of learning and if knowledge is not acquired, it cannot be shared or stored. All learning indices are associated with each other to some degree which proves the cyclic existence of OL. Furthermore, the organisational setting comes out to be a very important driver of OL, as it is highly correlated with the rest of the indices. For example,

Table 4. Descriptive statistics on learning index values

Index Variables	Minimum	Maximum	Mean	Standard Deviation
Individual learning	1.69	4.29	3.37	0.55
Learning from other parties	1.60	4.54	3.05	0.62
Organisational knowledge acquisition	1.67	5	4.21	0.86
Knowledge sharing	1.17	4.38	3.13	0.67
Storing of knowledge	1.00	4.71	3.24	0.86
Deployment of knowledge	1.00	5.00	3.26	1.22
Organisational setting	1.83	4.61	3.55	0.54
Overall organisational learning	1.42	4.65	3.40	0.67

"organisational setting" has been found to be substantially associated with "knowledge sharing" (0.647) and "knowledge storage," which is expectable as the organisational culture is the foundation on which knowledge management activities are constructed and sharing of knowledge is not possible without a supporting environment. There is a substantial positive association (0.647) between "knowledge sharing" and "learning from others," demonstrating the fact that if communication environment fosters knowledge sharing within a company, the effectiveness of communication with other parties is also high and visa versa.

General Findings about Impact of Company Characteristics on Organisational Learning

There are seven control variables defined within the research, namely: age of the company (years), total domestic turnover ($), total overseas turnover ($), total turnover ($), proportion of international works in total turnover (ratio), frequency of partnering with foreign companies (1 to 5 Likert scale), and frequency of partnering with foreign companies (1 to 5 Likert scale).

When possible associations are searched between the control parameters and OL indices, the Pearson correlation coefficients do not show strong levels of association between the variables. Some of the statistically significant correlations showing medium association (at the 95% level) can be summarised as follows:

- Most of the indices (knowledge storage, knowledge sharing, learning from others, and organisational setting) are positively correlated with total volume of works completed till date (total turnover expressed in $). Rather than the age, size (measured by total turnover) is found to be more associated with the learning indices.
- There is a moderate association between the "organisational setting index" and "total overseas turnover," implying that working in international markets may have an impact on the organisational culture, strategies, and structure of a construction company and may increase its capacity to learn.
- "Frequency of partnering with foreign companies" is moderately associated with "learning from other companies," which can be interpreted as: companies that have established partnerships with foreign companies learn how to learn from their partners and increase their OL ability.

Table 5. Correlations between learning indices and performance

Learning indices	Pearson correlation	Significance level (2-tailed)
Individual learning index	0.474	0.000
Learning from other parties index	0.508	0.000
Organisational knowledge-acquisition index	0.338	0.002
Knowledge-sharing index	0.553	0.000
Storing of knowledge index	0.570	0.000
Deployment of knowledge index	0.384	0.001
Organisational setting index	0.367	0.001
Overall organisational learning index	0.624	0.000

Findings about Impact of OL on Performance

The performance index, comprising two parts—mainly profitability and growth in contract awards—is assumed to reflect the performance of contractors compared to their competitors. The minimum, maximum, and average performance values computed in this research are 1.75, 4.56, and 3.42, respectively. Table 5 summarises the association between the OL indices and performance index. The abilities of knowledge storage, sharing, and learning from other parties have been found to generate significant differences in the performance levels of Turkish contractors. Also, the correlation found between the OL index and performance index is 0.624, showing a substantial positive association between the OL competency and performance.

Moreover, an independent samples t-test, in which basically the means of two samples are compared, has been carried out. The overall sample in this research is divided into two by taking the grouping variable as the OL index. Thus, the companies having an OL index value smaller than 3.40, which is the mean of the overall sample, have been grouped under one category whereas the companies having an OL index value greater or equal to 3.40 have been placed under another group. A significant difference has been found between the mean performance index values of these groups, and the null hypothesis that there is no difference between the means of these groups has been rejected at the 95% level. The results of hypothesis testing are given in Table 6. Findings demonstrate that OL ability is a significant performance driver for Turkish contractors.

Table 6. Independent samples t-test results

	OL	Mean	Significance
PERFORMANCE	>= 3.40	3.7030	0.000
	< 3.40	3.0947	

Shortcomings of the Proposed Model and Limitations of the Study

Although OL index is assumed to reflect the OL ability of companies, it is not without limitations. Following are some of the shortcomings of the proposed method:

1. **About the individual learning sources:** The OL ability is assumed to be high if the individual learning sources such as external seminars are utilised frequently. However, training of employees fails to contribute to OL if the newly acquired knowledge is not translated into practical knowledge and transferred into work processes.
2. **About knowledge repository:** Organisational knowledge consists of tacit and explicit knowledge. The organisational knowledge resides in the minds of employees and relations that employees have, and is embedded in the work processes that all fall into the tacit knowledge category. The knowledge-storage index can only measure the availability of explicit knowledge, which is codified by recording and communicating throughout the organisation. As argued by Kahkonen and Kazi (2003), knowledge repositories are information oriented and knowledge management systems are usually based on information processing. Even if they are knowledge bases, one should ensure that they are referred by people to facilitate their decision-making process. Whether the employees use the knowledge sources effectively, the stored knowledge turned into working knowledge cannot be measured by the proposed index directly.
3. **About all organisational learning sources and mechanisms:** The major assumption within the OL index is that if the knowledge is acquired, shared, disseminated, and stored, it turns into organisational knowledge. It is used to revise the current situation and leads to a change in the work processes. However, how the organisational knowledge is further used to create new knowledge and insights cannot be measured with the proposed index.
4. **About the organisational setting:** The culture of the organisation is hard to evaluate. Whether the OL culture exists within a company or not has been tried to be measured by the existence of necessary organisational structure and utilised strategies to increase OL. However, OL culture is mainly about mental models and mindsets. How the OL process is supported by mental models cannot be easily evaluated. Also, as Schein (1997) denotes, in complex systems, composed of

multiple inter-acting and inter-dependent subsystems, OL occurs in each sub-system and then efforts are taken to integrate them. Each sub-system has its own culture and the cultural setting of the organisation comprises those sub-cultures that have to be investigated, while considering cultural barriers and enablers for OL.

5. **Role of environment:** There has been a continuous debate between the theorists about whether OL is an adaptive process or a pre-planned process. If the learning process is assumed to be pre-planned, environment and organisation are considered to be two independent systems where organisation has no control over the environment, which implies that OL follows from stimuli stemming from outside the organisation. The proposed index is based on the idea that OL is a planned process. Environment has been indirectly incorporated into the model, assuming that the impact of environmental factors are reflected in the organisational setting. How the organisation interprets the environment, makes strategic choices, and adapts the changes are not covered within the model.

In light of the above discussions, OL index should be considered as a subjective measure based on quantifiable components of OL rather than a complete and objective measure. Moreover, the questionnaire results should be investigated by keeping in mind that findings establish association between OL ability and performance of construction companies rather than causation. Also, as noted before, the relation between performance and OL ability is a complex one, as benefits of OL competency are usually exhausted over time. Thus, there is a time lag between the current OL ability and the performance which has not been considered within the current research study.

Conclusion

In this chapter, a conceptual OL model and a corresponding method for the measurement of OL have been proposed so that the relation between the OL ability and performance of construction companies can be investigated empirically. Based on the conceptual framework, a questionnaire structure is proposed as a way to collect data about learning practices, and facilitate calculation of an OL index. The applicability of the model and measurement method has been demonstrated by a study conducted within the Turkish construction industry. The questionnaire results demonstrate that there are significant differences between the performances of construction companies when they are grouped according to their OL competency. Companies having higher OL competency perform better than the others having a lower OL ability.

However, OL competency is very hard to quantify due to the existence of tacit knowledge embedded in the work processes, services, and so forth; cultural aspects of the learning process; and the complexity of the information ecology. Finally, it must be stated that the proposed subjective reporting approach and simplified OL index have underlying assumptions and should not be treated as objective instruments that can perfectly quantify OL competency.

Practical Tips and Lessons Learned

- Companies willing to improve their OL should use a procedure to measure their current state of OL ability. The measurement of OL ability necessitates both evaluating the knowledge management abilities as well as the appropriateness of the company culture.
- The companies having higher OL ability are expected to perform better than the others. Thus, the time and financial funds reserved to increase OL should be seen as high return investments that will lead to long-term success.
- The case of Turkish contractors reveals some facts about OL which will be also valid for companies from other developing countries. Although knowledge is seen as a major source of competitive advantage, learning mechanisms are not utilised frequently and learning from other parties is not given enough importance. Even though mechanisms are used to create organisational knowledge, usually unavailability of knowledge-sharing and knowledge-storage mechanisms decrease OL competency. Also, necessary performance measurement systems and reward mechanisms are usually not in place to promote learning. The weaknesses of Turkish contractors may demonstrate potential areas of improvement for companies from other developing countries having similar characteristics as Turkey.

References

Alwani-Starr, G. (1997). *Effective learning networks: How organisations can benefit from learning new ways of working.* Construction Industry Research and Information Association. Issue 1.

Ataoglu, T. (2002). *An organisational learning framework as a driver of organisational performance of Turkish construction companies.* Unpublished MSc Thesis, METU, Ankara.

Birgonul, T.M., Dikmen, I., & Ataoglu, T. (2002). Organisational learning as a performance driver in Turkish construction industry. *Proceedings of the 18th Annual Conference of the Association of Researchers in Construction Management* (ARCOM) (pp. 647-656), 2-4 September, Northumbria University, UK.

Davenport, T. (1997). *Information ecology.* New York: Oxford University Press.

Dess, G.G., & Robinson, R.B. (1984). Measuring organizational performance in the absence of objective measures. *Strategic Management Journal, 5*(3), 265-275.

DiBella, A.J., & Nevis, E.C. (1998). *How organizations learn.* San Francisco: Jossey-Bass.

Dulaimi, M.F., & Ling, F.Y.Y. (2003). Opportunities to enhance organisational learning in Singapore's construction industry. In G. Ofori & Y.Y.L. Ling (Eds.), *Proceedings of the Joint International Symposium of CIB Working Commissions* (pp. 87-97), 22-24 October, National University of Singapore.

Egbu, C. (2000). Knowledge management in construction SMEs: Coping with the issues of structure, culture, commitment and motivation. In A. Akintoye (Ed.), *Proceedings of the 16th Annual Conference of the Association of Researchers in Construction Management* (ARCOM) (pp. 83-93), 6-8 September, Glasgow Caledonian University, UK.

Garvin, D. (1993). Building a learning organization. *Harvard Business Review,* (July-August), 78-91.

Hakansson, H., Havila, V., & Pedersen, A. (1999). Learning in networks. *Industrial Marketing Management, 28,* 443-452.

Holt, G.D., Love, P.E.D., & Li, H. (2000). The learning organization: Toward a paradigm for mutually beneficial strategic construction alliances. *International Journal of Project Management, 18,* 415-421.

Huemer, L., & Ostergen, K. (2000). Strategic change and organizational learning in two Swedish construction firms. *Construction Management and Economics, 18,* 635-642.

Kahkonen, K., & Kazi, A.S. (2003). Knowledge repository for construction projects: Hype or real solution? In G. Ofori & Y.Y.L. Ling (Eds.), *Proceedings of the Joint International Symposium of CIB Working Commissions* (pp. 87-97), 22-24 October, National University of Singapore.

Kale, S., & Arditi, D. (2002). Competitive positioning in United States construction industry. *ASCE Journal of Construction Engineering and Management,* (May/June), 238-247.

Kale, S., & Arditi, D. (2003). Differentiation, conformity and construction firm performance. *ASCE Journal of Management in Engineering,* (April), 52-59.

Knuf, J. (2000). Benchmarking the lean enterprise: Organizational learning at work. *ASCE Journal of Management in Engineering, 16*(4), 58-71.

Kululanga, G.K., Edum-Fotwe, F.T., McCaffer, R., & Price, A.D.F. (1998). Learning mechanisms for addressing improvement in construction companies. In W. Hughes (Ed.), *Proceedings of the 14th Annual Conference of the Association of Researchers in Construction Management* (ARCOM) (pp. 69-77), 9-11 September, University of Reading, UK.

Kululanga, G.K., Edum-Fotwe, F.T., & McCaffer, R. (2001). Measuring construction contractors' organizational learning. *Building Research & Information, 29*(1), 21-29.

Kululanga, G.K., McCaffer, R., Price, A.D.F., & Edum-Fotwe, F. (1999). Learning mechanisms employed by construction contractors. *ASCE Journal of Construction Engineering and Management, 125*(4), 215-223.

Kululanga, G.K., Price, A.D.F., & McCaffer, R. (2002). Empirical investigation of construction contractors' organizational learning. *ASCE Journal of Construction Engineering and Management, 128*(5), 385-391.

Love, P.E.D., Li, H., Irani, Z., & Faniran, O. (2000). Total quality management and the learning organization: A dialogue for change in construction. *Construction Management and Economics, 18,* 321-331.

Marquardt, M.J. (1996). *Building the learning organization.* McGraw-Hill.

Nonaka, I. (1996). The knowledge-creating company. In K. Starkey (Ed.), *How organisations learn.* London: International Thomson Business Press.

Orange, G., Boam, J., & Burke, A. (1998). A network system to facilitate organisational learning within the construction industry. *Proceedings of the 4th EATA International Conference on Networking Entities* (NETIES98), 15-16 October, Leeds Metropolitan University, UK.

Robinson, H.S., Carrillo, P.M., Anumba, C.J., & Al-Ghassani, A.M. (2001). Linking knowledge management strategy to business performance in construction organizations. In A. Akintoye (Ed.), *Proceedings of the 17th Annual Conference of the Association of Researchers in Construction Management* (ARCOM) (pp. 577-586), 5-7 September, University of Salford, UK.

Schein, E.H. (1997). *Organizational culture and leadership.* San Francisco: Jossey-Bass.

Senge, P.M. (1990). *The fifth discipline: The art & practice of the learning organization.* Currency Doubleday.

Senge, P.M. (1994). *The fifth discipline: Strategies and tools for building a learning organization.* Currency Doubleday.

Snyder, W., & Cummings, T. (1998). Organisational learning disorders: Conceptual model and intervention hypotheses. *Human Relations, 51*(4), 531-546.

Vakola, M., & Rezgui, Y. (2000). Organizational learning and innovation in the construction industry. *The Learning Organization, 7*(4), 174-183.

Chapter XI

Evaluating an Organisation's Learning Culture Using Learning Histories

Marcus C. Jefferies
University of Newcastle, Australia

Chen Swee Eng
University of Newcastle, Australia

Ralf Zenke
Department of Health and Human Services, TAS, Australia

Abstract

This research describes a recently developed approach to organisational learning known as "learning history". Learning history involves the analysis and reflection on project experiences and business processes. The learning history of a construction industry organisation is documented and the outcomes of the learning history are used to evaluate the organisation's methods of acquiring and managing knowledge. This learning history technique is applied in a practical setting to a case study of the Tasmanian State Public Health Sector, Australia. The findings of the learning history are used to evaluate the organisation's knowledge management in order to evaluate its "learning culture". The organisation developed new learning capabilities and learning history is a practical tool for organisational learning which can be widely used for knowledge transfer. Fostering knowledge management leads, indirectly, to more efficient and effective processes. In order to sustain the organisation's learning culture then, improvement of its organisational learning style, the learning process (knowledge management), and its organic learning style is necessary.

Introduction

The research approach to organisational learning described in this chapter reflects a recent development in this area, known as a *learning history*, the practical application of an organisational learning tool.

The learning history concept was developed in the U.S. at the MIT Center for Organizational Learning. The Center was established in 1990 to apply and test the concepts proposed in Senge's *The Fifth Discipline* (1990). This new organisational learning tool is applied in a practical setting, and the outcomes of the learning history used to evaluate the learning culture in the Tasmanian State Public Health Sector (Health Co.).

In terms of the research objectives, it is the intention of this chapter to describe, apply, and test an organisational learning tool in order to address the issues associated with promoting organisational learning. Subsequently, the learning history of a construction industry organisation is documented, and the outcomes of the learning history are used to evaluate the organisation's methods of acquiring and managing knowledge.

Learning Objectives

The reader will learn and understand the following objectives from this chapter:

1. Concepts of organisational learning and learning culture.
2. The concept of an organisational learning tool known as *learning history.*
3. The application and testing of the learning history tool within a construction organisation.
4. How an organisation's knowledge management can be evaluated in order to improve efficiency and practice.

The Concept of Organisational Culture

Despite being an important concept, there is debate as to the definition of organisational culture. For example, according to Schein (1985), organisational culture is seen as a pattern of basic assumptions invented, discovered, and developed by a given group as it learns to cope with the problems of external adaptation and internal integration that has worked well enough to be considered valid and taught to new members as the correct way to perceive, think, and feel in relation to those problems.

In contrast, Denison (1984) defined organisational culture as the underlying values, beliefs, and principles that serve as a foundation for an organisation's management system, as well as the set of management practices and behaviour that both exemplify and reinforce those basic principles.

Schein's definition of organisational culture is generally adopted because it involves assumptions, adaptations, perception, and learning.

However, there seems to be wide agreement that organisational culture refers to a system of shared meaning held by members that distinguish the organisation from other organisations (Becker, 1982; Schein, 1985). This system of shared meaning is a set of key characteristics that the organisation values. Robbins, Waters-Marsh, Caciope, and Millet (1994) suggest that there are 10 primary characteristics that capture the essence of organisational culture:

1. **Member identity:** The degree to which employees identify with the organisation as a whole, rather than with their type of job or field of professional expertise.
2. **Group emphasis:** The degree to which work activities are organised around groups rather then individuals.
3. **People focus:** The degree to which management decisions take into consideration the effect of outcomes on employees within the organisation.
4. **Unit integration:** The degree to which units within the organisation are encouraged to operate in a coordinated or independent manner.
5. **Control:** The degree to which rules, regulations, and direct supervision are used to oversee and control employee behaviour.
6. **Risk tolerance:** The degree to which employees are encouraged to be aggressive, innovative, and risk-seeking.
7. **Reward criteria:** The degree to which rewards such are pay increases and promotions are allocated according to employee performance, rather then seniority, favouritism, or other non-performance factors.
8. **Conflict tolerance:** The degree to which employees are encouraged to air conflicts and criticisms openly.
9. **Means-ends orientation:** The degree to which management focuses on results or outcomes rather than on the techniques and processes used to achieve those outcomes.
10. **Open-system focus:** The degree to which the organisation monitors and responds to change in the external environment.

Every organisation has its own unique culture, even though it may not have been consciously created. It will probably have been created unconsciously, based on the values of senior management, the founders or core people of that organisation. Over time individuals (particularly the organisation's leaders) attempt to change the culture of their organisation to fit their own preferences or changing marketplace conditions. This culture then influences the decision-making processes, it affects the style of management and what is determined as success.

"Organisational culture is the key to organisational excellence... and the function of leadership is the creation and management of culture." (Schein, 1985)

Managers must understand organisational culture if they want to manage or change it. Schein (1985) gives a flavour of the concept when he states that organisational learning, development, and planned change cannot be understood without considering culture as the primary source of resistance to change.

Subcultures and Subsystems

Robbins and colleagues (1994) say that most large organisations have a dominant culture and numerous sets of subcultures. A homogenous culture in an organisation is rarely the case in practice. Most organisations of any size contain many identifiable subcultures—the beliefs, values, and assumptions of which may compete with the dominant culture. A dominant culture expresses the core values that are shared by most of the organisation's members. Subcultures tend to develop in large organisations to reflect common problems, situations, or experiences that members face.

Organisational Learning

Organisations can be seen to learn as the collective patterns of behaviour amongst organisational members change and adapt to their environment. Individuals act as learning agents for the organisation by detecting and correcting errors in the organisational behavioural patterns, which in turn become embedded in the culture of the organisation's learning culture (Argyris & Schön, 1978). Nevis, Di Bella, and Gould (1999) define organisational learning as the capacity or processes within an organisation to maintain or improve performance based on experience.

organisational learning is a process that is embedded in the organisational culture, where individuals continuously learn how to learn together (the learning organisation) and as a result maintain or improve organisational performance. It is important to note at the outset the useful distinction between organisational learning (learning by individual and groups/teams in the organisation) versus the learning organisation (learning by the organisation as a total system) (Schein, 1999).

Critical to organisational learning is how knowledge is communicated. Successful organisations are those that consistently create new knowledge and are able to disseminate it widely through the organisation and embody it in new technologies and innovations. A critical part is the ease of access to all forms of knowledge. Levinson and Asahi (1995) suggest that the following steps are integrated into organisational relationships:

- becoming aware and identifying new knowledge;
- transferring/interpreting new knowledge;
- using knowledge by adjusting behaviour to achieve intended outcomes; and
- institutionalising knowledge by reflecting on what is happening and adjusting learning behaviour.

Organisational Culture and Learning Culture: A Relationship

It can be argued that the organisational culture determines the appreciation of a learning culture and has paramount influence on the learning culture's strength or weakness (i.e., performance). Learning culture is defined by Schein (1999) as the ability and willingness of an organisation (i.e., the organisational culture) to capture and utilise the knowledge and experiences of its people (i.e., organisational learning), and to use this capacity (i.e., learning organisation) to maximise organisational performance. This definition provides the context in which learning culture is seen for this research.

Schein (1999) identifies that the learning culture of an organisation is the consequence of the organisation's prior experience and learning, and is the basis of its continuing capacity to learn. The performance of the learning culture depends very much on the context of the organisational culture.

The Learning Process

Malhotra (1996) sees the key ingredient of the learning organisation as how organisations process their managerial experiences. Rhodes (1996) also places emphasis on management as the very key to control the organisational learning process. According to Inkpen and Crossan (1995), organisations rarely learn from projects because they do not have the mechanisms or intent that would allow knowledge to be captured. The essential prerequisite for organisations to institutionalise learning are systems that routinely and regularly collect intelligence from projects and partners. A learning history as a tool to capture knowledge from projects is an option to improve organisational memory.

Inkpen and Crossan's learning cycle (interpreting–integration–institutionalising) applies to all three levels of organisational learning and differs from Kolb's learning cycle (1984) as a model of individual learning. The phases of his learning cycle—hands-on experience, reflection, abstract conceptualisation, and active experimentation—have been tested extensively and have been shown to accurately model individual learning (Newcombe, 1997).

It can be argued that Kolb's four phases of individual learning are to a certain degree reflected in Inkpen and Crossan's learning cycles for all three levels in an organisation. Kolb's phases of "hands-on experience" and "reflection" may coincide with Inkpen and Crossan's "interpreting." The phases of "integration and institutionalising" may relate to "conceptualisation and experimentation." One can argue that once an organisation has integrated and institutionalised a lesson learned, it actively experiments with an abstract concept.

Organisational learning occurs at individual, group, and organisation level. A learning cycle applies to all three levels of organisational learning and has different phases:

- Individual level by interpreting/reflection
- Group level by integration/conceptualisation
- Organisational level by institutionalising/experimentation

The three levels are dependent upon each other. Without individual learning there is no group learning, without group learning there is no organisational learning.

Learning Styles

Mumford and Honey (1986) define Kolb's individual learning styles as:

- Activists who learn best where there are new experiences/problems/opportunities from which to learn:
  - Reflectors who learn best where they are allowed to watch/think/cogitate over activities.
  - Theorists who learn best from activities that are part of a system/model/theory.
  - Pragmatists who learn best where they have an opportunity to try out/practice techniques with coaching/feedback from a credible expert.

Newcombe (1997) considers that, regardless what learning style an individual has, new project experiences alone are not sufficient for organisational learning unless time is allowed for reflection; mentoring, which encourages conceptualisation; and a non-threatening atmosphere, which encourages experimentation.

Edum-Fotwe (1996) states that the concept of organic learning derives from theories of organisational learning. It relates to how organic systems (and individuals) such as corporate establishments imbibe information from their external and internal environments. According to Argyvis and Schön's (1978) classic analysis of organisational learning, learning mechanisms for organic systems can be grouped into adaptive learning (single loop) and generative learning (double loop/deutero).

Research Method

The learning history process entails project de-brief interviews with participants and a qualitative data analysis of their project experiences, which leads to the formulation of process improvement and risk reduction strategies based on the generic lessons learned. The dissemination of the learning history and its acquired knowledge within the organisation is a vital part of the organisational learning process and an invaluable process for knowledge management and transfer within the learning organisation.

The first step in the research of organisational learning is to collect the stories by means of interviewing individuals who have been involved in events of organisational change and learning. Using the interviewing or storytelling approach for a broad spectrum of organisational members is the main characteristic of a learning history. This research technique aims at giving voice to stories that are not heard in the traditional approach of organisational theory. The value of this approach is that it creates the opportunity for a reflective discussion and comparison of the diversity of meanings and themes. The learning history makes extensive use of participants' own narratives, as well as the researchers' assessments of the story (formulation of generic lessons and proposed process improvement strategies).

Case Study Project Background

The following learning history describes the very successful collocation project of a private hospital with a public health care facility in Hobart, Tasmania by Health Co., a major player in private health care provision in Australia. The Health Co. Private Hospital opened on 29 November 1999. The opening was the culmination of approximately two years of planning and development by Health Co., following the announcement of the collocation project by the state government in July 1997. A budget in excess of A$44million was allocated for the development, acquisition, construction, and fit-out of the facility. The design of the facility was completed by the in-house 'Building Development Team', and subsequently a 'Construct Only' contract was awarded to build a facility comprising a 150-bed acute hospital on 11,000 m2 over nine storeys. The construction period was nine months.

Results

Thirty project participants from all facets were interviewed. This includes the Health Co. core project team, relevant business unit managers involved in project work, team members from the corporate head office, and representatives from the Royal Hospital. Their individual experience is presented in the following structure and discusses a particular aspect of learning history:

- **Participants' Quotes and Position Name:** Quote or quotes relevant to the topic, reflecting the broad opinion of Participants.
- **Generic Lesson:** Statement of interview results, which show a regularity across the team or individual comment of importance of outstanding innovative nature, followed by a conclusive comment where applicable.
- **Reflective Improvement or Proposed Strategy:** If the generic lesson leads to a process improvement /strategy, it will be formulated here. Not all generic lessons will lead to a strategy. In general, generic lessons lead to a process improvement, but only a few will result in the formulation of a proposed strategy.

The Health Co. Private Hospital Collocation Project: Learning History and Proposed Process Improvement Strategies

Project Structure

- **Project Manager:** "It should have been identified as a project and, more importantly, handled as such as soon as Health Co. was announced preferred proponent."
- **Senior Management Team Member:** "We should have made the assumption that we are going to be successful and should have started the project then and developed project teams. We believed we had a very good tender document and we were confident that our bid was well constructed."
- **Unit Manager:** "There was a big gap after the tender was won. I thought that was a negative. Setting up how the project is going to be run and communicating it to the people could have happened then."

- **Generic Lesson:** The vast majority of the project team identified the time lapse between the announcement and "switching into project mode" as a lost opportunity for project preparation and effective time management.
- **Reflective Improvement:** Utilise this time to put an initial project scope document together involving key project personnel.

Pre-Project Workshop

- **Corporate Team Member:** "Right at the start we should have a project group discussion regarding expectations, roles, and responsibilities, project management principles, the importance of project planning, and timelines. We also should have a discussion of how we culturally handle this project, how we develop a team environment that is a true team environment, and to develop an understanding of each other. In this environment a number of people are new to each other, from corporate and locally, and haven't really worked that closely together. They are suddenly thrown together with tight timeframes and a huge project, working on different and new issues. You need to get that team right. I think that will benefit the company in the long term anyway."
- **Generic Lesson:** A pre-project workshop that addresses the above-mentioned issues was seen by all team members as a valuable step to improve future project performance.
- **Senior Management Team Member:** "There is no process in place whereby deficiencies in management expertise or experience are identified. There is course work you can do, usually in weekend modules. If you have an understanding about what you are required to do, you can approach it much more confidently and effectively. It is not easy to identify how to be a project manager and what this entails. Health Co. should invest a small amount of time and effort in encouraging team members to attend relevant staff development courses. That's not a major undertaking, but would improve performance. Team members without previous project experience would benefit from participating in a number of project management courses which outline the importance of various generic project management principles."
- **Reflective Improvement:** Hold a pre-project workshop and consider involving a number of people with recent significant project management experience for initial consultation. Outline Health Co.'s cultural approach and, if relevant, clarify the cultural difference/ethos between the public and private sectors. This is of particular importance in improving project performance and the outcome of service agreements. Set up a reporting structure and encourage a mindset of open communication. Encourage team members to regard a project as a "short-term business" that needs to be run in a professional way.

Resources - Key Appointments

- **Project Manager:** "There was an even longer gap between clarifying project management roles and business unit manager roles. That was very destabilising. A lot of people felt very uncomfortable that the business unit manager appointments took so long. A lot of people's enthusiasm waned and they became quite negative. If a business unit manager expresses discontent, this has a ripple effect on their staff. They become unsettled. This was quite negative to the process."
- **Business Unit Manager:** "I was frustrated in the beginning and was pushing to be involved, worried that nothing was happening as some of the services were quite complex and difficult to scope. I started to do some work on my own in anticipation. Project teams should have been formed earlier."
- **Generic Lesson:** It is necessary to find a fine balance between cost implications and early key appointments. All team members felt that early role ownership improves the attitude towards project issues. It encourages project thinking and planning and maintains enthusiasm. Uncertainty creates angst, and angst is less harmful earlier in the process than one month before opening.
- **Reflective Improvement:** Start the selection and appointment process for key positions as early as possible. Most team members consider six to nine months prior to project finalisation appropriate. This also enables project managers to delegate certain tasks and details to these key positions, which otherwise leads to increasing over-allocation of project managers and hinders their focussing on the "big picture." If cost implications preclude an early appointment, then the recruitment process for key positions must be finalised as soon as practical, with the commencement date established for the positions one month before opening. Performance criteria must be established for these new roles, which would be reviewed after an appropriate period. If expectations are not met, the consequences must be openly communicated at this point. This provides a clear indication about expectations early in the process.

Improved Interaction Between Project Managers and HR Manager

- **Project Manager:** "Human resources is such an important part of each project, and it is paramount that a dedicated management resource is applied to it. It is important to be coordinated centrally. In doing so, it highlights the absolute need for effective communication between the human resources activities of the individual project and the project managers. I think the process was about right, but the communication between the project manages and human resources component needs to be very good. You could expect that the project manager and the human resources manager have regular face-to-face meetings throughout the project about the status quo. This did not happen enough."
- **Generic Lesson/Reflective Improvement:** The above quote illustrates the common perception of the team that regular meetings regarding human resources issues

were insufficient. This could be improved by having regular updates during information forums and also having improved interaction during informal project meetings between the individual project managers and the human resources manager. This might require an additional resource for human resources (e.g., for interviews regarding basic organisational positions) to better utilise the human resources manager's role for facilitation of project meetings and the definition of appropriate processes.

Negotiation Process with Public Sector

- **Senior Management Team Member:** "When we developed the tender, the assumption was that we could provide enormous advantages for the public sector. This concept was a mistake and elevated the level of risk. There was a lack of understanding of what the expectations of the public sector were. Important issues of the private sector—cost and affordability—do not exist in the public sector. There was an unwillingness to consider that the public sector is as efficient (or even more) as the private sector."
- **Project Manager:** "There is such cultural difference even if you think we are the same. That's not only related to how the systems are funded. For negotiations with the public *sector, the best approach to make it successful is to put on the 'public* hat' for a moment and look at the negotiation process from their perspective."
- **Generic Lesson:** Try to understand the objectives of the public sector and show how Health Co. can facilitate this outcome. This makes negotiations more focussed. Gain and maintain insight in the public sector and update your knowledge about it.
- **Reflective Improvement:** Understand that altruistic practice may actually pay a dividend in the end. Do not concentrate exclusively on short-term profit; invest in unforeseeable future events (e.g., by contributing to training). Involve yourself in the community where you wish to operate and make profit.

Consultation with Stakeholders

- **Project Manager:** "It is important to have open and honest dialogue, otherwise it may have a negative impact on future negotiations. It is also important to involve people whom the service agreements will affect in the process of setting it up and not to underestimate any of the participants. Stakeholders need to take ownership of the whole process to avoid being dictated to. You also have to be very careful about releasing financial details of service agreements to clinical people in particular. They don't have the 'big picture' of the financial situation and interpret it negatively. That impacted on the relationship in this area."
- **Project Officer:** "We put a lot of time into setting up relationships and getting communication working between the two sides. We gave them (i.e., operational

working groups) a lot of encouragement about what it means to work together and we developed some ideals. If new players came in, they had a different background and understanding. We had so many people coming in all the time and they were not briefed. That's where the communication difficulty arose from."

Generic Lesson:

- Initial open and honest dialogue creates trust, which is necessary for future negotiations.
- Do not underestimate any stakeholders.
- Someone who ultimately has responsibility for service agreements needs to be involved in the development and have ownership of the whole process.
- The project controller clearly communicates to the team what each group aims to achieve before working together. This improves efficiency.
- New team members need to be briefed and expectations clarified in order to maintain good communication.
- Ensure that everyone involved understands that financial arrangements are part of the 'big picture' and should not make judgements in any way. If this is pursued early in the process, potential problems can be minimised.

Discussion

Effect of Learning History on the Organisation and Its Learning Culture

The concept of organisational learning was for the corporation, Health Co., elusive, but the openness of the relevant key manager with regard to this topic was positive. This enabled the research to progress with the learning history process. Previously, few, if any examples of learning in this organisation have been well documented.

According to key members of the organisation, the reasons for this lack of documented learning was twofold. Firstly, there was the unanswered question of how learning can be assessed without a link to specific performance objectives. It was seen that the nature of organisational conditions is their relationship to performance. Secondly, it was not clarified which demands in the organisation lead to pressures for assessment and how these pressures influence the phenomena (i.e., organisational learning) they are intended to measure.

This issue clearly relates back to the 'fear' of assessment. This required tireless efforts to outline the benefits of the learning history concept, which tries to avoid this demise by capturing valuable intellectual property of both positive and negative experiences.

Effect on Organisational Performance

The learning history had the effect that Health Co. realised that in today's information economy, the most valuable non-capital resource in a company is its knowledge. Most of a company's embedded knowledge is locked up in the heads of its employees. The learning history supported the goal of knowledge management and expertise location.

These concepts aim to unlock this knowledge before it "walks out the door" and to place it in the hands of those individuals who are making 'on the ground decisions' (i.e., knowledge transfer). This negates or diminishes the need to engage external consultants to retrieve this knowledge when needed at a later stage. This positive financial impact on the bottom line was a convincing and contributing factor for the company.

Health Co. began to understand the value of non-capital resources and the emerging tendency that organisations are being increasingly valued on their intellectual capital as opposed to physical assets. The core objective of this process was to develop strategies that will enable Health Co. to build on tangible benefits in future projects. The collated material of this project has been used to devise strategies which minimises the occurrence and impact of negatives, thereby minimising risk and improving risk allocation and risk management. The strategies also maximise and build upon identified positive features.

Effect on the Organisation's Learning Culture

This project's de-brief and analysis captured valuable intellectual property, which will enable Health Co. to retrieve this corporate knowledge to continually gain capabilities and improve competitiveness by broadly diffusing these experiences and new innovations within Health Co. Without that approach, the learning effort would have stopped with the end of the collocation project, and organisational experience and actionable knowledge would have been lost. Individual perspectives of both the project's achievements and shortfalls were joined together coherently. Health Co. learned what happened, why it happened, and how to translate it into more effective action for future projects.

Participants in this learning effort commonly stated that they underwent a transformation. As they developed capabilities together, gained insights, and shifted their shared mental models (often without being aware), they changed their assumptions about work and the value of good personal and organisational interrelationships. The organisation's senior management is aware that this viable learning effort may not produce tangible results for several years, but its most important results may include new ways of thinking and behaving.

The value of the learning history came not so much from the document itself, but from the consultation process that was built around it. Reflective interviews gave participants a chance to talk openly and expressively about their experiences which is a significant source of learning in itself. The intention of this learning concept is to pass on knowledge necessary for new teams to initiate their own learning cycle. Health Co. obtained "actionable knowledge"— that is, knowledge that covers tacit issues such as relationships and working habits, which were brought to the surface, examined collaboratively, and most importantly, communicated. Such knowledge provides the key for repeating success and avoiding failures.

Health Co.'s Organic Learning Style

The analysis of the project processes as described in Health Co.'s learning history suggested a remarkable predominance of adaptive over generative learning mechanisms. This doesn't come as a surprise, as Argyris and Schön (1978, 1996) state that organisations are good at single-loop learning, but poor at double-loop learning, which requires a more extensive rethink of underlying norms, policies, and objectives.

The reason for the imbalance between adaptive and generative learning might be twofold. Firstly, Health Co. being a more adaptive organisation, as the researcher's study suggests, focuses on incremental improvements, often based on the past track record of successes. It can be argued that Health Co.'s learning history, being a retrospective analysis of a successful project, fits into this category. Secondly, the fundamental assumptions underlying the existing ways of doing work were generally not questioned. It seems that this was not necessarily required. Change does not have to occur just for the sake of change, particularly if established procedures work well. For double-loop learning to occur, there must be a change in the individual, group, or organisational theory-in-use. This is unlikely to occur during a high-pressure project with tight timeframes such as the one studied.

The advantage of a retrospective learning history is that any shortcomings due to adaptive learning can now, during the "breathing time" after the project, be detected and reacted to by adoption of generative learning principles. This means the retrospective rethinking of existing ways of doing work.

Health Co.'s Organisational Learning Style

The results of the previous analytical comparison of Health Co.'s Learning Organisation structure are quite remarkable. They show an even distribution of distinctive facilitating factors and learning orientations through the whole spectrum of the learning process cycle (i.e., knowledge acquisition, dissemination, and utilisation). This distinctive characteristic of Health Co.'s learning cycle makes it particularly suited to organisational learning, particularly double-loop learning.

The knowledge utilisation stage of the learning cycle indicates Health Co.'s tendency to incremental/corrective learning. This is an indication that the organisation tends to adaptive learning, which is also shown in the previous learning culture evaluation (see organic learning style).

This finding, however, seems to be in contrast to the above claim that the learning history process encourages generative learning and Health Co.'s indication to implement the resulting proposed process improvements. This discrepancy may be based on the fact,

Figure 1. Concept and benefits of learning history

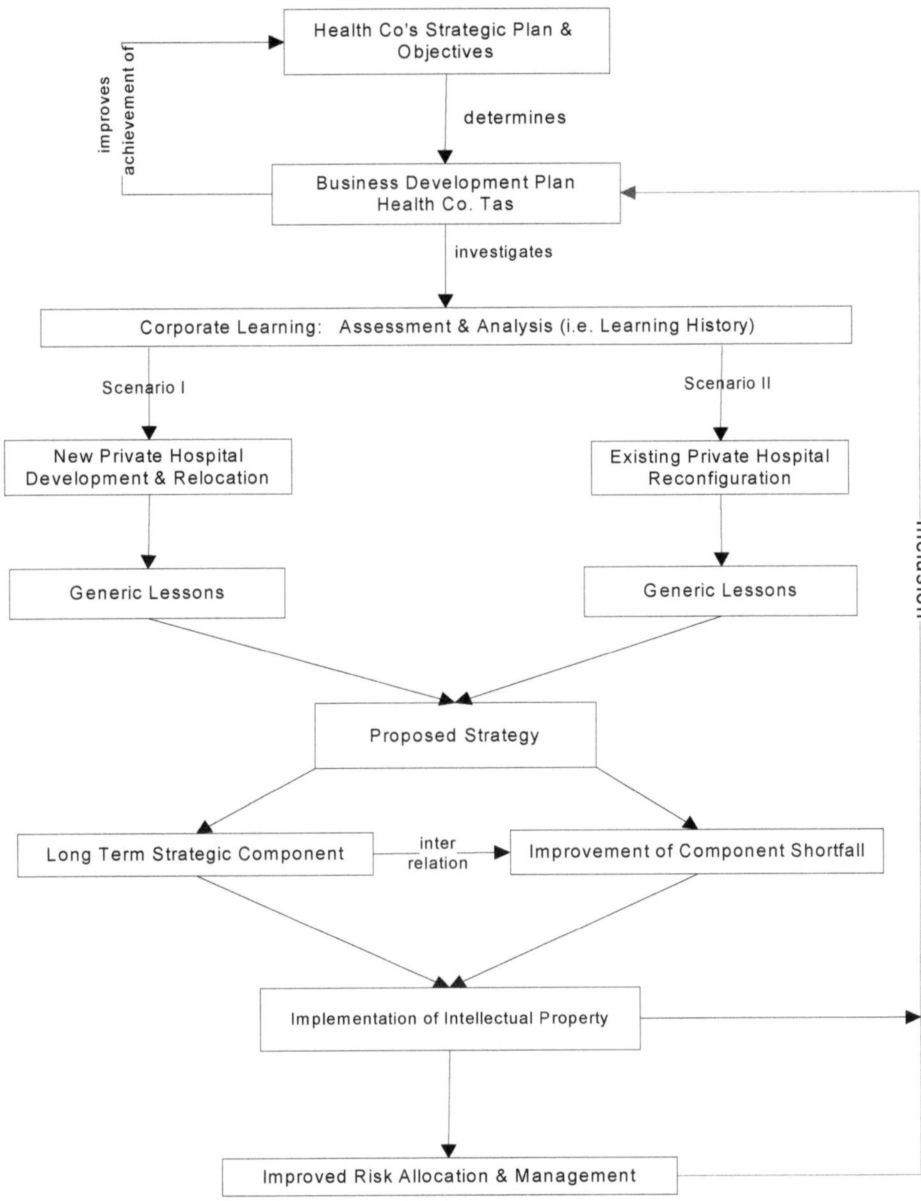

as suggested by data, that no specific facilitating factor, which determines effectiveness and efficiency for learning to occur, is assigned to the knowledge utilisation stage.

Health Co.'s Learning Process

The processes studied seem to place an emphasis on individual skill development, involved leadership, and systems perspectives. This may be explained by an underlying correlation between knowledge specialisation and information flow between organisational units, which are interdependent. Functional diversification occurs at Health Co. because different types of expertise are necessary to execute distinct activities in processes, which leads to knowledge specialisation. Nevertheless, there is a connection between these organisational units, that is, the corporate and local teams who share needs and goals and use this specialised knowledge to address problems and find solutions caused by processes of this systemic relationship.

Therefore, at Health Co., there seems to be a strong correlation between knowledge specialisation and knowledge exchange/transfer. This occurs because of the high number of knowledge workers with expertise in specialised fields. The current trend, not only at Health Co., towards knowledge specialisation is likely to compound this situation in the future, as new knowledge and new knowledge specialisation areas are created every day. This fact alone underlies again the importance to manage this acquired knowledge (knowledge management) constantly with appropriate organisational learning tools such as a learning history.

Strategic Benefits of Learning History

In general the learning history had several effects. Firstly, and perhaps most importantly, it built trust. People who believed their opinions have been ignored in the past felt validated by the presence of those opinions in the document. Project participants who felt isolated realised they were not alone in their efforts to improve both their and Health Co.'s future. The learning history appeared to be especially effective in allowing individuals to raise issues that they wanted to talk about, but had not had the courage to speak out loud. The document with its anonymous commentary from participants does this for them in a safe environment.

This project has proven that the single most critical factor for success is the quality of human interaction within the organisation, which depended on the humility and openness of their leaders who direct the effort.

Figure 1 shows a flowchart developed to provide senior management with an overview and outline of the learning history concept and its long-term strategic benefits for the organisation.

Conclusion

The learning history process can be seen as a set of interrelated activities of organisational learning carried out to improve project process functions by capturing both general and often specialised bodies of knowledge held by organisational members who perform those functions. Health Co.'s collocation learning history may not have produced measurable results as yet, but it has produced a theory of how the organisation got there.

The outcomes of the research are threefold:

1. An organisation developed new learning capabilities.
2. The new organisational learning tool (i.e., the learning history) can be used widely.
3. Improved processes that can be tested and can lead to replicable outcomes.

Fostering knowledge communication, transfer, and management will to lead to an indirect way to more efficient and effective processes, even if no "visible" improvements accrue from the process improvement attempts. In other words, a tangible result like the learning history often signifies an intangible gain.

Practical Tips and Lessons Learned

The following issues have been identified as tips and/or lessons learned:

- Organisational learning, and the concept of learning history, must be clearly understood by all stakeholders so that any 'fear of assessment' can be overcome.
- A learning history tool builds trust and facilitates successful communication.
- A successful organisation will encourage quality human interaction.
- Organisational leadership must be open in order to facilitate the learning history tool and encourage communication.
- An organisation's most valuable non-capital resource is its 'knowledge', which must be captured, and subsequently transferred, before it is lost.
- A learning history tool adds value far beyond producing documentation, in that the process itself, particularly reflective interviews, gives stakeholders the opportunity to discuss their experiences and therefore enables a form of knowledge capture.
- Acquired knowledge must be constantly managed in order to benefit from the increasing trend of organisations that consist of high numbers of knowledge workers with specialist expertise. Applying learning history is one approach to successfully manage this knowledge.

- Using organisational learning tools, such as learning history, allows for the successful fostering of knowledge capture, transfer, and management, and ultimately leads to process efficiency and effectiveness.

References

Argyris, C., & Schön, D.A. (1978). *Organisational learning: A theory in action perspective.* Reading, MA: Addison-Wesley.

Becker, H. (1982). Culture: A sociological view. *Yale Review, 71,* 513-527.

Denison, D. (1984). Bringing corporate culture to the bottom line. *Organisational Dynamics, 13*(2), 5-11.

Edum-Fotwe, F. (1996). Advancing construction management research through organic learning styles. In S.E. Chen, R. Kenley, & M. Skitmore (Eds.), *International electronic conferences 384 1997/1998* (pp. 23-28), AUBEA, Melbourne.

Inkpen, A., & Crossan, M. (1995). Believing is seeing: Joint ventures and organisational learning. *Journal of Management Studies, 32*(5), 595-618.

Kolb, D. (1984). *Experimental learning.* Englewood Cliffs, NJ: Prentice-Hall.

Levinson, N.S., & Asahi, M. (1995). Cross-national strategic alliances and inter-organisational learning. *Organizational Dynamics,* (Autumn).

Malhotra, Y. (1996). Organizational learning and learning organizations: An overview. Retrieved from www.brint.com/papers/orglrng.htm.

Mumford, A., & Honey, P. (1986). *The manual of learning styles.* Honey, Maidenhead.

Nevis, E., Di Bella, A., & Gould, J. (1999). Understanding organisations as learning systems. *Proceedings of the Society for Organisational Learning and International Consortium for Executive Development Research,* Lexington, Massachusetts, USA.

Newcombe, R. (1997). Procurement as a learning process. In S. Ogunlana (Ed.), *Profitable partnering in construction procurement CIBW92* (pp. 285-194). London: E and F.N. Spon.

Robbins, S.P., Waters-Marsh, T., Caciope, R., & Millet, B. (1994). Organisational behaviour—concepts, controversies and applications. Sydney: Prentice-Hall.

Schein, E.H. (1985). *Organisational culture and leadership.* San Francisco: Jossey-Bass.

Schein, E.H. (1999). *Organisational learning: What is new?* MIT Sloan School of Management. Retrieved from learning.mit.edu/res/wp/10012.html.

Senge, P.M., Roberts, C., Ross, R.B., Smith, B.J., & Kleiner, A. (1994). *The fifth discipline fieldbook: Strategies and tools for building a learning organisation.* New York: Doubleday.

Chapter XII

Uncertainty and Information in Construction: From the Socio-Technical Perspective 1962-1966 to Knowledge Management – What Have We Learned?

Alan Wild
Consultant, Birmingham, UK

Abstract

This chapter questions the possibility of knowledge management in construction, other than among leading organisations handling a restricted population of projects. Socio-technical research in UK coal mining from the 1950s and construction from the 1960s and other relevant studies are compared. The extent to which the tacit order, instability, and diffuseness of construction may, practically, undermine knowledge management is explored. Knowledge management is compared to other methods in terms of the stability required for it to be effective. It is demonstrated that this stability is not usually available in construction, whose projects and processes are subject to an unusually wide range of disturbances.

Introduction

This book offers a socio-technical perspective on knowledge management in construction. This chapter discusses the original socio-technical research on construction, the Building Industry Communications Research Project (BICRP) by the Tavistock Institute of Human Relations (TIHR) from 1962-1966 (Boyd & Wild, 2003; Wild, 2004). Located between TIHR studies of UK coal mining and concepts of contextual turbulence in 1965, BICRP archives and publications demonstrate the state of construction, revealing the tacit order as deeper and wider than assumed. Problems of information, uncertainty, and indeterminacy pushed socio-technical ideas to their limit.

The knowledge management discourse assumes accessibility of the tacit order through structured inquiry. Are the categories of sense making; communications; information management; intellectual property and capital; innovation, communities of practice, and knowledge capture through IT latent within TIHR ideas? As a then leading analysis for re-framing managerial practice, these compare to knowledge management, which may also reach the limits of analytical penetration in the tacit order of construction. The chapter discusses TIHR work; 1960s studies of organisation; reflective practice and knowledge management, and uncertainty and information in the tacit order of construction and mining.

Learning Objectives

By the end of this chapter, readers will

1. appreciate the intrinsic instability of construction projects and
2. appreciate the effects of this instability on 'methods of management'.

Tavistock (TIHR) Studies of Mining: Knowledge and Learning; Recursiveness and Reflexivity in a Traditional Working Class Community of Practice

TIHR studies of mining are organisation theory classics. In the transition to the 'long-wall method', the responsible autonomy of the previous craft mining process was lost. This aligned methods, organisational and legal constraints within the face-to-face, self-managing workgroup; its context of time, space, and darkness; the absence of intermediate management and the moral intensity of the community, which buffered the operational core of collieries, and stabilised the 'organisational ground'. Long-wall technology never realised its production engineering potential. It required an organised, large working group to match the "...structure of the long face and the shift sequence...(creating) difficulties of communication and working relationships...the scale...of the task transcend(ed) the limits of simple spatio-temporal structure..." completion of a job in one place at one time. Engineering difficulties exacerbated the differentiation and interdependence of the workforce. Production required "...one hundred percent performance at each step...working to some extent against the threat of its own breakdown...," creating a tendency to crisis management. Stability implied "...a common skill of a higher order than that required...to carry out, as such, any of the operations belonging to the production cycle..." (Trist & Bamforth, 1951).

Adaptation of the craft system, assisted by the sociologically bounded community and stable political support, created the 'composite long-wall'. This early example of 'learning organisation' and 'community of practice' recorded improved efficiency, safety, and welfare (Trist, Higgin, Murray, & Pollock, 1963). The composite long-wall method was developed from the tacit and local knowledge of the pits without knowledge managers (Heller, 1993).

> *"At the time the long-wall method developed, there were no precedents for the adaptive underground application of a machine technology. In the absence of relevant experience in the mining tradition, it was... inevitable that... culture borrowing...should have taken place. There was also no psychological or sociological knowledge in existence at that time which might have assisted in lessening the difficulties." (Trist & Bamforth, 1951)*

The 'composite long-wall system' resolved tensions between uncertainty and norms of rationality (Thompson, 1967). Consensus in the local industry permitted attainment of the design principle (Clark, 2000) for coordination by mutual feedback through multi-skilling, self-regulation, and redundancy in people (Emery & Trist, 1973), reducing coordination costs. Transferring the concept (Trist et al., 1990) required a learning process involving reflexivity and recursive-ness:

> *"...the capacity of actors to modify present action by imagining its consequences based on previous experiences..." and "...the socially accomplished reproduction of sequences of activity and action because the actors involved possess a negotiated sense that one template from their repertoire will address a new situation with improvisation through time as events inside processes." (Clark, 2000, p. 69)*

Improvisation implies finite capacity for change (Clark, 2000), given the pluralism of social structure in organisations (Burns, 1966) and the re-construction of the role systems of mining.

The Evolution of Tavistock Research and Social Thought

TIHR studies in the 1950s confronted problems of communication, learning, information, and uncertainty. Later projects including the BICRP (Higgin & Jessop, 1963, 1965; Crichton, 1966) added wider multi-organisational and longer time horizons as uncertainty and complexity rose in contextual and task environments (Emery & Trist, 1973). BICRP archives clarify the socio-technical theory of construction evolved (Boyd & Wild, 2003) to cope with these problems. Construction exposed the limits of socio-technical ideas, which focus on the inter-penetration of the psychological, social, and technological in work roles and culture. Roles vary in their definition, integration with, and contradiction of other aspects of organisation. People carry the environment as culture to and from work roles, inducing instability and acting as change agents. This personal quality of culture imparts social vitality.

Interactions of individuals and groups involve potentially competing systems including technologies as cultural artefacts (Trist, 1990). Roles constrain through designated rights, obligations of status, mutual expectations, and the array of roles in the role-set. Misalignment generates role conflict, overload, and stress (Merton, 1958). Project teams change through time, implying cultural diversity, unstable roles, and tacit coordination (Stringer, 1967) within "...unpredictable (but inventable) configurations of people and technologies..." (Groak, 1994).

Mining and construction compare in their male workforce and spatio-temporal difficulties. Contextual instability makes the population of construction projects (Groak, 1994) a set of shifting figures on a shifting sociological ground (Trist, 1976). The organisational ground of projects is unstable. Production "...that is subcontracts, is a set of shifting figures in a world of stratified instabilities, sociologically derived at first and then self-induced" (Wild, 2003a).

The BICRP 1962-1966: An Emerging Theory of Construction

Higgin and Jessop (1963) used communications and role diagnostically, revealing instability and variability across project role systems and teams. Unsettled definitions of jobs and responsibilities, threat of losses, and absence of agreed rules for relationships induced offensive and defensive behaviour by representative bodies:

> *"As the social aspects of... communications... embody the fabric of the building team structure, it would be necessary... to learn something of the way in which the teams get built up from among interdependent, autonomous units... communications in the building team are up against built-in difficulties before they start." (Higgin & Jessop, 1963, p. 37)*

External economic, social, and technical forces beyond the control of any party weakened internal coordination. OR analysis had linked communications and roles to the institutions, interests and value systems of construction, and their diverse configurations in projects (Higgin & Jessop, 1963). This diagnosis of a mismatch in *"...the network of technical and social relationships forming the structure of the building team"* (BICRP Archives, Research Committee, 18 November 1963) was confirmed at two conferences in spring 1963. The research proceeded from this, seeking consensual adaptation and improvement, through appropriate redefinition of roles and responsibilities derived from a formal view of the intrinsic technical relatedness of operations. Design was indeterminate. Construction operations could be foreseen reasonably accurately and had improved in efficiency through critical path and related methods. The phases required different approaches. Signing of contracts changed roles as team members became subject to constructor not architect coordination. The research focussed on acute problems of relationships and communications with complex clients and multiple participants in design and construction processes:

- Building the brief, including sketch plans
- The structure and relationships of the design team beyond this initial stage
- Contractor assumption of responsibility from the design team leading to a planning process
- The conduct and control of the building operations (BICRP Archives, Research Committee, 18 November 1963)

The sequence of briefing, exploration, decision, re-appraisal, and revision in the design team, along with the communications network and reasons for later client intervention were relevant. Operations involved people engaged at the professional and commercial interface where *"...many of the tensions created by the instability of the environment were manifested"* (BICRP Archives, 18 August 1964).

Scheduling had implications for relationships and communications and was affected by design decisions. Construction operations could be improved given unified control and communications processes. The research would develop critical path methods in design using OR concepts: computer logic; viewing design as a progression of decision 'states', analogous with certain types of statistical system; the 'theory of games' and 'information theory':

> *"... 'sharper' techniques are necessary...which will more closely identify the complex of inter-related decisions necessary for developing a design...reached in various organisational frameworks by a design team of different groups communicating information in different ways at different stages of development..."* to formally optimise *"...an overall result from a set of value variables, which, at this stage, seems to be the essential operation in developing the brief and in the contractors' building himself into the process." (BICRP Archives, Research Committee, 18 November 1963)*

Non-technical aspects of operations originated in institutional roles. This division of responsibility and the personal, group, social, and economic interests implied:

> *"...the strength of these attachments and...possibilities for change in them, which will allow development...Merging the two lines of inquiry should produce an integrated concept of role. The connecting concept for this merging will probably be that of distinctive competence, which is at once both a technical and a social definition of a role, or perhaps a combination of roles." (BICRP Archives, Research Committee, 18 November 1963)*

In August 1964 the researchers extended indeterminacy to construction planning and control. Construction planning was apparent. They were trying to understand briefing, design, and control to generate the information required for construction operations. A constrained prototype model had emerged from house design. The constraints were:

- Incomplete control over the operational environment translates into instability.
- There is uncertainty in the supply chain over quality, availability, and delivery.
- Formal techniques imply stability, hence their poor performance in practice.
- Attempts to use (and justify) them lead to inter-role tensions.
- Informal (un-verbalised) techniques allowing greater adaptability are applied—what are they and can they be formalised and improved?
- Single pieces of information are generated from many sources, often insufficiently related; confusion results and responsibility is unclear.
- The hidden agendas of site meetings and how these relate to instability in construction control processes.

They were asking about the purpose:

> *"... any form of communication serves as a contribution to the operations of the building process as a whole...Only when the operations involved and the responsibilities of those undertaking them have been understood, is it possible to know what form an appropriate network of communications...should take." (BICRP Archives, 18 August 1964)*

On the 29th of January 1965, a more complex, abstract re-interpretation, essentially Crichton (1966), emerged. There were five interacting systems:

1. *Operations:* Action necessary to design and construct, divorced of people.
2. *Resource Controllers:* Architects, QSs, Contractors, and so forth

3. *Directive Functions:* Briefing, designing, design costing, constructional control
4. *Adaptive Functions:* Variations, extemporisation, buildability, design ahead of consents, cut and load, claim in anticipation, post-contract working drawings, exploitation of errors and crises, re-deployment, use of substitutes, re-design, and on-site design. There were immense differences between the directive functions and the adaptive functions, and these led to ill-will and confusion and the establishment of the fifth system.
5. *A 'Psycho-Social System':* Which reflected these differences

Rational accounting was unavailable (Higgin & Jessop, 1963; Crichton, 1966). The construction process was not self-contained as the client:

> *"...continued to influence action throughout and after the briefing stage and the variation order recognised this fact. The lack of formal links between the different directive functions was exemplified by three different costing systems: cost planning in terms of elements, by trades in the bill of quantities and by certificates etc. in final accounts. Effective comparison between these three was not possible. " (BICRP Archives, 29 January 1965)*

But a wider problem existed. No one really knew what was going on. Detailed information on the processes and purposes of communications was lacking (BICRP Archives, 18 August 1964):

> *"Dr. Higgin elaborated on the normal method of Construction Control. The General Contractor set up agreed time targets (a Programme). This depended on variables not within the control of the Builder, the nature of the contract between the General Contractor and the supplier; the effect of previous negotiations between the Architect and Sub-contractor or supplier and the nature of the Nominated Sub-contractor. The myth of certainty was essential for work to start. The function of contingencies was to reconcile formal/ informal systems, and the new contract tried to adapt to the variability of the system. In these conditions, a critical path network became a record, not a planning technique, and rigidity was a common defence against chaos." (BICRP Archives, 29 January 1965)*

Andrew Darbyshire, Chairman, stated that his Steering Committee:

> *"...had always worried about the acceptability and the usefulness of the research from the point of view of the sponsors... They might thus have come to expect too much from the project; but they had come to realise very strongly what immense gaps there were in our knowledge and understanding of the building industry." (BICRP Archives, Research Committee, 15 June 1965)*

The Tacit Order of Construction in the 1960s

Turbulent environments demonstrate an intricate enmeshing of diverging causal strands (Emery & Trist, 1965). In construction, contextual and task environments, communications, information, uncertainty, and institutionalised role conflicts inter-penetrated so far, only the most abstract theory was available. The innovative operational research offered required significant development (Friend, 1997) to penetrate the complexity and dynamism of decision processes, align values, and avoid conflicts. The depth, complexity, and impenetrability of the tacit order of this fragmented system characterised by self-interest is asserted. Assumption of the 'myth of certainty' for a project - the transition of so many with different variants of personal knowledge dependent upon verbal transmission through a range of diverse, transient project teams- required tacit coordination within an unstable order.

TIHR presented the problem of knowledge and information as culture. Tacit knowledge is carried by individuals and groups inducing change (Trist, 1990; Trist et al., 1963). In mining, the composite long-wall system was elicited from local and tacit knowledge by a community, bounded by isolation, moral intensity, and political and economic support. In construction diffuseness, diversity, indeterminacy, and fragmentation flow from shifting social priorities and impatience. (Higgin & Jessop, 1963) Dynamic processes confront the unique configuration of many social knowledges and their coordination in the unstable temporal structures (Wild, 2003a) of projects. Tacit knowledge is contextual, but contexts vary in continuity and the potential for knowledge capture.

Uncertainty, Interdependence and the Sociology of Values

The BICRP is a critical milestone in the analysis of construction. Crichton (1966, p. 19) summarises problems of uncertainty and information played out in communications:

> *"It was found... in one particular case study that each time a design decision was taken, it set in train a chain of consequences which could and did cause the initial decision to be changed, a clear example of how decisions and actions depend on one another. Since the full implications of any decision or action can seldom if ever be forecast with absolute accuracy, a communications system which assumes that they can will simply not work"*

Projects may be driven towards their origins in the indeterminacy and diffuseness of design at the professional and commercial interface, especially susceptible to value conflicts and environmental instability. This is 'reverse time compression': contingencies stack up against schedules (Wild, 2003b). External, shifting social forces and conservative ideologies acting in fragmented project role systems transformed (Higgin & Jessop, 1963) serial into reciprocal interdependence. This reflects tensions between information, uncertainty, and norms of rationality. Pooled, serial, and reciprocal interdependence require standardisation, planning, and mutual adjustment. Each contains the others, implying increasing costs of coordination and time. The most formally efficient structure works back from reciprocal to pooled interdependence (Thompson, 1967). A valid operational decision sequence implied comprehensive attention and feedback (Crichton, 1966). Computer metaphors assumed serial interdependence, requiring disavowal in the face of tacit coordination (Stringer, 1967), that is 'adaptive functions'. Such organisational pluralism required attention to a:

> *"...sociology of values, or perhaps a calculus of values, which is more than just experiential...the connections between systems must be realised in terms which are categorically different from those applying within systems. They relate to the value systems of the individuals and of the culture... any preference ordering between systems (which must be a moral ordering) has to be in accordance with fundamentally different criteria from those which obtain within systems (which can be a technical ordering)." (Burns, 1966, p. 248)*

Bresnen (1991) demonstrates project integration, attained through contingent alignment of a client interest, seeking allies in a context of shifting client roles with the strategy of a management contractor. At the temporal and organisational boundary of design and contracting to which TIHR researchers attached importance, an alignment of values occurred. It was sufficient to steer the project to completion, confirming the TIHR focus on optimisation through this route. Strategic intent was reflected in project staffing by the contractor and client attention at board level. Management contracting was innovative, requiring new roles and behaviour including re-negotiation of previously adversarial relationships. This led to knowledge exchange in a re-design imposed by a budget cut. Despite tensions, the contextual organisational features facilitated collaboration.

Management contracting emerged after the BICRP. Higgin and Jessop (1963) recognised, in relation to the 'package deal', that procurement method is contingent on a wider envelope of accountability. Techniques are insufficient onto themselves, requiring conditions of stability to work (Beardsworth, Keil, Bresnen, & Bryman, 1988; BICRP Archives, 18 August 1964). Strategic choice - evolved from the BICRP (Friend, 1997) - reflects structured exploration of uncertainties over environments, governing values, and information. Green (2001) argues for this, deconstructing value and risk management, which:

> *"...do not have any meaning in isolation of the way they are enacted and people enact value management in many different ways. There is no established causal link between the use of value management and a resulting better product...Furthermore there is no consensus on which techniques constitute value management."*

Uncertainty and information underpin transaction cost economics (Winch, 1989, 2000, 2001). Rationality and efficiency lie in reduction through hierarchy or markets of information costs. Uncertainty lies in information processing capacity and bounded rationality. Parties to transactions experience different levels of information. Computational difficulties around configurations and interactions in complex problem areas generate uncertainty. The coordination costs of reciprocal interdependence are equivalent to information costs in transaction cost economics (Clark, 2000). Opportunism, driven by self-interest in a fragmented system (Higgin & Jessop, 1963) and the effects of impacted information, are experienced as different uncertainties and communication difficulties.

Construction, Situations and Reflective Practice

Construction is a world of potential situations (Wild, 2004) characterised by complexity, instability, uncertainty, uniqueness, and value conflict (Schon, 1983):

> *"...the presence of a great number of organisational interests creates a pull towards under-boundedness within construction projects whose teams rarely attain stability and this must be recognised as their normal condition." (Boyd & Wild, 1999)*

Bresnen (1991) exemplifies non-emergence of a situation as strategic intents fortuitously converged into collaboration. Different inter-organisational and power dynamics, including the treatment of subcontractors, facilitated communications and knowledge exchange.

Sub-contracting and indirect management induce a game of wider discretion and instability due to unfamiliarity of people. Implicit strategies reduce uncertainty from project transience affecting power dynamics, surveillance, and performance expectations. Long-term networks with teams of transferable, subcontracted core workers buttress market relationships improving performance and quality. Market variations influence power and dependence. Problems of time and unpredictable activity inhibit management continuity. Spatial constraints clash with parallel events in critical paths inducing a conflicted, improvised production process. Time compression, which concentrates activities more densely into limited space, induces directive management. Roles, authority, and the expectations of site managers are clear in a fragile environment at the mercy of the weather, deliveries, labour and information problems, and accommodating multiple demands in temporally limited contexts of short-termism, a fragmented division of labour and subcontracting; pressures from the client, firm, and the value of the job are also considered (Beardsworth, Keil, Bresnen, & Bryman, 1988).

Construction as a system of Situations (Wild, 2004) requires the judgements of practitioners in constructing and setting problems as a condition for problem solving (Schon, 1983). This counters the recursive drift of managerial solutions into technical rationality, which:

> *"...responds to Situations through: selective inattention, simplification for situational control and...residual categories...the Reflective Practitioner proceeds through: Appreciative Systems; Over-arching Theories; Role Frames and Media, Languages and Repertoires." (Boyd & Wild, 1999)*

Expert consultancy frequently replicates pre-packaged (Mumford, 1980) techniques dependent on stability (BICRP Archives, 18 August 1964) and susceptible to institutionalised fragmentation (Wild, 2001). Appreciative Systems (Vickers, 1965) meld judgements of facts and values, returning to the sociology of values; assimilating socio-technical theory to Schon (1983), given technological change as a source of instability (Schon, 1971) and transaction cost economics given problems of uncertainty and information.

Construction, the Tacit Order and Knowledge Management

Construction is striking in its cohesive fragmentation (Wild, 2002a, 2002b, 2001) and recalcitrance (Reed, 1989) to official reform efforts. It is a precarious social field, "...constrained...to systemic reproduction..." (Archer, 1995) in which "...conflict may be both endemic and intense...without causing any basic structural change" (Lockwood, 1964, p. 249).

Tacit coordination (Stringer, 1967), a distinctive organising process, involves knowledge exchange, and a 'negotiated order' between roles with a limited time of service to a project.

TIHR metaphors of 'computer logic' (BICRP Archives, Research Committee, 18 November 1963) anticipate the use of IT in knowledge management. Techniques of exploration are emphasised to provide optimisation models. This 1960s discourse did not offer discursive penetration (Clark, 2000) of the shifting complexities of construction. Skills in improvising industry linkages and technology fusion generate opacity in the project. This:

> *"...defines ad hoc whatever supply of services, finance, information and products are possible and necessary...the project induces its own demand chain, its needs and resources, its own process and consequential processes and its own specific organisation...dominance of the project changes ideas*

about how we organise highly responsive networks of skills, where we locate research and development effort, on what we focus for productivity improvements...It all gives renewed significance to Hillebrandt's potent and fundamental research question: What do we mean by the capacity of construction?" (Groak, 1994)

There are no intrinsic reasons for supposing that appreciation of construction is more easily attained today. Knowledge management is sophisticated, but construction is more complex than in the 1960s, implying the importance of the tacit order as theories-in-use (Argyris & Schon, 1974). Client experience is critical (Atkins & Wild, 2000; Sabbagh 2000). Why should a client, who has never procured construction previously, see knowledge management as a strategy for reducing his uncertainty about the services offered? "How often does a client find that had he known as much at the beginning of a project as he knows at the end, he would have made quite different decisions?" (Higgin & Jessop, 1963 p. 75).

Post-war reports on British construction reflect a template (Wild, 2004) of problems and rational managerial solutions backed by uncritical research (Connaughton, 2000). Until 1963 they lacked an analytical framework for the wider context of construction operations "...and had not been able to do more than canvass best opinion and agree [to] general precepts on that basis" (Higgin & Jessop, 1963 p. 35).

Policy is always pursuing a shifting, complex, and opaque world. An impressive array of client and public resources backs present attempts at construction 'reform'. The slow response to current initiatives (Murray, 2003) echoes the BICRP. Aylwyn Lewis, BICRP Secretary, wrote to the Research Committee chairman (1 July 1965):

"...I do not think that I have really changed the two basic views with which I started the Project. These were:

1. *that a combination of sociology and operational research was necessary to tackle the problems of communication between the autonomous units that make up the building team;*
2. *that change in these relationships would only come about in a controlled manner if the research was integrated with a representative but progressive cross-section of the industry who would sponsor and promote its findings." (BICRP Archives)*

Construction apparently replicated the finite capacity for change of organisations (Burns, 1966) exacerbated by production as projects. Crichton (1966) argued that this stability had advantages but changed too slowly for 'modern' needs. Post-war reports (Murray & Langford, 2003) reflect a treadmill of societal impatience (Higgin & Jessop, 1963) and political criticism, a reflexive, double-edged rhetoric of modernisation as its own theme (Beck, 1986) with proliferating change and research agendas. C.E.D. (known as Bertie - BICRP Archives) Wooster, Director of Building Management, Ministry of Public Building and Works, exemplifies this rhetoric:

> *"To the extent that society is not satisfied with the performance of the industry (already overloaded by about 10 per cent, and who is satisfied with our buildings and towns?), and to the extent that the industry does not match up to the task facing it, in the next ten years, it is failing to meet its objectives; and the failure can only be due to inadequate resources or mis-application of those we have." (Wooster, 1964)*

Yet technology fusion (Groak, 1994) implies a capacity for innovation despite the apparent disorganisation of construction. Does this enable change by the tacit transmission of ideas and practices? Does knowledge require formalisation in a managerial system to be made effective? There had been the:

> *"...remarkable recovery of the building materials...and the construction industries from the war period ...their flexibility in meeting new demands on their services in the past fifteen years; the introduction of new materials, increased mechanisation and new techniques; the steady rise in output, and the avoidance of major industrial disputes." (Emmerson, 1962, p. 4)*

Does this imply interaction of organisational conservatism, misinterpreted as fragmentation, with technical innovation and tacit R&D inducing disjointed incrementalism (Hirschman & Lindblom, 1962) in projects? In this world taken for granted is innovation and learning denied by 'assumptions of ordinariness' (Trist et al., 1990)? Are paradoxes resolved as mutual adjustment, partly driven by power? Concurrently with BICRP there were strikes on the Horseferry Road and Barbican projects. Variations and instability in design provoked disputes over incentive pay, earnings, and related areas (Clegg, 1972). Is this also the metropolitan leading edge?

Construction in Society and Society in Construction

> *"We have...been surprised at the long list of the sorts of people with whom communications, usually of a consulting nature, will be undertaken - chemists, physicists, engineers of all kinds, doctors, financiers, sociologists, other clients, meteorologists, and many others." (Higgin & Jessop, 1963, p.18)*

Briefing reveals the socio-technical complexity of buildings. Design intensity (Clark, 2000) is a source of potential uncertainty, instability, and value conflicts in a process, which quickly attains the character of an extensive conference of both sides. A requirement for optimisation emerges here and later in construction from sources related to the client, whose complexity was poorly appreciated. Construction failed to account for and was impatient with such social forces. A general awareness and a capacity to engage with the client, developing appropriate techniques to deal with interdependent decisions, were required. Internal diversity implied an unusually wide range of interests, institutionally conditioned by uncertainty, including that from other projects, creating an ideal context for conflict and intrinsic difficulties for reforms. The system partly determined the behaviour of all the parties (Crichton, 1966).

Schon (1971) provides a comparison. In 1963 the U.S. Government focussed on 'backward' industries, whose R&D lagged behind their contribution to GNP and were composed of a few large firms and lots of small ones, low profit margins, and little investment in technology. American construction was considered fragmented. Only $1 million had been spent on product development in 1962. The report concluded that the government should stimulate building research. The dynamic conservatism of the industry, as an organised constituency, killed off a proposed public R&D programme. The proposals had clarified the networks of a complex coalition of shared interests, livelihoods, and a 'theory-in-use' (Argyris & Schon, 1974) relating prevailing technologies to individual and collective values. Without a clearly visualised future stable state, the existing system would have to be pulled apart. This is a view of construction as embedded (Clark, 2000) in society. London is a dynamic construction forcing-house (Ackroyd, 2001), explaining the metropolitan character of the 'progressive cross section'. Construction reform without public participation would lack viability:

> *"...no single characteristic...will be appropriate to measure the efficiency of a building process. We must measure the total performance by a yardstick of total performances... In the final analysis the community only gets the building industry it deserves. Therefore the client/user/community complex also needs to associate itself with the industry in...coordinating...further large-scale research." (Crichton, 1966, pp. 59 & 61)*

In the context of a politically driven impatience (Higgin & Jessop, 1963) with a wider stagnation, could society attain a viable consensus over the opaque social field of construction disturbed by national planning and political change? (Wild, 2002a, 2002b). The Barbican and Horseferry Road disputes reflected, partly, the disruption of "what sometimes passes for consensus in Britain...the artefact of a tenuous balance of opposing forces" (Birnbaum, 1962, p. 283).

A similar sociological concern pervades Boyd and Wild (1999). They assert the equivalence of construction and organisation development. OD involves those who do the work in diagnosing the efficiency and effectiveness of their work organisation, rather than offering a recipe or formula (French & Bell, 1990). The organisational iceberg reflects a Pareto ratio of the 'formal' and 'informal'. Construction is multi-organisational (Stringer, 1967). The multiple interactions of multiple 'informal systems' imply construction as a 'sociological iceberg' explaining self-induced uncertainty (Stringer, 1967). Beyond this there are objections to the formal-informal systems dichotomy. TIHR researchers (BICRP Archives, 18 August 1964) suggested a synthesis of these, transcending their own assumptions. However:

"...the informal system has served as a receptacle for (matters) which were seen as irrelevant to the formal organisation or incompatible with its purpose...all that had defeated Taylorite attempts at human engineering...." (Burns, 1966, p. 235)

Managers and workers devise multiple interpretations of an organisation and institute mutually sustained working relationships. Multiple organisational goals emerge and a "...different conceptual framework, from that of the formal organisation of management..." (Burns 1966, p. 235).

Rational models imply instrumentalism towards goals and valued future states. This reflects misplaced concreteness and ignores the problem of reconciling the unity of aims in the model with the empirical fact of disunity of purpose within organisations (Burns, 1966). Does under-boundedness (Boyd & Wild, 1999) to society proliferate disunities of purpose in projects? Can demand side conditions of power and organisation insulate the project? Tate Modern (Sabbagh, 2000) was doubly insulated. Stewardship for a protected building avoided internal value conflicts over aesthetics, leaving the 'technical' problem of optimising display space. The clients' establishment constitution and experience rendered the project hegemonic (Lukes, 1974), structuring information to set aside bounded rationality, align values around project prestige, and suppress information uncertainty. Such concerns provoked Green (2001) to comment on one of the many proliferating boundaries (Wild, 2002c) between construction and society:

"The process of risk management...depends...upon...the need to maintain a viable political consistency within the client organisation. It is notable that there is as yet no framework that embraces both the notion of technical risk with the less tangible uncertainties that characterise the strategic interface between construction projects and client organisations...."

Conclusion (and a Provocation to a Debate)

"...Is it somehow 'natural' that we appear to find the challenge of solving new problems (or of finding apparently fresh solutions) more seductive than plugging away at the old?" (Connaughton, 2000)

The BICRP anticipated the discourse of knowledge management. Will this permit us to learn more about construction? Is this relevant when the problem may be to learn to manage projects differently? Successes will be presented. Will they confirm that a new perspective induces these until it reaches its limits, as did socio-technical ideas when exposed to construction in the 1960s?

The diffuseness of construction requires a significant tacit order. Knowledge management assumes this is accessible to structured inquiry (Clark, 2000). Are there constraints such as spatio-temporal fragmentation (Wild, 2003a) appreciable from socio-technical research? Does integration require contextual conditions not readily present in the under-bounded (Boyd & Wild, 1999) world of construction? Is knowledge management a 'fresh solution', a renewed combination of sociology and operational research, an old wine poured in a new bottle by a new generation of 'experts'? As 'technique' is it as dependent on stability as other formal methods reported in the BICRP? Can it operate undisturbed by existing strategic recipes, networks, professional values, and personal constructs (Boyd & Wild, 1996)? The BICRP was assimilated to a construction project. Why should or how could knowledge management avoid being drawn into the turbulence and communication difficulties of construction like previous initiatives? Will it be routinised as a technique and fall away in effectiveness as construction is still improvised to completion in a project-based world of intrinsic unmanageability? (Wild, 2004, 2002c).

> *"If so many have preached the unification of the UK construction industry for so long with so little effect, how could we succeed? Is it indeed a lost cause? Have we been barking up the wrong tree all these years? Is it our destiny to build a fragmented future on a fragmented past and make the best of it?" (Andrews & Darbyshire, 1993, p. 37)*

Andrews (1984) and Connaughton (2000) on the managerial - the socio - and Koskela (2003) on production - the technical side of the socio-technical equation - replicate Darbyshire's statement from June 1965 concerning the need for:

- "...a general theory of the organisation and management of construction." (Andrews, 1984)
- "...consistency and coherence in the development and accumulation of knowledge necessary for a useful, coherent theory of construction management...." (Connaughton, 2000)
- "...integrated prescription for construction renewal...Our present theoretical understanding is not necessarily solid enough. We need to develop fully the theoretical foundation or first principles of production in general and of construction especially." (Koskela, 2003)

In the problem of 'society in construction and construction in society', expectations outrun the capacity of construction to deliver, generating an eternal performance gap. Putative theories and methodologies of reform are drawn into the turbulence of construction as a cohesively fragmented social field. Knowledge management awaits its own partial demise. In contradistinction to 'integration', a theory of improvisation may explain the production of the built environment in under-bounded project worlds of non-recurrence and re-configuration.

Post-war reports (Murray & Langford, 2003) are symptoms of the instability and indeterminacy of construction. Hence the focus of management attention and research can be shifted from control to the steering of projects as negotiated orders (Reed, 1989). The difficulties of construction can be grounded on the qualities sought from the constructed world, which people inhabit, work in, and pass through. What is constructed and how is ideological. Powerful and cohesive clients get the buildings they want. Society, indifferent to other aspirations, gets the constructed order it deserves. This refracts what is there; what can be built; what we are prepared to pay for, and the resulting constraints upon us. In principle, construction is impossible, given the potential for value conflicts in and around the process. Dilemmas are resolved tacitly within the project. Within a mobile, masculine confraternity creativity is re/de-constructed as yarns. That projects are completed is a profound tribute to those who do the work, both because and in spite of the situations within which they labour. This, the real achievement of construction, truly constitutes the management of knowledge.

Practical Tips and Lessons Learned

Construction projects are streams of disturbances punctuated by crises and occasional episodes of calm. The skills of project management (Boyd & Wild, 1999) include:

- Coping with discontinuity at start up
- Eliciting organising behaviour: consulting to the uncertainty in the client
- "Est": working in left- and right-brain modes
- Handling network centrality
- Holding and sharing power, laterally and vertically
- Negotiating and re-negotiating: contracts, roles, and expectations
- Skills of collaboration: managing the uncertainty in the project
- Spanning boundaries
- Steering through complexity
- Working with procedures as diagnostic devices

Acknowledgment

This chapter is dedicated to Professor John Andrews who, in the last two years of his life, took an active interest in my attempt to re-interpret construction. He died in November 2003.

References

Ackroyd, P. (2001). *London, the biography.* London: Vintage Books.

Andrews, J. (1984). Comment at the *CIB W-65 International Symposium on Organisation and Management of Construction,* University of Waterloo, Ontario, Canada, July.

Andrews, J., & Darbyshire, A. (1993). *Crossing boundaries*. London: CIC.

Archer, M. (1995). Social integration and system integration: Developing the distinction. *Sociology, 30*(4), 679-699.

Archives of the BICRP. (1962-1966). London: Royal Institute of British Architects.

Argyris, C., & Schon, D.A. (1974). *Theory in practice: Increasing professional effectiveness.* San Francisco: Jossey-Bass.

Atkins, J., & Wild, A. (2000). Automotive facilities procurement and the new construction paradigm. *Proceedings of the Sixteenth Annual Conference of the Association of Researchers in Construction Management*, Glasgow Caledonian University, 6-8 September.

Beardsworth, A.D., Keil, E.T., Bresnen, M., & Bryman, R. (1988). Management transience and subcontracting: The case of the construction site. *Journal of Management Studies, 25*(6), 603-625.

Beck, U.C. (1986). *Risk society: Towards a new modernity.* London: Sage Publications.

Birnbaum, N. (1962). Great Britain, the reactive revolt. In Birnbaum, N. (1971). *Toward a critical sociology.* New York: Oxford University Press.

Boyd, D., & Wild, A. (1996). Engaging with personal constructs to improve construction projects. In A. Thorpe (Ed.), *Proceedings of 12th Annual Conference of ARCOM,* Sheffield, September.

Boyd, D., & Wild, A. (1999). Construction projects as organisational development. *Proceedings of the 15th Annual Conference of the Association of Researchers Construction Management,* Liverpool John Moores University, 15-17 September.

Boyd, D., & Wild, A. (2003). Tavistock studies into the building industry. In M. Murray & D. Langford (Eds.), *Construction reports 1944-98.* Oxford: Blackwell.

Bresnen, M. (1991). Construction contracting in theory and practice: A case study. *Construction Management and Economics, 9,* 247-263.

Burns, T. (1966). On the plurality of social systems. In Burns, T. (Ed.). (1969). *Industrial man*. London: Penguin Books.

Chrichton, C. (1966). *Interdependence and uncertainty: A study of the building industry.* London: Tavistock Publications.

Clark, P.A. (2000). *Organisations in action. Competition between contexts.* London: Routledge.

Clegg, H.A. (1972). *The system of industrial relations in Great Britain.* Oxford: Blackwell.

Connaughton, J. (2000). How can construction management research matter? Keynote Address of the *16th Annual Conference of the Association of Researchers in Construction Management.* Glasgow Caledonian University, 6-8 September.

Emery, F.E., & Trist, E.L. (1965). The causal texture of the environment of organisations. *Human Relations, 18*(1), 21-32.

Emery, F.E., & Trist, E.L. (1973). *Towards a social ecology.* London: Plenum Press.

Emmerson, H. (1962). *Survey of problems before the construction industries: A report prepared for the Minister of Works.* London: HMSO.

French, W.L., & Bell, C.H. (1990). *Organisation development* (4th edition). Englewood Cliffs, NJ: Prentice-Hall International.

Friend, J. (1997). Connective planning from practice to theory and back in: The social engagement of social science, volume 3. In E. Trist, F. Emery, & H. Murray (Eds.), The socio-ecological perspective. Philadelphia: University of Pennsylvania Press.

Green, S.D. (2001). Towards an integrated script for risk and value management. *International Journal of Project Management, 7*(1), 52 -58.

Groak, S.(1994). Is construction an industry? *Construction Management and Economics, 12,* 287-293.

Heller, F. (1993). Another look at action research. *Human Relations, 46*(10), 1235-1242.

Higgin, G., & Jessop, N.K. (1963 & 1965). *Communications in the building industry.* London: Tavistock Publications.

Hirschman, A.O., & Lindblom, C.E. (1962). Economic development, research and development, policy making: Some convergent views. *Behavioural Science, 7,* 211-222.

Koskela, L. (2003). Is structural change the primary solution to the problems of construction? *Building Research and Information,* 31(2), 85-96.

Latham, M. (1993). *Trust and money.* London: HMSO.

Lockwood, D.A. (1964). Social integration and system integration. In G.K. Zollschan & H.W. Hirsch (Eds.), *Explorations in social change.* Boston: Houghton Mifflin.

Lukes, S. (1974). *Power: A radical view.* London: Macmillan.

Merton, R.K. (1968). *Social theory and social structure.* New York. The Free Press.

Mintzberg, H. (1989). *Mintzberg on management: Inside our strange world of organisations.* New York: The Free Press.

Mumford, A. (1980). *Making experience pay.* London: McGraw-Hill.

Murray, M. (2003). Re-thinking construction: The Egan report 1998. In M. Murray & D. Langford (Eds.), *Construction reports 1944-98.* Oxford: Blackwell.

Murray, M., & Langford, D. (2003). *Construction reports 1944-98.* Oxford: Blackwell.

Reed, M. (1989). *The sociology of management.* Hemel Hempstead: Harvester Wheatsheaf.

Sabbagh, K. (2000). *Power into art. Creating the Tate Modern, Bankside.* London: Allen Lane.

Schon, D.A. (1983). *The reflective practitioner: How professionals think in action.* New York: Basic Books.

Schon, D.A. (1971). *Beyond the stable state.* London: Temple Smith.

Stringer, J. (1967). Operational research for multi-organisations. *Operational Research Quarterly, 18*(2), 105-20.

Thompson, J.D. (1967). *Organisations in action.* Englewood Cliffs, NJ: Prentice-Hall.

Trist, E.L. (1976). Engaging with large-scale systems. In P.A. Clarke (Ed.), *Experimenting with organisational life.* London: Plenum Press.

Trist, E.L. (1990). Culture as a psycho-social process. In E. Trist & H. Murray (Eds.), *The social engagement of social science, volume 1: The socio-psychological perspective.* Philadelphia: University of Pennsylvania Press.

Trist, E.L., & Bamforth, K.W. (1951). Some social and psychological consequences of the long-wall method of coal getting. *Human Relations, 4,* 3-38.

Trist, E.L., Higgin, G.W., Murray, H., & Pollock, A.B. (1963). *Organisational choice.* London: Tavistock Publications.

Trist, E.L., Higgin, G.W., Murray, H., & Pollock, A.B. (1990). The assumption of ordinariness as a denial mechanism. In E. Trist & H. Murray (Eds.), *The social engagement of social science, volume 1: The socio-psychological perspective.* Philadelphia: University of Pennsylvania Press.

Vickers, G.A. (1965). *The art of judgement.* London: Chapman and Hall.

Wild, A. (2001). Construction projects as teams or situations: Criticising the techniques of communication approach. *Proceedings of the 17th Annual Conference of the Association of Researchers in Construction Management,* University of Salford, 7-9 September.

Wild, A. (2002a). False coherence and cohesive fragmentation in the post-war corporatist coordination of construction. *Proceedings of the Eighteenth Annual Conference of the Association of Researchers in Construction Management,* University of Northumbria, 2-4 September.

Wild, A. (2002b). The management college that never was. *Proceedings of the Eighteenth Annual Conference of the Association of Researchers in Construction Management,* University of Northumbria, 2-4 September.

Wild, A. (2002c). The unmanageability of construction and the theoretical psycho-social dynamics of projects. *Engineering, Construction and Architectural Management, 9,* 345-351.

Wild, A. (2003a). Iron triangles and holy trinities: Time or temporality in construction management. *Proceedings of the Nineteenth Annual Conference of the Association of Researchers in Construction Management,* University of Brighton, 3-5 September.

Wild, A. (2003b). The working party report to the Minister of Works: The Phillips Report on building 1950. In M. Murray & D. Langford (Eds.), *Construction reports 1944-98.* Oxford: Blackwell.

Wild, A. (2004, March). Re-interpreting the building industry communications research project. *Construction Management and Economics, 22*(3), 303-310.

Winch, G.M. (2001). Governing the project process: A conceptual framework. *Construction Management and Economics, 19,* 799-808.

Winch, G.M. (1989). The construction firm and the construction project: A transaction cost approach. *Construction Management and Economics, 7,* 331-45.

Winch, G.M. (2000). Institutional reform in British construction: Partnering and private finance. *Building Research and Information, 28,* 141-155.

Wooster, C.E.D. (1964). Building management in the '70s. *The Structural Engineer, 42*(3), 79-85.

Section III:

Emerging Models and Solutions

Chapter XIII

The Culture of the Knowledge Advantage (K-Adv): A Holistic Strategic Approach to the Management of Knowledge

Derek H.T. Walker
RMIT University, Australia

Tayyab Maqsood
RMIT University, Australia

Andrew Finegan
Charles Darwin University, Australia

Abstract

This chapter introduces a concept 'the knowledge advantage' (K-Adv) that describes a model in which knowledge leadership provides the setting in which an organisation's knowledge vision can be developed and activated. This is supported by a people infrastructure that allows people to effectively create and share knowledge. This infrastructure is further supported by an information and communication technologies (ICT) infrastructure. This chapter also provides examples of how the K-Adv can be developed, and several tools are illustrated for this purpose. These include using a soft systems methodology (SSM) approach as part of developing the knowledge leadership vision and how the K-Adv can be used for benchmarking at both a course and fine-grained level. The contribution that this chapter makes is to provide some strategic and practical tactical tools that companies can use to develop their competitive advantage.

Introduction

Mobilising the necessary intellect for developing and maintaining a firm's core competences requires high levels of sensitivity to the needs, motivation, and power (in terms of available energy) of humans to unleash their creative energies. This demands a change in perception of the way that firm's maximise their capacity to deliver outcomes from that embedded in an extractive resource-based view of the firm.

Sound business strategy for organisational sustainability is inextricably linked to knowledge management, innovation diffusion, change management, leadership, and building organisational competencies. This chapter presents a well-supported framework for achieving a knowledge advantage that could be of pivotal value to organisations that presently lack an understanding of how leadership, ICT, and workplace culture underpinned by supportive organisational processes can have a major impact upon their becoming a learning organisation that effectively manages its knowledge assets. For those organisations with high-level strategic foresight that enhances their competitiveness from a knowledge perspective, this model provides a useful checklist for application as a benchmarking model.

The approach proposed in this chapter is to concentrate on the concept of a knowledge culture as delivering competitive advantage and examine how a knowledge advantage can be culturally achieved through holistic management of knowledge. The chapter is structured as follows. First we explain the knowledge advantage (K-Adv) concept and then provide an overview of the model to be explored. This is followed by a brief discussion of the implications of commitment on deploying the K-Adv. We then discuss how the K-Adv model can be deployed, and we discuss soft systems methodology (SSM) as a tool for identifying stakeholders and thence knowledge that they can contribute. We follow this with discussion of a second tool, how benchmarking can be used with the K-Adv model, and we continue with a discussion of a generalised maturity model that can be used to provide a course-grained benchmark of the maturity stage of an organisation in deploying its K-Adv. We summarise the chapter with conclusions and recap the chapter's content.

Learning Objectives

Readers of this chapter should be able to:

1. Understand the need for leadership in defining and strategising a knowledge vision, so that organisations can start the process of developing a competitive advantage based upon corporate and individual stakeholder knowledge.
2. Appreciate, from the K-Adv model presented in this chapter, that the K-Adv is founded upon a People/Cultural infrastructure supported by ICT.
3. Understand how the model can be applied in a practical way to manage knowledge by organisations in their specific setting.

Background

According to a Greek myth, King Midas wished for everything he touched to be turned to gold. Unfortunately, eating food, drinking water, and even having human contact became impossible because all touched objects turned into gold. A purely resource-based theory of the firm, with its aims to harness competitive advantage to maximise the firm's profits, misses a vital element of the necessary resources—the human mind and the knowledge contained therein. This chapter provides a model developed from a three-year research project to investigate how construction organisations can better develop competitive advantage through their access to, and use of, knowledge. We argue that this delivers them a competitive advantage that we refer to as their *knowledge advantage* (K-Adv).

The research draws upon findings from several PhD studies associated with the CRC research project in terms of a better understanding of ICT diffusion. It is interesting that the Australian Standards Association has also been developing a framework for knowledge management in the construction industry, and the work reported upon in this chapter has been able to be communicated to that association—they have independently developed similar ideas of a model that recognises drivers and inhibitors for innovation, and the management of knowledge also focuses upon the same core elements outlined in this chapter, namely ICT as an enabler; leadership to provide vision and identify priorities; and supporting organisational processes, knowledge content, and people (culture) (Walker, 2003). Thus, the model presented in this chapter is further supported in its general findings from a range of emerging literature, and has both institutional support and has been supported by several important industry groups in Australia, including government public works departments, a major first-tier construction contractor, and a leading global construction engineering consultant.

The model comprises three elements facilitating delivery of a K-Adv:

- knowledge leadership that provides the organisational support, backing, championing, and vision to create strategies and implement them;
- a well-functioning and supportive ICT infrastructure to enhance communication, coordinate problem-solving activities that generate knowledge in new contexts, and transfer of both explicit (easier to achieve) and tacit (highly complex to achieve) knowledge; and

- a supportive and facilitating people infrastructure to focus in particular on the highly problematic tacit knowledge as well as transferring explicit knowledge.

Identification of these elements is not new and the need for these are generally agreed upon by firms that undertake sophisticated strategic planning; however, for many less strategically mature organisations, the model presented provides useful ideas with which to improve their strategy development and deployment maturity level.

The K-Adv Concept Explained

An organisation's K-Adv is its capacity to liberate latent creativity and innovation potential through effective management of knowledge both from within its organisational boundaries and its external environment (Walker, 2004). The K-Adv requires a coordinated approach in addressing leadership actions to establish and deploy a vision of what the K-Adv means to the organisation, to support the people infrastructure necessary to effectively use knowledge in their business activities, and to provide the necessary enabling ICT infrastructure to do so.

Creativity is a function of three components—expertise, intrinsic motivation, and creative thinking skills. Creativity lies at the intersection of all three components (Amabile, 1998, p. 78). Expertise is the person's knowledge of a particular domain in question. Maslow (1943) first promoted a theory of a hierarchy of motivation factors stretching from physiological, security, affiliation, and esteem to self-actualisation. Intrinsic motivation is the desire and passion that wells up from within a person to contribute to making a real difference, rather than get extrinsic rewards such as physical or security needs. Creative thinking skills relate to flexibility, imagination, and nimbleness of mind. Innovation is central to a forward movement that depends on trying something different or completely novel, and testing theories about how the innovation could or should affect an outcome against reflection of the experience of the experiment. This requires a 'safe' environment where it is OK to experiment and make mistakes—as long as lessons learned are internalised, hopefully turned from tacit knowledge to explicit knowledge, or at least shared (Nonaka & Takeuchi, 1995; Nonaka, Toyama, & Konno, 2001).

The literature on business strategy has much to say about competitive advantage and the struggle for supremacy over competitors. For example, Zack discusses the process involved in developing a knowledge strategy (Zack, 1999, p. 134). He describes an organisation's knowledge in terms of *core, advanced, and innovative knowledge. Core knowledge* is the minimal knowledge required to stay in the game. *Advanced knowledge* enables a firm to be viable relative to its competitors. It may have similar scope and quality of knowledge to its competitors, but a specific differentiated knowledge that places it in a niche market situation differentiates this knowledge as being of an advanced nature. *Innovative knowledge* allows it to lead its industry segment(s) and significantly differentiate itself from competitors. When core knowledge matches the competition, the organisation becomes a viable competitor. However, when one organisation has advanced knowledge relative to its competitor, it places that competitor in a laggard position. When an organisation has innovative knowledge relative to its competitor, then the competitor is at risk because, as well as having a differentiation competitive advantage, the leading organisation may well have advanced innovative practices that lower cost and thus provide cost advantage.

The K-Adv Model: An Overview

Figure 1 illustrates the K-Adv model. The central and focal point of the concept is knowledge leadership. This is linked to the ICT and people infrastructures that help turn the idealised knowledge advantage vision into reality. These three components or attributes dynamically interact to shape a preferred future.

If we first concentrate on the knowledge leadership element, we see that strategy and its enactment comprise two sub-elements—envisioning and vision realisation. Knowledge leadership recognises that a K-Adv is realised through people and their creative energies, and knowledge that is grounded in their individual experience and ability to interpret and re-interpret meaning from experience. Thus, a knowledge vision depends upon people; moreover, it depends upon a range of people from both within and external to any organisation. An important part of the K-Adv is an ability to envision a preferred

Figure 1. The K-Adv model

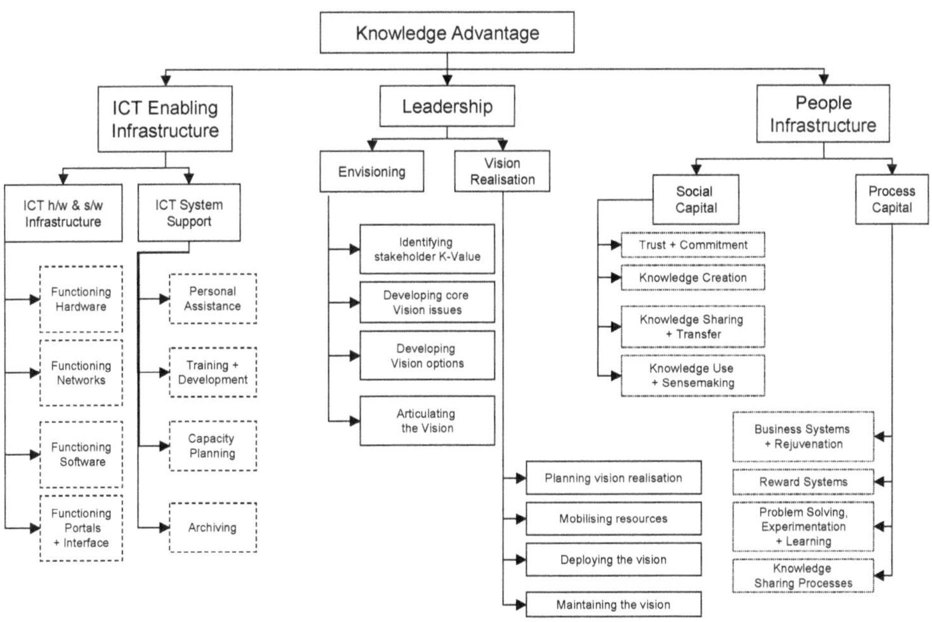

Figure 2. Knowledge leadership attributes

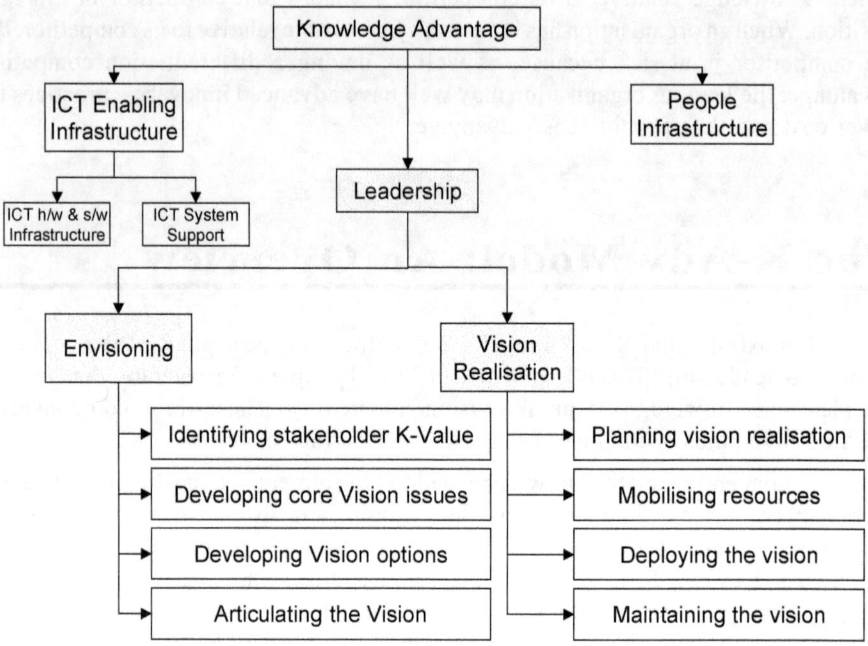

future knowledge strategy through the identification and value of useful stakeholder knowledge. That depends upon first identifying and understanding stakeholder environments, which naturally leads to identifying stakeholders and the knowledge that they possess.

Figure 2 illustrates the identifying stakeholder K-Value as the first of four sub-elements or attributes. The second attribute is developing core vision issues; this is derived from understanding the stakeholder's (knowledge) value proposition.

A useful tool that can be used for gaining these insights is undertaking a soft systems methodology (SSM) study (Checkland, 1999; Fernie, Green, Weller, & Newcombe, 2003). This helps to unearth deeper issues surrounding a study's problematic issues. We tested this approach through a pilot project and discovered that erstwhile hidden stakeholders can be identified and that core knowledge issues can be better understood from the development of 'rich pictures' as part of the study (discussed in further detail later in this chapter). This led to a better understanding of stakeholder knowledge value proposition as well as identification of knowledge assets (Walker, Finegan, & Maqsood, 2003). From analysis it is possible to develop knowledge vision options that have been filtered, prioritised, and validated by targeted stakeholder groups. Finally, the knowledge vision must be articulated—making the vision clear, explicit, and well understood using communication channels best suited to the target audience.

The realisation sub-element/attribute follows a typical project management approach. A detailed planning phase is then necessary, followed by mobilisation of identified needed resources. This requires negotiating acquisition of these resources, attracting the

Figure 3. People infrastructure attributes

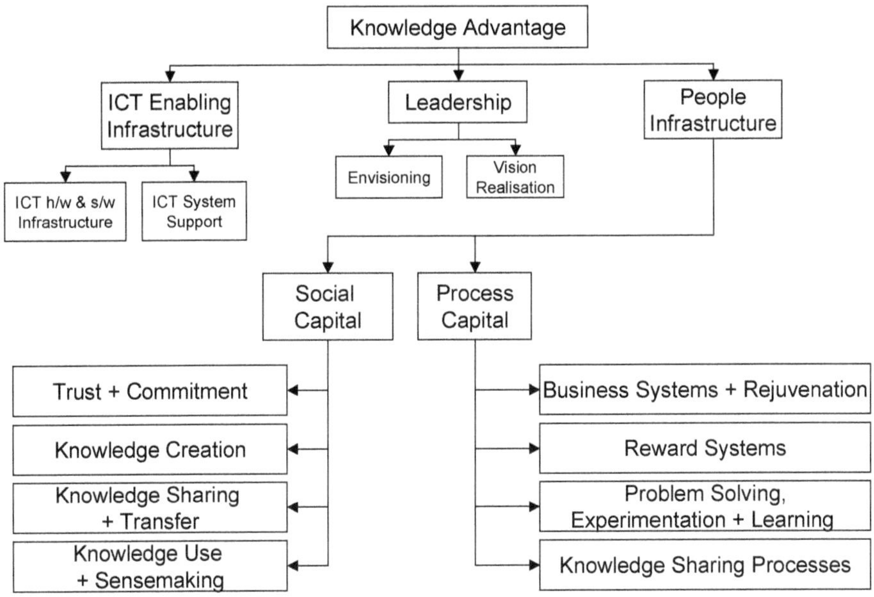

requisite talent (people with skills and behaviours needed) and the procurement of all identified resources necessary to enact the vision. The knowledge vision can then be deployed through leading the team(s) charged with its deployment, coordination of resources including people, and of course developing appropriate monitoring and control systems for the vision realisation project. An often-neglected aspect that must be addressed is maintaining the knowledge vision and this requires feedback, leadership support, continual communication, and relationship maintenance. The leadership realisation sub-element/attribute requires both an ICT and people-supporting infrastructure. ICT is necessary because in the construction industry, the project-based nature of work undertaken results in teams of people who need to communicate and share information and knowledge, being dispersed geographically. ICT, through groupware technologies, facilitates virtual teams to create, share and transfer, and store and retrieve both information and knowledge. ICT also provides various media channels for people to use for communication and joint problem solving. Realising a knowledge vision also requires a strong and viable people infrastructure that maximises advantages that ICT offers and management systems, procedures, and styles to unleash creativity and flexibility of thought and action.

Figure 3 illustrates the two sub-elements/attributes for the people infrastructure. Social capital provides the goodwill and commitment towards the K-Adv vision and its deployment. To gain the necessary trust and commitment that is required, the organisational values and culture need to support the creation of social capital. Not only must the organisation's values and culture be clear and consistent, but also it should provide close synergies with individual's values for there to be coherence and embeddedness that delivers a desire in individuals to want to realise the knowledge vision. Knowledge creation is a fundamental human activity that follows the SECI process (Nonaka & Takeuchi, 1995): (<u>**S**</u>) socialisation of tacit knowledge; (<u>**E**</u>) externalisation to make tacit knowledge explicit; (<u>**C**</u>) combining explicit-exiting explicit knowledge with the organisational stock of knowledge; and (<u>**I**</u>) internalisation, which is re-absorbing knowledge in its richer and expanded form. Knowledge sharing and transfer involves networks of people, and so this needs to be configured in some way; for people to want to join knowledge networks, they must anticipate deriving some value and they must have a *desire* to share knowledge as well as have the *capacity* to share knowledge. Knowledge use and sensemaking are inextricably interlinked. This requires engendering some ambiguity and creative chaos, redundancy of resources to allow people to think and reflect, requisite variety of stimuli and channels of communication, and a capacity for reflection and curiosity. These sub-elements build the necessary social capital required to realise the K-Adv.

The people infrastructure necessary for the K-Adv also requires an organisation to have a responsive and supportive environment. This can be provided through organisational processes, routines, and procedures that form the necessary process organisational capital. Figure 3 illustrates the need for business systems and rejuvenation that provides organisational structure to support: flexibility and creativity; a management style and empowerment strategy that stimulates people with a challenge, but that then supports them in their finding a way to address the challenge; designing group and individual interaction to promote sharing knowledge; and a decision-making approach that sup-

Figure 4. ICT infrastructure

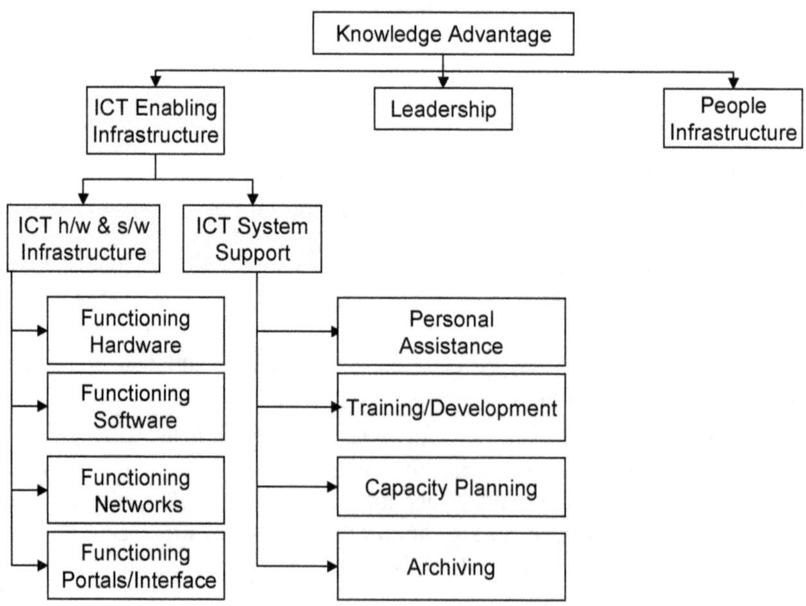

ports flexibility. There is also need for a reward system that is strategic and clear in its goals, and has an individual-to-team motivational fit and individual-to-team cultural fit, so that the rewards are matched to maximise impact of developing social capital. Additionally there is need for a problem solving, experimentation, and learning philosophy that: promotes experiential learning, has an appropriate response to failure to build social capital, links research and development with training and development, and encourages adaptation and re-invention of the organisation to capitalise on social capital and the K-Adv. Finally, a knowledge-sharing process needs to facilitate cross-levelling of knowledge throughout the organisation, to make valuating knowledge explicit, and to support cultural awareness so that the rich potential of diverse perspectives can be accessed and capitalised upon.

The people infrastructure required to deliver a favourable K-Adv requires supporting mechanisms that allow knowledge creation, sharing, and transfer. This calls for a suitable technological infrastructure (see Figure 4) that allows people to seamlessly share communication without loss through translation, and the transcription of explicit information and knowledge for later reuse and reference. People also need groupware application technologies to virtually link themselves when problem solving and coordinating their knowledge-related activities. To do all this, both hardware and software needs to be supportive. Hardware needs to be available, matched to meet current needs, be functional to do the required job, and be dependable and reliable. Software needs to comprise groupware application with interoperable features to enable digital information to be easily and rapidly transferred and seamlessly shared. Networks that effectively connect business units and organisations may span the supply chain with sufficient bandwidth capacity and shared protocols to enable information and data to be effectively shared and transmitted both upstream and downstream. Most ICT application-mature organisations have functioning portal interfaces that provide facilities to ease gaining ready access to valuable data, information, and knowledge sources; relevant and pertinent helpful content and scope of information and knowledge that can be accessed via a portal interface; and the relevant authorities and security to ensure that only those who are authorised gain access to knowledge. The ICT 'system' must deliver what is needed in a user-friendly way and also be secure with appropriate protective firewalls against unauthorised entry to confidential information and knowledge.

An ICT infrastructure will not function without the necessary level and quality of support. Figure 4 illustrates four sub-elements/attributes for ICT support; two of these relate to an ICT people-infrastructure. Personal assistance from non-programmed resources includes mentoring, person-2-person help (in a variety of forms that provide response to individual queries, programmed resources with staff that can respond in a pre-determined way to help them, as well as a traditional Q&A or systemised way of responding to typical queries), as well as a communities of practice approach that provides peer-2-peer help or informal assistance with colleagues helping each other when requested. Training and development prepares and sustains ICT users pursuing a more formal response that is provided through non-programmed mentoring and mutual adjustment support arrangement between people working to help each other, programmed internal training and development arrangements, and also external access to training and development. ICT system support also is provided by the organisational response of capacity planning manifested by demonstration of the organisation's planning cycles for both ICT infrastructure and support services based upon the current and future needs of the knowledge workers dependant upon this infrastructure. Further, archiving information and creating well-maintained, effective knowledge depositories provides critical ICT infrastructure support. Legacy systems and data/information needs to be managed to empower knowledge workers to maximise their effective use of ICT to develop and maintain the K-Adv.

Implications of Commitment on K-Adv Deployment

A key implication, evident from the model presented, is that it requires high levels of commitment from both the organisation (or business unit concerned) as well as from participants being empowered to release their creative energies to realise the K-Adv. But what exactly does commitment mean in this context? Also commitment applies equally to the organisation being committed to the workforce, as well as workers being committed to their host organisations. Notions of commitment are provided by Meyer and Allen (1997), who developed a typology of three forms of commitment, each with a different set of drivers that power this intrinsic force.

The highest level of commitment, *affective* (the want to force), is manifested by a deep-seated drive that aligns and deeply bonds the goals of the individual concerned with the organisation. Because these goals are indistinguishable, they are completely internalised by the individual. *Normative* (ought to) commitment relies on a sense of duty, purpose, and group identity (as opposed to individual identification). In such conditions this translates into a desire to 'not let the side down' or 'doing your bit/fair share'. *Continuance commitment* (need-to) relies on the investment sunk-cost that an individual has invested and the psychological calculation that is made when conducting a cost/benefit analysis about whether to retain commitment or to withdraw commitment. Thus, commitment is inextricably bound up in organisational behaviour and its impact upon the individual. Mere compliance lies at the lower interface of *continuance* commitment, in which the individual does only that which has to be done. Many management processes require only compliance to minimum-level standards even though the management rhetoric is focussed upon commitment.

The way that raw forces of commitment (to be used or abused) are deployed explains the dangers of brainwashing, the reality-rhetoric gap, and the failure to maintain a consistent and empowering attitude to deal with talented knowledge workers who have much to offer (Green, 1999; Green & Lenard 1999). This is why the issue of leadership is so pivotal to the K-Adv through cultural change—it sets the scene and affects the likely drivers and barriers to releasing creative energies of people.

Application of the K-Adv Model

The K-Adv model serves a useful purpose of providing a road map for strategic planners in the construction industry to align their corporate goals with maximising the organisation's competitive position through the K-Adv framework. Management efficiencies could save considerable costs by reducing wasted management and/or administrative energy through unleashing people's creative energies. Greater focus can be placed on the main game rather than side-shows, of fixing problems rather than fixing the system (Argyris & Schön, 1978). By enhancing competencies through the K-Adv, improved products, services, and management approaches would likely emerge that would provide the organisation with a differentiation advantage.

In this section we will concentrate on just one important aspect of K-Adv leadership, relating to stakeholders and use of tools to apply the concept of the K-Adv. Three tools will be discussed. One tool is the soft systems methodology (SSM), used to analyse problematic situations that help to unearth valuable knowledge about stakeholders and thence knowledge about what these people may know that can contribute to an organisation's K-Adv. The other two tools relate to techniques and tools for benchmarking an organisation's K-Adv performance. We stress that we are merely illustrating these tools as an example of three of many tools and techniques that can be applied to using knowledge about an organisation's K-Adv that can yield productive results in it realising competitive advantage.

Tool 1: Soft Systems Methodology (SSM) Used for Stakeholder Identification

Discussion about what a stakeholder is and the degree to which these people should be considered in business and social interactions has sparked much debate centred on, for example, an instrumental philosophy (Jones & Wicks, 1999). They argue that (positive or negative) outcomes will become evident if the interests of certain groups (as political constituents with a legitimate voice) are taken into account. The argument follows that to maximise benefit, stakeholders should be given the opportunity to influence plans and outcomes by voicing their concerns about their contributions to planning and monitoring outcomes. The usefulness of this view is that it provides a framework for prediction.

The management literature stresses the need for a customer focus in moving beyond the narrower commercial interest of selling goods to not only learn from the client by finding out how to satisfy their needs, but also by working with them to discover ways of doing so (Prahlad & Ramaswamy, 2000; Lederer & Hill, 2001; Ulwick, 2002). In supply chain management theory for example, upstream contributors to the value chain view those downstream as their customers to whom they should be adding value (Womack, Jones, & Roos, 1990; Mathews, 1999; McGeorge & Palmer, 2002). Thus from a value chain perspective, it is wise and pertinent to consider how a broad range of stakeholders might affect the development of a K-Adv. Briner, Hastings, and Geddes (1996) argue that organisations need to consider both the community interest and those who support

Figure 5. Example of a rich picture from Walker et al., 2003

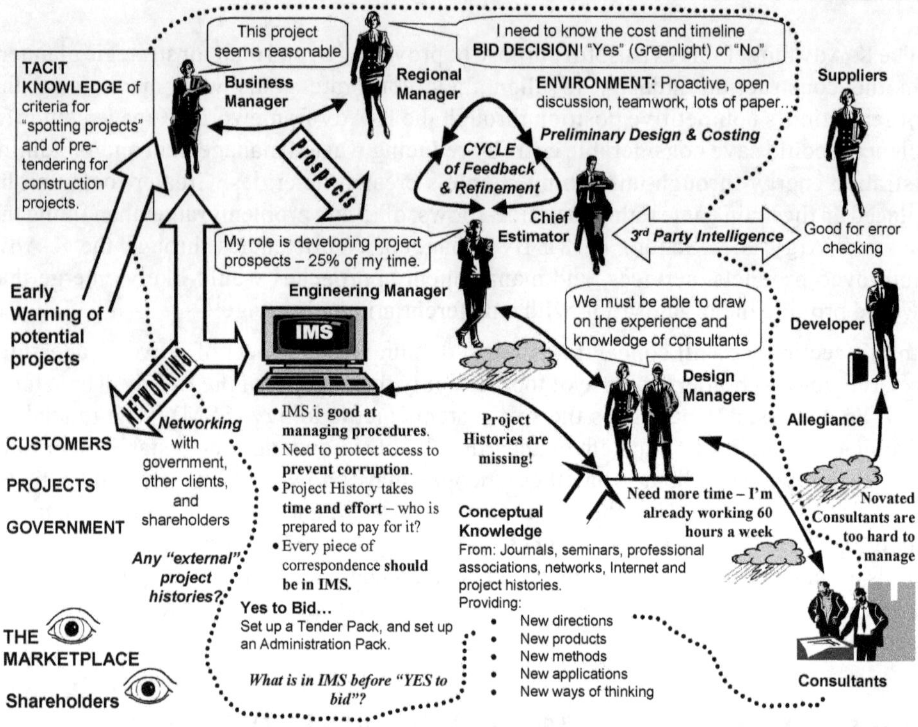

project teams. More lately there has been an increasing focus on developing a family friendly workplace as a means of winning the war for talent by attracting the best and brightest minds (Michaels, Handfield-Jones, & Axelrod, 2001). All this has relevance to developing a K-Adv and supports the need for visionary leadership as a key driver of a K-Adv through creating the best environment for knowledge to flourish.

The traditional systems approach to problem solving is generally based on problems being solved by a fragmentation approach—one stage at a time. This is appropriate in complex and highly structured situations that can be well defined. In information systems, this is formalised in the system development life cycle (SDLC) and the alternative method of prototyping. However, complex and poorly defined systems often conceal interesting hidden sub-text issues that are difficult to readily unearth. Understanding these contextual issues requires a joint exercise of sensemaking being embarked upon to fully understand the situation environment and dynamics.

Soft systems methodology (SSM) works within a seven-step framework. These steps are not necessarily sequential, and generally the first four steps are undertaken with a number of overlapping iterations. In Step 1 the situation or problem is identified in an unstructured form as a problematic *situation*. From a stakeholder identification problem illustration perspective, the *situation* is identifying business opportunities that are worth submitting a proposal on. The choice of whether to respond to a tender or request for information on a project (or potential project) requires a great deal of consideration, because the cost of responding to a proposal is often high and the chances of successfully winning the opportunity necessarily more modest.

In Step 2 the problem is expressed. This is where explicit knowledge must be effectively unearthed through a SECI process. Using an SSM approach, we interview as many participants in the *situation* as is practical who can explicate their tacit knowledge about the *situation*. This is made explicit through a *rich picture* medium. *Rich pictures* are deceivingly trivial at first sight because of their apparent child-like interpretation of a *situation*. This format, however, conceals a sophisticated way of inclusively garnering impressions and interpretations of experiences, feelings, and manifestations of driving and inhibiting forces that create the situation dynamic—stories that help in the sensemaking process so vividly described by Weick (1995).

The rich picture illustrated in Figure 5 is included here as an indicative impression of the style adopted rather than important content. This rich picture comprises a narration of a story as told by interviewed participants from their perspective and worldview. People are shown in comic-like sketches to de-sensitise personal accounts and to provide a holistic interpretation. Flows of dialogue and action are also represented with key issues highlighted, in this case in rain clouds. Relationships between key elements and perceived sub-systems are also indicated together with their links.

This rich picture represents a connective human communication channel that expresses the situation through an elicitation process (resultant from interviews or possibly surveys). Often unstructured interviews are conducted where respondents are encouraged to express their unease in graphic means. The principal is to unearth sub-textual information and knowledge because those interviewed generally have valid tacit knowledge to offer that is difficult to explicate in more conventional means.

Step 3 comprises the interpretation of the rich picture into a root definition. This is where high perceptive skills are called upon to take the rich picture and offer a more systemic and formulaic summary. This is the formulation of a "good" Root Definition, and it is decisive to the creation of the conceptual model in Stage 4. A Root Definition is tested against a group of elements known by the mnemonic CATWOE, that defines a checklist for:

- **C**ustomer (beneficiary or victims of the situation)
- **A**ctors (those directly affecting the situation)
- **T**ransformation process (what is happening in terms of inputs being transformed into outcomes in this situation)
- **W**eltanschauung (worldview of participants - the underlying narrative that addresses the question, "Why bother with this situation of endeavour?")
- **O**wner (the entity most affected by the particular situation)
- **E**nvironment (what lies outside the situation)

The Root Definition of the chosen system is expressed in statements that incorporate the points of view that make the activities and performance of the systems meaningful. The

CATWOE provides the analyst with a framework for ensuring that all points of view and interest are considered in the knowledge elicitation process. It should be a concise description of a human activity system which captures a particular view of it as a transformation process—INPUT/TRANSFORMATION/OUTPUT.

Step 4 is the development of an account of what must be done to achieve the transformation described in the Root Definition. This is generally illustrated as an activity model and uses an eclectic approach to use whatever techniques may be available so that 'hard' systems may be used in the form of flow charts, simulations, animation, statistical or mathematical models, or even theatre/performance. By comparing what is perceived to be the way things happen, including subtext and the full picture with the conceptual

Table 1. Root definition and CATWOE

ROOT DEFINITION—Pre-Tendering	**Customer:** Regional Manager (RM)
A system owned by the pre-contracts team who, together with the Chief Estimator and the Design Managers, takes prospective projects from the Business Manager, together with knowledge, processes, and technology, and prepares preliminary understanding of the project and cost estimates. This is used to assist the Regional Manager in assessing the feasibility of making a tender bid. This must be undertaken within short timeframes and with expert assistance from consultants. This is taking place in a very competitive environment where the "fit" to our business objectives and corporate goals, cost, and the timeline are all important.	**Actors:** Engineering Manager, Chief Estimator, Design Manager, Pre-Contracts Team, Business Manager **Transformation:** Knowledge, processes and technology, together with details of prospective projects, are used to prepare an understanding of the project and a cost estimate for assessing the feasibility of a tender bid. **Weltanschauung (world view):** To assess the feasibility of making a tender bid, we (RM) need a good understanding of the project—does it fit our corporate objectives—and cost and timeline details. **Owner:** Pre-Contracts Team **Environment:** Competitive, quality, cost and time critical, community and corporate goals

Figure 6. Conceptual model of pre-tendering process

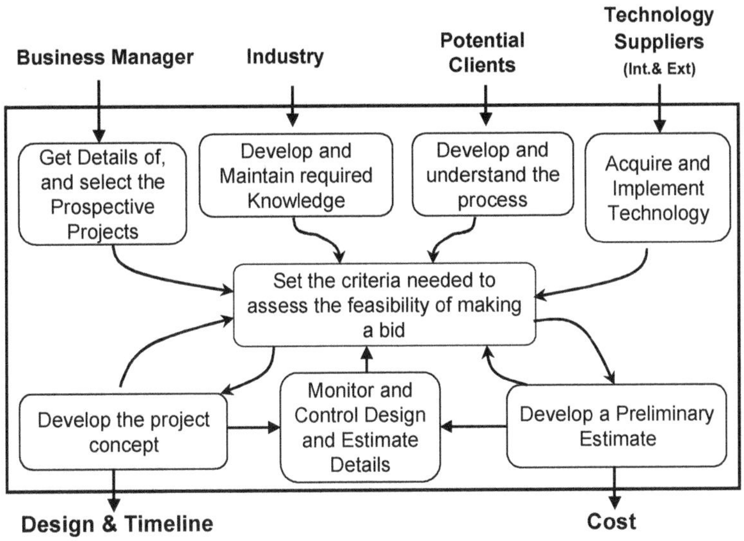

model, you can reveal a lot of interesting questions to be addressed, assumptions to be re-visited, and dysfunctional behaviours/actions that need to be remedied in Step 5. This step provides the reality check for Stage 4, but also throws forth a challenge to the owners of the situation, to rethink and re-analyse underlying assumptions for a more creative and fulfilling outcome to be reached.

This leads to Step 6 in which specific recommendations are made and plans for their implementation formulated. This may trigger changes to organisational structures, changes to procedures, and/or changes to organisational culture. The format of these recommendations requires careful considerations because as Schein (1993) has argued, change triggers two types of anxiety. Anxiety 1 is the fear associated with an inability or unwillingness to learn something new because it appears too difficult or disruptive. Anxiety 2 is the fear, shame, or guilt associated with *not* learning something new. The valence of Anxiety 2 needs to be greater than Anxiety 1, so this implies that communication strategies for the need for change need to form an important part of any set of recommendations.

Finally in Step 7, action is taken to make changes and/or restart the process using feedback loops to test and monitor the changes. In this way it can be seen that SSM is a reflective learning process as well as an action learning approach to problem resolution (Schön, 1983; Argyris & Schön, 1996).

In this section we demonstrate SSM as a viable approach for addressing difficult situations that employs a *bricolage* approach involving a creative, adaptive, and flexible response to problems. The bricoleur is a thinker who makes creative use of situations to find innovative fixes to problems, and as Weick argues, "What makes for skilled bricolage is intimate knowledge of resources, careful observation, trust in one's intuitions,

Figure 7. Summary of the SSM approach to identifying stakeholders

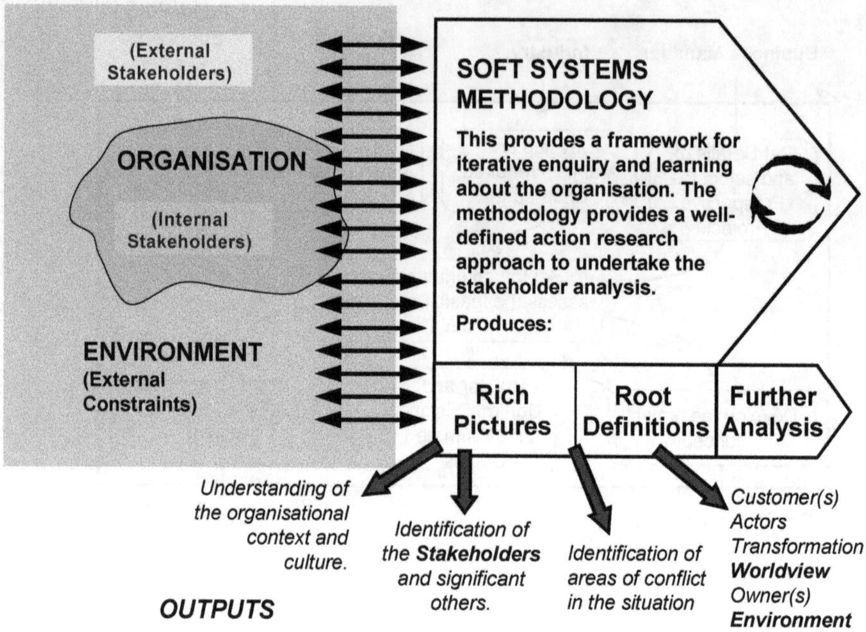

listening, and confidence that any enacted structure can be self-correcting if one's ego is not too heavily invested in it" (Weick, 2001, p. 63).

The example illustrated above was taken from the analysis of a pre-tendering system that identifies knowledge assets employed in business development through identifying appropriate projects for possibly tendering on. This involved identifying stakeholders with knowledge to contribute to this process and mapping their potential contribution of knowledge (Walker et al., 2003). It is an appropriate example to use in this chapter as it involves understanding stakeholder environments as well as identifying various players or stakeholders in this pre-tender process that would not be identified using a more structured and 'hard' systems analysis approach.

That particular analysis provided a root definition and model that identified a conceptual model that identifies the nature and role of stakeholders to the situation under study. Table 1 and Figure 6 illustrate the model derived from systems thinking and other forms of analysis of the situation as unearthed through the rich pictures.

This SSM approach in general presented in Figure 7 indicates that this is a powerful tool for understanding stakeholder environments and identifying the stakeholders themselves. This approach has many advantages over a 'hard systems' or more purely analytical approach such as using a work breakdown structure (WBS) type approach to identify tasks involved in a process and, from that, participants or stakeholders. This is because the more rigidly systematic approach tends not to allow the 'voices' of shadow or not immediately obvious participating stakeholders to be heard.

Tool 2. Benchmarking for Use in Strategic Planning

The K-Adv model is used as a basis to develop a matrix that provides a fine-grained way to map the K-adv maturity level of the organisational unit being studied. The dimensions of this matrix are logical components of the sub-elements of the model. For example, in Figure 2 the sub-element 'Identifying stakeholder K-value' was identified. In practical terms, what does this mean? Extensive review of the literature reveals three characteristics or components of the K-Adv Leadership element:

- understanding stakeholder environment so that an appreciation of the nature and identity of potential knowledge sharers may be gained;
- identifying stakeholders so that they can be targeted as potential sources of useful knowledge for sharing and exchange; and
- identifying what these stakeholders know so that the knowledge vision can be developed to appeal to their value proposition, as well as from the point of view of the organisation under scrutiny.

Once these characteristics have been identified, they can be analysed against a maturity scale of organisational commitment to the K-Adv principles in order that this tool may be used for benchmarking either against other business units or for a single business unit benchmarking itself at different points in time. The proposed maturity hierarchy comprises scales originally developed by Rogers (1995), modified by Carlopio (1998, p. 5), and then further modified to reflect associated stages of readiness in using the K-Adv model. The five levels of implementation maturity relate to: an *inactive* state of mere *awareness* of the question; a *pre-active* state where some form of *initiation* of action is taking place to address the issue; an *active adoption* of a solution strategy; a *pro-active* state where the action is *accepted and is being adapted* to best suit the individual context; and an *embedded* state where the adopted solution is built into the fabric of the organisation's culture, including its *review and continual refinement* and periodic overhaul to meet the prevailing context.

Once the matrix has been developed for each sub-element, the cells can be populated with descriptors that provide a useful guide in assessing the maturity level of an organisation for each maturity level of each characteristic of each K-Adv sub-element/attribute, as illustrated in Figure 8. By previously undertaking a SSM study on representative aspects related to stakeholder identification, the meaning and relevance of the descriptions becomes more achievable. For example, in the situation studied above and described in the rich pictures and other representation of the root definition, CATWOE, and conceptual model illustrated in Figure 6, could lead to the evaluation of a particular business unit or organisation.

The K-Adv guide explains each of the sub-elements/attributes described in Figures 2, 3, and 4, and each element of these has been further categorised on a matrix with five stages of K-Adv maturity scaled against the sub-elements/attributes. For example the envisioning sub-element/attribute 'identifying stakeholder K-Value' was described in terms of understanding stakeholder environments, identifying stakeholders, and iden-

Figure 8. Example of K-Adv benchmarking gap analysis

Maturity	Performance Characteristic Understanding stakeholder environments	Identifying stakeholders	Identifying what stakeholders know
How can the K-vision be improved, by →	Understanding the environments where stakeholders come from.	Understanding who these stakeholders are.	Understanding what stakeholders know and to whom benefits would accrue from their knowledge.
Inactive AWARENESS	Only considering the financial environment perspective.	Identifying the paying customer/clients	Only being able to identify how their knowledge might bring tangible financially beneficial to the organisation's or project's $ bottom line.
Pre-active INITIATION	Beginning to consider risk and opportunities more broadly so that environmental and social risk can be quantified.	Identifying internal and external customers/clients within the organisation.	Being able to identify how their knowledge might bring tangible and intangible benefits to the organisation's/project's $ bottom line.
Active ADOPTION	Considering in both quantitative and qualitative terms, the environmental and/or social environment to reduce these risks in financial terms.	Identifying internal and external customers/clients across the supply chain.	Being able to identify how their knowledge might bring a limited range of 3BL benefits to the organisation or project.
Pro-active ACCEPTANCE ADAPTATION +	Being able to develop an understanding of all three 3BL environments on their own terms.	Identifying the most significant stakeholders with an interest in knowledge about the financial element of the 3BL plus one of the other two elements.	Being able to identify how their knowledge might bring financially and a full range of 3BL benefits to the entire supply chain.
Embedded ROUTINISATION + INFUSION	Being able to fully comprehend the knowledge needs and supply of expertise that various players have in contributing qualitative and quantitative knowledge about their 3BL environments	Being able to identify and classify the most significant stakeholders by 3BL environments.	Being able to identify full 3BL benefits that each stakeholder's knowledge could contribute to the global community.

tifying what stakeholders know. A question naturally is raised as to how this aspect of creating a K-Adv vision (envisioning) can be improved by ensuring that each of these components are properly addressed (see Figure 8).

Rigorous benchmarking measurement can be undertaken between organisations or business units, or a time sequenced benchmarking exercise could be undertaken to plan for K-Adv development. Gap analysis can then be undertaken, and this provides the roadmap for analysis; strategic planning for taking action (change) can form the basis for developing a K-Adv vision.

Figure 8 presents a matrix comprising four columns and seven rows (including Row 1, the headings). Column 1 provides the five levels of implementation maturity in response to the maturing question being addressed. Columns 2, 3, and 4 contain text as a scenario that best describes the maturity for each of the five levels for the maturity question being asked. In this case the question (Column 1, Row 2) is "How can the K-vision be improved by…" for the performance characteristic of:

- understanding the environments where stakeholders come from (Column 2, Row 2);

- understanding who these stakeholders are (Column 3, Row 2); and
- understanding what stakeholders know and to whom benefits would accrue from their knowledge" (Column 4, Row 2).

In the course of undertaking an appropriate investigation approach, the current and desired (at some future time) maturity level can be determined from the descriptors populating the matrix cells best describing a scenario for each maturity level. It may become clear for example that the current situation indicates that the financial bottom-line value is viewed as the only valid way of looking at a stakeholder. Thus "only considering the financial environment perspective" (Column 2, Row 3) would be the current position for the "understanding stakeholder environments" sub-element or attribute. Discussion with key decision makers in that organisation may reveal that the preferred position at some future time "T" could be "being able to develop an understanding of all three 3BL[1] environments on their own terms." Thus the tools of SSM - with its rich pictures, root definition, and further analysis through developing a conceptual model that can be compared with the actual observed way that the issue under study operates - can provide one powerful tool for rigorously making judgements and assessments necessary to benchmark the K-Adv maturity matrix. The gap between current and desired future levels of maturity for the question asked for each of the three sub-elements of 'identifying stakeholder K-Value' can be measured. Using this process of benchmarking helps the organisation compare one business unit with another, one company with

Figure 9. Model illustrating organisational learning and interaction with external sources of innovation (Maqsood, Finegan, & Walker, 2003)

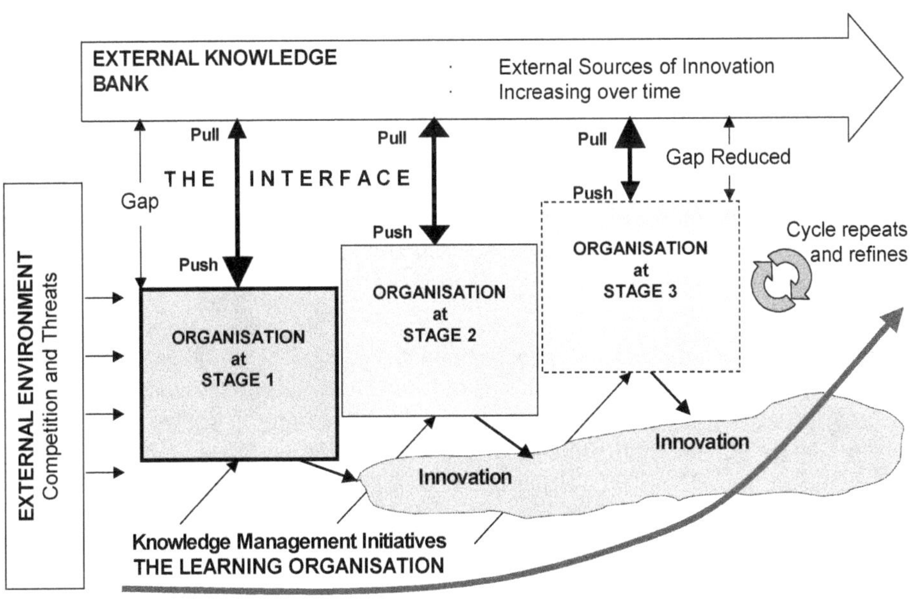

Figure 10. Detailed picture of three stages of organisational transition

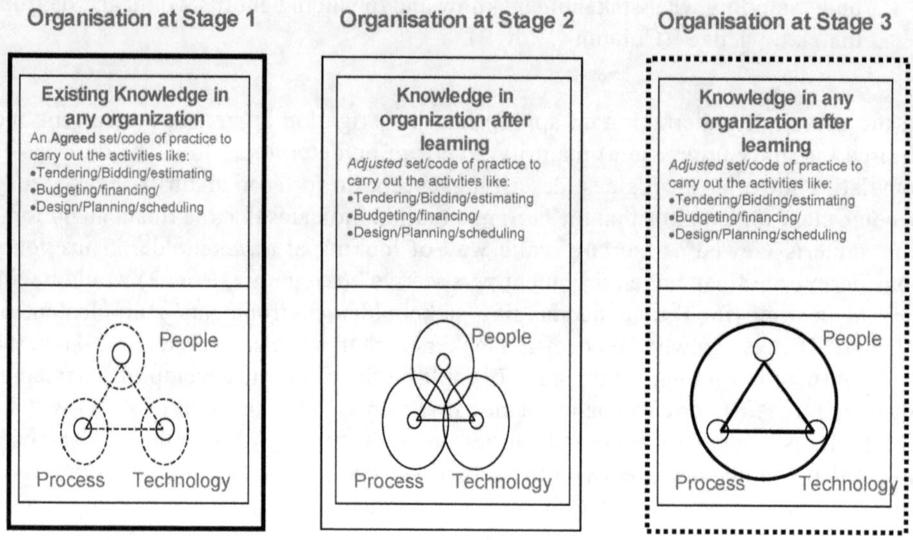

another, or to prepare a strategy plan to move from one maturity level to a higher one. Thus, the K-Adv concept provides a powerful tool for gaining competitive advantage through fine-grained benchmarking.

Tool 3 - Generalised Maturity Level Assessment for Absorptive Capacity

Cohen and Levinthal (1990, p. 128) define absorptive capacity as the ability of a firm to recognise the value of new external information, assimilate it, and use it for commercial ends. An alternative to Tool 2 above that provides a more coarse-grained generalised level (indicated on a three-level maturity scale) benchmarking tool is presented below. This may be used to consider how the organisation or business unit appears on a maturity scale related to the extent of its absorptive capacity (its ability for pulling knowledge from external sources or having these sources push knowledge into the unit) and the degree of rigidity or imperviousness of the organisation (or business unit) to sharing knowledge with external knowledge sources.

The model presented in Figure 9 is based upon analysis of the literature and a series of preliminary interviews in the construction industry. It explains the transformation of an organisation over time by illustrating organisational learning. It shows three transformation stages that are indicative of the transformation process that is actually a continuous process. These are further detailed in Figure 10.

The Figure 9 model could be applied, for example, to analyse the degree of understanding stakeholder environments, identifying stakeholders, and identifying what these stakeholders know that may be of value to the organisation's K-Adv. While it provides a more general alternative benchmarking tool to the populated matrix approach illustrated in Figure 8, it still needs rigorous case study data to be presented to justify any assessment given using this model.

The Figure 9 model distinctly shows a typical organisation and sources of knowledge external to it, referred to as an external knowledge bank. This knowledge bank or repository may consist of academic institutions and research centres dedicated to the production of new scientific knowledge. Knowledge may include new innovative processes, products, and technologies, as well as training and preparing people to utilise these processes through formal education. In addition, this knowledge bank also consists of externally existing knowledge and flows because of social interaction by the organisation with other organisations in a supply chain. The external knowledge bank is always increasing with time.

In this model, the interface is conceptualised to exist between the organisation and external knowledge. This interface operates under two main influences visualised as "pulling" and "pushing" forces. The external knowledge sources may exert an unsolicited *push* force towards the organisation to adopt new knowledge. A *pull* force refers to the drawing-in force of the organisation actively attempting to obtain knowledge from sources external to it. The distance between the external knowledge source and the organisation is an indication of the gap that exists between external knowledge and its adoption by the organisation - a gap between academic research and actual practice.

Figure 10 illustrates what is happening within the organisation at each stage where there is a transition from impervious thick boundaries isolating the organisation from its external knowledge environment with people, process, and technology similarly segregated and isolated within impervious boundaries with integrated people, process, and technology.

Stage 1: Before Transformation

This is the state of an organisation before or at the commencement of undertaking a knowledge management initiative. In this state, the innovation sources push external knowledge into the organisation. Inside the organisation, there is specific code of practice and organisational procedures to carry out its role depending on its position in the supply chain. People using various processes and technologies carry out their organisational duties. A high degree of segregation exists between people, processes, and technologies, and each organisational part has a specific area of influence.

A Stage 1 the organisation is able to enhance its productivity, improve its decision making, and keep its competitive advantage. At the start of the transformation, the internal knowledge bank will be very weak and less detailed, and will improve over time. The internal knowledge bank will also act as a source of feedback for external sources of innovation, providing details of the effect of innovation for further refinement and new developments. At this stage the feedback mechanisms are very weak, showing less collaboration and interaction with the external sources of innovation.

Stage 2: Transformation in Process

This stage is where the organisation has improved its absorptive capacity and has learned to do so through knowledge management initiatives. In this state, the culture of the organisation is becoming less of a barrier. The quest for learning becomes stronger. This results in an increase in pulling forces that the organisation exerts on the external sources of innovation that brings new knowledge into the adopting organisation. Academic institutions and research centres do not have to strongly push their new knowledge at this stage. Through learning, the organisation improves and streamlines its processes and routines. People change their attitudes, and under strong leadership become motivated to learn, adopt, and utilise the knowledge available. The area of influence grows and there is a reduction of segregation. This is the state where people are learning and seeking to adopt the officially deemed useful knowledge.

Stage 3: Ideal Transformed State

This stage represents a state where the organisation has reached an enhanced maturity. The culture of the organisation becomes highly adaptable and offers virtually no resistance to any flow of knowledge inside the organisation. Knowledge readily flows from outside sources into the organisation through strong pulling forces that feed the learning needs. The organisation will have gained momentum over time in learning, and its appetite for knowledge will become insatiable. No more segregation remains between people in organisations, processes, and technologies. This state represents high levels of integration between these three organisational components. Knowledge management becomes the integral part of the organisation routines, and continually assists in developing and maintaining internal knowledge repositories. The feedback mechanisms become solid and strong, and the gap between external knowledge sources decreases remarkably. Under this state, the productivity levels of the organisation improve profoundly.

Figure 10 explains how the model in Figure 9 shows that as the organisation starts to learn and transform itself into a learning organisation under the influence of knowledge management initiatives, knowledge from external sources of innovation would start flowing in it under forces visualised as 'pull' rather than 'push'.

We suggest that the K-Adv is intricately built upon the concept of organisational learning. Learning is always associated with better outcomes. Having learned lessons means that the 'wheel is constantly reinvented' and that the 'same mistakes are not to be repeated again'. Since Argyris and Schön (1978) and Senge (1990) introduced the ideas of organisational learning and the learning organisation, researchers and practitioners alike are striving to devise strategies to inculcate these ideas in practice.

Conclusion

This chapter provides insights into another way of viewing how the construction industry (or perhaps any other enterprise) could orient itself towards its market. Currently, many organisations look at the resources they have available to them - financial, physical, and human - and they respond to market demands, often reactively with a hand-to-mouth approach that could be argued to be largely responsible for the construction industry culture of conflict, poor service to clients, and a declining attractiveness to talented potential employees (Latham, 1994; DETR, 1998; Department of Industry Science Resources, 1999). A different approach is needed that allows firms to free themselves of innovation barriers and strengthen the natural drivers of commitment that resides in most human beings undertaking enjoyable work that can fulfil their needs for relevance and self-actualisation.

The K-Adv provides such a model, and this chapter has provided an introduction that is holistic in nature because it links the guiding leadership required for the people infrastructure that needs to be supported by effective ICT. A practical example of how it could be used was provided to illustrate the power of the concept. The concept has also been validated with industry partners through the Cooperative Research Centre in Construction Innovation in Australia and was well received. During late 2004 to early 2005, it will be further and exhaustively pilot-tested to develop a detailed business plan guide for its general application as a consulting tool.

The chapter progressed from a discussion of what the K-Adv is, to how it may be used for benchmarking, which in turn leads to strategic planning on ways to develop a competitive advantage through product/service differentiation. The use of SSM was introduced as an example of how a K-Adv maturity matrix can be used for benchmarking purposes, as well as how it can be used for mapping knowledge assets. This was followed by a discussion of how benchmarking using the K-Adv maturity matrix can be achieved for undertaking gap analysis that would drive a process of strategic planning for improving an organisation's K-Adv. An alternative model was presented for representing the knowledge maturity level for an organisation that can also be used for gap analysis and can also trigger strategic planning for an improved K-Adv. Thus three techniques (SSM, benchmarking, and gap analysis based on a push-pull model of organisational knowledge maturity) were offered as possible tools to be used with the K-Adv model as a way in which organisations can visualise their knowledge culture as a first step in developing a knowledge advantage.

Practical Tips and Lessons Learned

- The K-Adv model recognises that there are three essential elements to creating a culture change in the construction industry with knowledge as its core sustainable competitive advantage asset—knowledge leadership, supported by a people-infrastructure that in turn is supported by an effective ICT infrastructure.

- The challenge is for the construction industry to embrace its commitment to a style and form of leadership that not only empowers its knowledge workforce, but also positively supports and underpins them with appropriate infrastructure, the means to unleash innovation, and creativity.
- Three tools illustrated below can be used to help facilitate use of the K-Adv.
  • Soft systems methodology is a highly useful diagnostic tool to unearth contextual knowledge about elements of the K-Adv.
  • Benchmarking can be undertaken using the K-Adv matrices for each K-Adv subsection, and these can be used to undertake gap analysis that in turn can lead to development, formulation, and enactment of strategy.
  • The push-pull model of organisational learning forms a useful basis for a course-grained benchmarking tool to measure absorptive capacity.

Acknowledgments

We acknowledge the Cooperative Research Centre in Construction Innovation for its support. The research study work reported upon here was part of a broader study into ICT diffusion and knowledge management under CRC Project 2001-004 (2B)—Delivering Improved Knowledge Management and Innovation Diffusion.

Endnote

1 3BL refers to the concept of triple bottom line, where performance is sought not only on financial criterion, but also on environmental and social sustainable measures. For more detail on 3BL concepts, see Elkington, J. (1997). *Cannibals with forks*. London: Capstone Publishing.

References

Amabile, T.M. (1998). How to kill creativity. *Harvard Business Review, 76*(5), 76-87.

Argyris, C., & Schön, D. (1978). *Organizational learning: A theory in action perspective*. Reading, MA: Addison-Wesley.

Argyris, C., & Schön, D. (1996). *Organizational learning II: Theory, method, and practice*. Reading, MA: Addison-Wesley.

Briner, W., Hastings, C., & Geddes, M. (1996). *Project leadership*. Aldershot, UK: Gower.

Carlopio, J. (1998). *Implementation: Making workplace innovation and technical change happen*. McGraw-Hill.

Checkland, P. (1999). *Systems thinking, systems practice*. Chichester, UK: John Wiley & Sons.

Cohen, W.M., & Levinthal, D. (1990). Absorptive capacity: A new perspective on learning and innovation. *Administrative Science Quarterly, 35*(1), 128-152.

Department of Industry Science Resources. (1999). *Building for growth – an analysis of the Australian Building and Construction Industries* (p. 88). Canberra, Australia: National Building and Construction Committee (NatBACC) for the Government of Australia.

DETR. (1998). *Rethinking construction, report*. London: Department of the Environment, Transport and the Regions.

Elkington, J. (1997). *Cannibals with forks*. London: Capstone Publishing.

Fernie, S., Green, S.D., Weller, S.J., & Newcombe, R. (2003). Knowledge sharing: Context, confusion and controversy. *International Journal of Project Management, 21*(3), 177-187.

Green, S. (1999). Partnering: The propaganda of corporatism. *Journal of Construction Procurement, 5*(2), 177-186.

Green, S., & Lenard, D. (1999). Organising the project procurement process. In S. Rowlinson & P. McDermott (Eds.), *Procurement systems: A guide to best practice in construction* (pp. 57-82). London: E&FN Spon.

Jones, T.M., & Wicks, A.C. (1999). Convergent stakeholder theory. *Academy of Management Review, 24*(2), 206-221.

Latham, M. (1994). *Constructing the team, final report of the government/industry review of procurement and contractual arrangements in the UK construction industry*. London: HMSO.

Lederer, C., & Hill, S. (2001). Seeing your brands through your customer's eyes. *Harvard Business Review, 79*(6), 125-133.

Maqsood, T., Finegan, A., & Walker, D.H.T. (2003). A conceptual model for exploring knowledge channelisation from sources of innovation in construction organisations: Extending the role of knowledge management. *Proceedings of the 19th ARCOM Conference* (pp. 613-621), Brighton , UK, 3-5 September.

Maslow, A.H. (1943). A theory of human motivation. *Psychology Review, 50*, 370-396.

Mathews, J. (1999). Applying partnering in the supply chain. In S. Rowlinson & P. McDermott (Eds.), *Procurement systems: A guide to best practice in construction* (pp. 252-275). London: E&FN Spon.

McGeorge, W.D., & Palmer, A. (2002). *Construction management new directions – second edition*. London: Blackwell Science.

Meyer, J.P., & Allen, N.J. (1997). *Commitment in the workplace – theory, research, and application*. Thousand Oaks, CA: Sage Publications.

Michaels, E., Handfield-Jones, H., & Axelrod, B. (2001). *The war for talent*. Boston, MA: Harvard Business School Press.

Nonaka, I., & Takeuchi, H. (1995). *The knowledge-creating company*. Oxford: Oxford University Press.

Nonaka, I., Toyama, R., & Konno, N. (2001). SECI, *Ba* and leadership: A unified model of dynamic knowledge creation. In I. Nonaka & D. Teece (Eds.), *Managing industrial knowledge – creation, transfer and utilization* (pp. 13-43). London: Sage Publications.

Prahlad, C.K., & Ramaswamy, V. (2000). Co-opting customer competence. *Harvard Business Review, 78*(1), 79-87.

Rogers, E.M. (1995). *Diffusion of innovation*. New York: The Free Press.

Schein, E.H. (1993). How can organisations learn faster? Lessons from the Green Room. *Sloan Management Review, 34*(2), 85-92.

Schön, D.A. (1983). *The reflective practitioner – How professionals think in action*. Aldershot, UK: BasiAshgate ARENA.

Senge, P.M. (1990). *The fifth discipline – the art & practice of the learning organization*. Sydney, Australia: Random House.

Ulwick, A.W. (2002). Turn customer input into innovation. *Harvard Business Review, 80*(1), 91-97.

Walker, D. (2003). *Interim Australian standard on knowledge management – AS 5037(Int)-2003.* Melbourne, Australia: Standards Australia.

Walker, D.H.T. (2004). The knowledge advantage (K-Adv) concept. *The knowledge advantage (K-Adv): Unleashing creativity and innovation* (p. 200). Brisbane: I-Con.

Walker, D.H.T., Finegan, A., & Maqsood, T. (2003). Using a soft systems methodology approach to knowledge elicitation – an Australian case study. *Proceedings of CIB W55, W65, and W107 – Knowledge Construction Conference* (Volume 1, pp. 209-220), National University of Singapore, 22-24 October.

Weick, K.E. (1995). *Sensemaking in organizations*. Thousand Oaks, CA: Sage Publications.

Weick, K.E. (2001). *Making sense of the organisation*. Oxford: Blackwell Publishers.

Womack, J.P., Jones, D.T., & Roos, D. (1990). *The machine that changed the world – the story of lean production*. New York: Harper Collins.

Zack, M.H. (1999). Developing a knowledge strategy. *California Management Review, 41*(3), 125-145.

Chapter XIV

Knowledge Emergence and Adaptive Management: An Exploration on the Co-Production of Project Needs and Requirements by Client-Specialist Groups

Michael Whelton
University of California, USA

Ari Pennanen
University of Tampere, Finland

Glenn Ballard
University of California, Berkley, USA

Abstract

Project definition is the project phase that identifies the needs and values of project stakeholders, and develops appropriate design solutions to satisfy them. Project purposes are constructed through the interconnected relationships of stakeholder needs and values, and the project constraints. Projects comprise diverse stakeholders whose needs and values often conflict given that environments have limited project resources to fulfill all stakeholder interests. The research adopts the perspective that project definition is a learning process that requires the shared understanding of stakeholder needs and values, in order to resolve them into a collective statement of project purpose. This chapter describes the significance of the project definition as a project delivery phase and outlines a project definition model for knowledge creation. The research identifies that project managers must be able to learn about the needs and values of stakeholders, and adapt the process accordingly to solve those needs. A workplace planning system is described and a case study is documented to support the proposition that project definition is a complex adaptive process, and that knowledge about project purpose emerges from group collaboration.

Introduction

Within the broad knowledge management spectrum of research, this work is concerned with the collective group process within the project setting. The project is of great importance as it forms the collective knowledge work space for the project participants. In particular, this research is concerned with group knowledge - its creation, sharing, and management within project definition activity.

This research is concerned with understanding effective management principles of group processes in the project definition phase of construction projects. We are particularly concerned with organizations that perceive the need for a physical facility to support their strategic and operational goals. Project definition refers to front-end planning and design activity for facility projects requiring capital investment.

We define project definition as the first phase in project delivery consisting of three modules: determining purposes (stakeholder needs and values), translating those purposes into criteria for both product and process design, and generating design concepts against which requirements and criteria can be tested and developed.

The primary goal of the project definition process is to define the customer's project purpose. Without properly understanding what customers are trying to achieve, designs are not likely to meet those needs. Purpose is a term to signify intent and is the primary driver in value generation. Developing project purposes is a dynamic process and changes to each purpose throughout the course of project definition. Customer and stakeholder needs transition through various states of definition and various levels of commitment by stakeholders. We propose that purpose emerges from collaborative group interaction, that is, it is considered to be an emergent product of the group creation process.

This research addresses the value generation process, notably the value identification and value proposition phases, which are primary activities in project definition. The research focus is on managing the active knowledge creation of "project purpose" by project groups comprised of client (strategists and operations) and industrial specialists. Management typically works within a co-production model (with client groups and specialists) of project development. The research is particularly centered on project managers and the way they dynamically manage changing purposes in the early phase of project definition. Customer and stakeholder needs transition through various states of definition and various levels of commitment by stakeholders. We propose the use of an adaptive management framework to manage such dynamic complexity. We seek to explore adaptive management techniques that steer purpose creation and change. Facilitating group learning is a key management capability in this phase.

Based on exploratory research the authors have proposed a collaborative framework (Whelton, Ballard, & Tommelein, 2002) to manage project definition groups. The authors advance the development of this framework with industrial fieldwork. The objectives of this research are twofold. The first research objective identifies and describes management's role and responsibility with respect to value generation. Particular emphasis is placed on creating collaborative conditions for value generation.

Secondly we consider that management of project purpose is a dynamic learning process. We adopt the perspective of complexity and dynamic change. Management capacity to adapt is based on its ability to recognize and understand the dynamic behavior of the group learning process and the emerging purposes created. The research highlights the importance of management steering and the iterative process of re-framing the problem, as new information is created by the problem-solving group.

This chapter describes a case study of a workspace planning and management system developed to support strategic and operational planning of capital facilities. The case illustrates the management steering process and the value generated in the project definition phase. We test the proposition that project definition is an adaptive process and client purpose-based knowledge is an emergent outcome of the collaborative process.

Finally, a set of characteristics of the management system are proposed to assess its capability to create collaborative and adaptive conditions for project definition development.

Learning Objectives

The reader should learn the following from this chapter:

1. Understand the significance of project definition
2. Identify barriers to the successful creation and management of group knowledge
3. Understand the main developments on process knowledge in this project delivery phase
4. Understand the project definition process as a group learning process
5. Understand workplace planning processes and management steering
6. Understand how management adapts the group process to support the knowledge creation process

Strategic Significance of Project Definition

Project definition refers to front-end planning and design activity for facility projects requiring capital investment. This activity is frequently referred to as front-end definition, front-end loading, pre-design, or pre-project activity. Once a client identifies a perceived need for a facility, a project definition process is implemented to develop the project need into tangible product-process requirement specifications and concepts, or alternative client solutions. Project definition is the process prior to final investment decision making. The process usually covers the preparation of project proposal, project initiation, design, and appraisal. The Construction Industry Institute (CII) (1995) defines pre-project planning as the process of developing sufficient strategic information for owners to address risk, and decide to commit resources to maximize the chances of a successful project. The objective of project definition is to maximize successful project realization through the production of strategic information for the owner and the development of project implementation solutions.

A primary objective of projects is to generate value. In Koskela's (2000) Transformation-Flow-Value theory of production, value generation is viewed as a process where value for the customer is created through fulfillment of his requirements. Womack and Jones (1996) state that value can only be defined by the ultimate customer, and it (value) is only meaningful when expressed in terms of a product or service, or both. The supplier/provider role is partly defined in terms of delivery of value to customers. However, producers engage in production in order to accomplish their own purposes/create value for themselves. Further, producers' perceptions of customer value tend to be influenced by their professional and personal backgrounds.

Project stakeholders include anyone who has a stake in the project. They can include both customers and producers, along with other interest groups. Stakeholder values and their value judgments are the dominant influence on how purpose transitions from needs to requirements. Value systems influence the way customer needs are construed and represented. Hence, value generation is not a simple process and requires subtle management action in order to be successful. The opportunity to create customer and stakeholder value through collaboration has been well recognized in the practice of value management. Barton (2000) views value management as "a structured, facilitated process in which decision makers, stakeholders, technical specialists, and others work collaboratively to bring about value-based outcomes in systems, processes, products, and services." Within this discipline, there is considerable attention towards understanding how to elicit and manage client values. Liu and Leung (2002) propose multiple dimensions of value by which they perceive individual value systems to develop.

Murman and Allen (2002) propose a framework for value creation consisting of three phases: value identification, value proposition, and value delivery. The framework is based on the premise that value should be delivered only after identifying value and constructing robust value propositions. The first phase, value identification, involves identifying stakeholders and what they consider to be of value, also understood by negotiation researchers as 'interests'. Once initial needs and values are identified, the process moves to the proposition phase where the needs of stakeholders come together. The propositional phase creates collective purpose for the project. It identifies value dependencies and differences between the stakeholders. Latent needs and hidden stakeholders may also be identified. This phase ultimately seeks to create stakeholder alignment and collective commitment.

Common Barriers to the Creation of the Knowledge

Barrett (1999) and Kamara and Anumba (2001) highlight the management problems associated with project definition. They identify common industry barriers to effective project definition processes, which include:

- inadequate involvement of all the relevant parties,
- insufficient time allocated to project definition,
- inadequate consideration of the owner perspectives by the project team,
- poor communication among the parties, and
- inadequate management of changes in project requirements.

These management issues often result in the misrepresentation of collective project needs and values, group indecision, bounded rationality, and lost opportunities to innovate and create value for the facility owners and project stakeholders.

The management of project definition has received attention from various industry and academic research initiatives over the past decades. Initiatives for improvement occur across various bodies of knowledge, but primarily include the areas of strategic facility planning, organizational decision paradigms, process re-engineering and modeling, client briefing, client requirements processing, design methodology, and architectural theory. Table 1 provides a summary of these initiatives.

A Working Model for Project Definition

We define project definition as the first phase in project delivery consisting of three modules: determining purposes (stakeholder needs and values), translating those purposes into criteria for both product and process design, and generating design concepts against which requirements and criteria can be tested and developed.

Project definition can be perceived as a learning process so to understand the variables associated with project definition. Learning as discussed by Argyris (1999) occurs when

Table 1. Project definition research disciplines and primary approaches (Whelton, 2004)

Research Discipline	Notable Contributors	General Approach & Contributions
Strategic Facility Planning and Workplace Design	Teicholz (2001)	Proposes a range of strategic issues that real estate and facility management pursue to support client organizational strategy.
	Lambert, Poteete, & Waltch (2000)	Develops workplace strategies that include: workplace cost reduction, workplace efficiency improvement, workplace productivity improvement, and employee satisfaction and retention within the work environment.
	Horgen, Joroff, Porter, & Schon (1999)	Highlights the importance of workplace design in organizational performance. Collaborative process model of workplace design consists of integrating space, organization, technology, and finance.
Stakeholder Paradigms & Perspectives	Woodhead (1999)	Identifies the range of paradigms and perspectives owners use for decision making in the pre-design phases of capital projects including: the capital investment paradigm, the cost-benefit analysis paradigm, the financial paradigm, the strategic paradigm, the marketing paradigm, organizational perspectives, management perspectives, the property development paradigm, the planning permission paradigm, and the preliminary design paradigm.
Project Scope Management	Construction Industry Institute (1999)	Shows that the Project Definition Rating Index (PDRI) can be effectively used to improve the predictability of project performance. The PDRI tool allows the project team to quantify, rate, and assess the level of scope development on projects prior to beginning development of construction documents.
Client Briefing	Barrett & Stanley (1999)	Critiques process issues in client briefing activity. Proposes key solutions including: client empowerment to inform, educate, and make decisions; management of project dynamics; appropriate user involvement; appropriate information and visualization techniques; and appropriate team building.
	Green & Simister (1999)	Uses a social constructivist approach to business process re-engineering in client briefing. Investigates the use of soft systems methodology (Checkland & Scholes, 1999) as an aid to client briefing.
	Smith (2000)	Develops a methodology for developing strategic needs analysis through group facilitation and strategy analysis software.
Value Management	Green (1994)	Develops a methodology (SMART) to support stakeholder learning workshops and strategic choice.
	Barton (2000)	Views value management as "a structured, facilitated process…." Develops insights into soft value management through facilitated group processes. Concludes that hard value analysis methods are inappropriate for complex project initiations.
Process Protocol Development	Kaglioglou, Cooper, & Aouad (1999)	The Generic Design and Construction Process Protocol (GDCPP) specifies a set of stages for project definition activity, namely: *Pre-project stage:* demonstrating the need, conception of need, outline feasibility, substantive feasibility study, and outline financial authority; and *Pre- construction stage*: outline conceptual design and full conceptual design.
Requirements Engineering	Kamara, Anumba, & Evbuomwan (2000)	Defines client requirements processing as involving the definition, analysis, and translation of client requirements into solution-neutral design specifications. The model sub-divides into three main stages: define client requirements, analyze client requirements, and translate client requirements. These stages sub-divide further into activities and utilize appropriate information gathering tools, decision support tools, and quality assessment tools (e.g., Quality Function Deployment) to develop solution-

Table 1. Project definition research disciplines and primary approaches (Whelton, 2004), cont.

	Eodice (2000)	Identifies notable relationships with requirements and needs: "A need is an identified desire for the product which has been formally expressed and accepted by the group…An advocate is an individual or group that assumes responsibility for taking whatever action is necessary to implement a specific need…A requirement may be defined as any pairing of a specific need with a specific advocate, or a series of advocates, that leads to implementation of the need." Claims that the confusion with need and requirement can be avoided if one adopts the view that when a need becomes constituted into the final product design (or specification), only then does it become a requirement.
Product Development	Bruce & Cooper (2000)	Establishes protocols for capturing front-end knowledge.
	Hook & Farry (2001)	Specifies a requirements management process of Validation and Verification (V&V): validation of need (Is the need necessary?) and verification of implementation (Can the need be implemented through a viable solution?).
	Ulrich & Eppinger (2000)	Promotes empathic methods for user needs analysis along with systematic process guidelines when transforming needs and values into design requirements. The action of requirements engineering requires process quality criteria such as clear and timely communication, accurate and unambiguous written documentation, clarity through language use and writing, ownership, and traceability to stakeholder or agent. Metrics regarding requirement conformance, correctness, completeness, and clarity are central to these approaches.
	Thomke & Nimgade (2000)	Describes the IDEO product development knowledge creation process. Empathic user methods, rapid prototyping, and iterative group learning are central themes of the development process.
Design Methodology	MacMillan, Steele, Austin, Kirby, & Spence (2001)	Approaches the process of project definition through understanding design methodology in the conceptual design phase. Develops and verifies a structured framework to support interdisciplinary design. Proposes a generic model based on processes, tasks, and activities. Argues that after preliminary evaluation, the tool can lead to improved integration of interdisciplinary design, improved collaboration, and improved process understanding.
Architectural Programming	Cherry (1999)	Provides problem-solving methods and approaches to architectural programming for design.
	Peña & Parshall (2001)	Develops a programming method to establish client and project values to allow designers to respond with alternative solutions to defined problems. Defines programming as a process of five steps: 1) establish goals; 2) collect and analyze facts; 3) uncover and test concepts; 4) determine needs; and 5) state the problem.
	Hansen & Vanegas (2003)	Advocates the automation of the programming process through the use of information technology support. Develops a Web-based application to capture project definition information.

Figure 1. Project Definition Learning Model (Whelton & Ballard, 2002a)

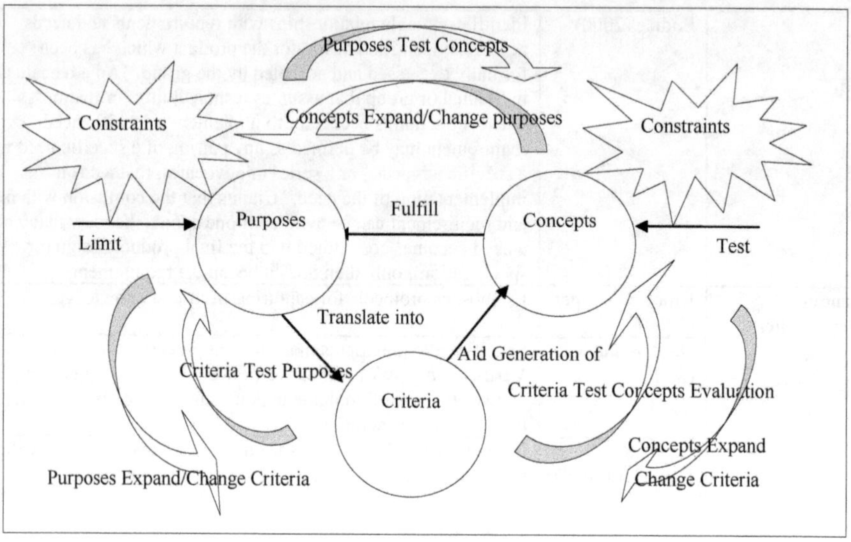

an organization achieves what is intended, that is, when there is a match between intentions and outcomes, and also when a mismatch is identified and corrected and turned into a match. The extended process results in double-loop learning (understanding the governing problem variables and altering actions) to determine how the original project goals and design criteria were set and established. Single-loop learning may focus on changing actions without a focus on the governing variables.

To relate this learning model to project definition activity, Figure 1 illustrates a set of learning cycles to test project purposes, criteria, and concepts. Single-loop learning occurs when the project definition group solves the initially stated project purpose. Governing variables may include the initial problem formulations represented in client purpose, team assumptions, stakeholder needs, and project constraints. Double-loop learning is advocated through appropriate testing of all project definition variables.

Capacity to Manage Project Constraints

The project definition process is significant in that it offers opportunity for the client and the team to identify constraints. Constraints are relationships that are maintained or enforced in a given context (Mayer, Painter, & Lingineni, 1995). Constraint statements describe the relationships between objects and processes. Constraints can be considered to act as enabling and limiting in a given context. Goldratt (1990) defines a constraint as "anything that limits a system from achieving higher performance versus its goal." By thoroughly identifying a constraint, clients and specialists can judge whether the constraint should be maintained or decide whether the constraint should be removed if possible to create value. The constraint may cause undesirable consequences or no longer support the organization's goals. There needs to be sufficient capacity in a project definition management system to reveal constraints and propose means of working with such constraints.

Origins of Purpose

In the domain of capital facilities design and construction, understanding the strategic and operational needs of the customer organization creates a basis for understanding project purpose. Smith (2000) argues that "when the strategic analysis of needs has been rigorously and conscientiously pursued, then it should result in a clearer view of the goals of the organization, a better definition of its real needs, and the strategic decision should recommend the best means to achieve those corporate goals."

Needs as the Basis for Creating Purpose

Altschuld and Witkin (2000) define a need as "a measurable discrepancy between the current and desired status for an entity." With a specific organization in mind, for example a healthcare service or an educational service, Level 1 addresses the needs of the organization's primary customers. Level 2 needs pertain to the organization's groups and individuals tasked with providing services to the customers in Level 1. Level 3 needs are those surrounding resources that support Levels 1 and 2. These needs may be facilities and support systems. Level 3 is the needs level within which construction professionals normally operate. Construction professionals may only see their role as supporting Level 3 needs, but Levels 1 and 2 are the primary organizational needs that must be satisfied. A holistic approach to needs is advocated to incorporate all levels of needs. Loss of focus on Levels 1 and 2 can result in poor performance of the project definition process.

The uncertainty associated with needs may be high, especially for large, multi-faceted, client organizations. Needs change over time in uncertain environments. Needs analysts may find it necessary to look at varying timescales so to identify the implications of needs and their dynamic changes (Nutt, 1993). The interdependency exists not only across levels of need, but also over time. For example, a healthcare organization has many unknowns regarding future needs. Healthcare administrators are tasked with fulfilling short-term needs, that is, what is needed now, yet in the long term, healthcare services are expected to change in terms of both the customer (patient) demographics and the technological service innovations that support medical care. Hospital facilities typically have long development cycles from inception to delivery. The requirements of one client group may be only considered as needs to another group. Managing choice and consensus as to what the project purpose is can be difficult without open process transparency and group commitment.

Table 2. Language constructs synonymous with managing project purpose

Purpose -Variables	Generic Definition
Need	A pressing lack of something essential
Latent Need	(Need) present, though not now visible, obvious, or active
Want	To have or feel need; something wanted: need or desire (a perceived need)
Aspiration	A strong desire to achieve something; an object of such desire
Desire	To express a wish for; request
Taste	Individual preference; inclination, critical judgment, or appreciation
Value	A relative worth, utility, or importance
Expectation	The act or state of expecting; ANTICIPATION; basis for expecting; ASSURANCE "they have every expectation of success"
Drive	An urgent, basic, or instinctual need
Intent	Having the mind, attention, or will concentrated on some end or purpose
Purpose	Something set up as an object or an end to be attained; INTENTION
Issue	A matter that is in disputed between two or more parties
Perspective	A mental view or prospect; the interrelation in which a subject or its parts are mentally viewed
Position	A firmly held point of view or way of regarding something
Preference	The act of preferring; the state of being preferred; the power or opportunity of choosing
Satisfaction	The fulfillment of a need or want; the quality or state of being satisfied
Advocacy	The act or process of advocating or supporting a cause or proposal
Function	The action for which a person or thing is specially fitted or used or for which a thing exists; PURPOSE
Goal	The end toward which effort is directed; AIM
Objective	Something toward which effort is directed; an aim, goal, or end of action
Priority	A preferential rating
Value Judgment	A judgment assigning a value (as good or bad) to something
Utility	Fitness for some purpose or worth to some end
Meaning	The thing one intends to convey—significant quality
Inquiry	The act or an instance of seeking truth, information, or knowledge about something—and yielding results, benefits, or profit
Definition	To determine or identify the essential qualities or meaning of
Criteria	A project criteria is a standard upon which a design solution adheres to
Demand	An act of demanding or asking, especially with authority
Constraint	A constraining condition, agency, or force
Concept	A "design concept" is a solution to a perceived problem or need
Reflection	To express a thought or opinion resulting from reflection
Usability	Capability of use; a design is judged on its fitness for use or in other words: "the extent to which it realizes the purpose of those for the whom the artifact is being produced"
Validation	Purpose requires validation: Is it truly what is sought?
Verification	The act or process of verifying: Can purpose be realized in a design solution and the project constraints?

March (1983) raises the idea of pre-existence of purpose. The language terminology that describes purpose can include: "values," "needs," "wants," "goods," "tastes," "preferences," "utility," "objectives," "goals," "aspirations," and "drives." The language of project purpose is often ambiguous, and therefore understanding that language is a prerequisite to developing a model of purpose creation. The terms needs identification, value statements, and requirements processing are often used synonymously. Table 2 shows the main language constructs observed in project definition conversations.

Central to the role of management is the capacity to establish a working language for managing multiple voices in the project definition conversation and dialogues.

Management Capability to Develop Project Purpose

Further theoretical understanding is necessary with respect to organizational problem solving and capacity to develop project purposes. The ability of project planning organizations to develop project purpose is quite often inhibited by their capability to perceive the environment around them and act effectively. Innes and Booher (1999) argue that "consensus planning" (a form of collaborative activity) is not only about producing agreements and plans, but also about experimentation, learning, change, and building shared meaning. Innes and Booher (2002) identify capacity in governance systems: "A governance system with capacity can learn, experiment, and adapt creatively to threats and opportunities. It is characterized by regular interaction among diverse players who solve problems or complete complex new tasks by working together."

An individual, organization, or system with capacity is one which is constantly learning and evolving. Capacity has to be built at four levels in a community: within members; within their relationships; within their organizational structure; and within the programs they sponsor.

In the realm of organizational innovation research, capacity theory is recognized as dynamic capability. Dynamic capability as defined by Teece (2002) "is the ability to sense and then seize new opportunities, and to reconfigure and protect knowledge assets, competences, and complementary assets and technologies to achieve sustainable competitive advantage." Achieving dynamic capability requires effective organizational sensemaking, understanding, and strategic and tactical action planning.

Dove (2003) uses the concept of agility to formulate dynamic capability. Agility is a balance between the physical ability to act (response ability) and the intellectual ability (knowledge management) to know what to act upon. It is the ability to apply knowledge effectively. Response ability relies on organizational relationships, processes, and communities of practice to access knowledge within various social and technical repositories embedded within the organization.

The notion of dynamic management capability is central to taking advantage of emerging opportunities to create value in the project definition phase.

Managing the Emergence of Purpose in Complex Socio-Technical Systems

Groups tasked with developing workplace strategies for organizations are faced with conditions of continuous change and uncertainty. Client organization life cycles, business (product and services) life cycles, and facility life cycles are in continuous states of change. We assume that project definition activity is a social process. Organization of this complex system is perceived to occur through networks of communications, conversations, and dialogues among system agents (stakeholders). This social system is perceived to be complex in detail and dynamically complex in behavior.

Complex systems are described in terms of both combinatorial, or detail complexity and dynamic complexity. A system may be made up of a large number of parts that have many interactions. Complexity frequently takes the form of hierarchy. A hierarchic system is composed of interrelated sub-systems. Complex properties exist in terms of parts and laws of interaction. The whole is usually more than the sum of the parts, and inferring the properties of the whole is more difficult (Simon, 1996). Complexity often results in features called emergent properties, which are properties of the system, though not of its parts (Axelrod & Cohen, 1999). Organizational systems have high levels of dynamic complexity (Sterman, 2000). Dynamic complexity arises from the interactions among the agents associated with the system over time. These dynamic influences can impede organizational learning and system performance.

Within the context of project definition, we identify the concept of emergent purposes based on the synergetic interaction of the customer, management, specialists, and other relevant stakeholders. We associate emergence with dynamic systems whose behavior arises from the interaction of its parts and cannot be predicted from knowledge about the parts in isolation (Corning, 2002). A key issue with understanding purpose is the synergy that a collaborative process creates. According to Corning, "*Synergy refers to the cooperative effects that are produced by two or more particles, elements, parts, or organisms—effects that are otherwise not attainable.*" A shared problem space or forum facilitates stakeholder conversations and dialogue. These spaces facilitate the development of needs and these needs transform into project requirements through some facilitated process. Purpose is seen to emerge once stakeholders progress to consensus.

Stakeholder worldviews are expressed in multiple paradigms and perspectives, which are in turn supported by strategies, methods, and tools, specific to that stakeholder background, knowledge, and experience. Therefore the research is directed towards understanding the management of the dynamic conversations that result among client representatives, project management, and Architecture-Engineering-Construction (AEC) specialists as they each bring their own perspectives to bear. Dialogue serves as a group communication mechanism that allows *value identification* to occur. Purpose is generated in the ongoing dialogue between stakeholders. As client entities and diverse industry specialists collaborate to produce individual and collective purposes, breakdowns of many types are to be expected given individual worldviews. Scharmer (2001) states that "reflective and generative dialogues" are minimum conditions for knowledge emergence and in our case for effective value identification and value propositioning.

Knowledge management researchers (Brown & Duguid, 1998) propose an architecture for organizational knowledge. It identifies social strategies to promote the spread of knowledge between communities. These strategies include "translation," "brokering," and "boundary objects." Translators are individuals who can frame the interests of one community in terms of another community's perspective. Translation requires trust of both communities to avoid value loss. Brokers by contrast are involved in participation. They are a feature of overlapping communities and often facilitate the flow of knowledge between communities.

Figure 2. Distributed knowledge agents in project environments (Whelton, 2004)

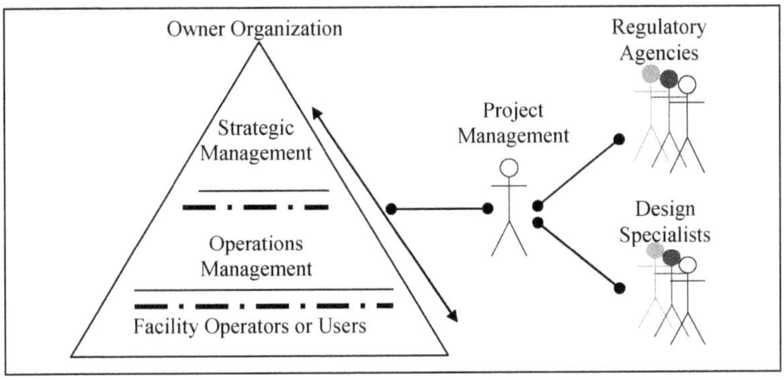

Knowledge agents interact through boundary objects. Boundary objects "forge coordinating links" between communities. These can be physical objects, technologies, or techniques shared by the communities. These objects serve to identify what is common and distinct about another community (Carlile, 2002). Such objects help clarify interest and attitudes, unearth hidden assumptions, and encourage reflection and learning loops. Problem-solving methods and tools serve as a primary object for such relations to occur and develop. Common language and appropriate transparency are necessary prerequisites for an effective group problem-solving space.

As illustrated in Figure 2, the planner needs to access the client organization and other specialists that contribute to the workplace definition. Such specialists include expertise in the client knowledge domain and expertise in the construction industry domain. The project manager is perceived to act as an adaptive agent and has capacity to activate distributed specialist knowledge and activate client-based knowledge both vertically and horizontally in the organizational structures.

Mapping Adaptive Management Action

Within the context of project management theory, the pre-dominant perspective of project management as planning, executing, and controlling has received criticism in relation to managing the definition of projects. One particular criticism pertinent to this research is centered around the constant process of re-defining the project (problem definition) well into the project's development (Melgrati & Damiani, 2002). Directive management techniques are suitable for systems that display low levels of complexity and change. Within the context of project definition activity, purposes are more difficult to establish in environments of high complexity and change.

Project management typically faces many challenges when operating in a complex system. Information is distributed within the project organizational network. Information may not be readily available for decision making. The information available may be understood as unimportant for decision making and requires the collaboration of the stakeholder groups to agree on. The outcome or solution is not predictable based on initial information. Quite often the initial information becomes less important in making judgments, as new information and increased shared understanding of the issues develop. The ability to find an optimum solution is difficult as there are multiple solution paths. Consensus and agreement on a "good enough" solution is often an appropriate means of progressing on in the project development. Therefore the role of project management as a facilitative agent in the group process is vital for project success.

A more adaptive style of management is necessary in complex environments. Adaptive management of project purpose is a dynamic learning process. Management's capacity to adapt is based on its ability to recognize and understand the dynamic character of the group learning process and the emerging purposes. The notion of adaptation is important for management to understand when operating in highly complex project environments. Ruitenbeek and Cartier (2001) view adaptive as the means by which groups learn from actions in one period, so that they may modify their actions in future periods. The essence of adaptive management is that managers can learn as actions unfold and as new issues emerge (Graham & Kruger, 2002). If we consider that multiple needs and values exist when defining the purposes of the project, control may be understood as effective facilitation. By facilitation we mean promoting interaction, participation, relating, and dialogues. Management may find itself continually iterating and redefining the basic premises of purpose, based on constant feedback and incremental learning cycles.

Case Study: Workplace Planning Management System

The characteristic of this adaptive management action requires further understanding. The following case study (a workplace management system) is understood as a knowledge management architecture that supports knowledge creation for project definition. The management system demonstrates adaptive capacity to develop collaborative project purpose statements in complex client organizations.

Haahtela Oy, a project management services company, has developed a strategic and operational workplace planning service for real estate clients. The Haahtela workplace planning system creates information to support the user viewpoint, the viewpoint of client management, and those of the real estate and building professionals. The system provides "a functional model by which spatial resource needs can be planned" (Pennanen, 2004). This planning is based on the organization's operational needs and by using operational measurement units. The developed operational model produces new information and new viewpoints which when reviewed will assist the organization in strategic workplace management.

The modeling process and management process combined seeks to facilitate strategic workspace management that supports the strategy of the business or service organization. From the viewpoint of the user, needs for spatial resources arise from the time of management planning of the operational environment and the geometric requirements.

From the company management perspective, the needs for spatial resources arise from those operations that management has chosen to realize the organization's strategic aims.

The planning concept is developed through a dialogue process between strategic and operational management and facility users. The process is supported by a management steering model and a workplace planning procedure. The steering model is based on feedback-loop control where the feedback is a description of current or desired state (use of resources) due to wants and wishes. The state is measured by:

- defining a bill of activities together with the client;
- defining temporal and geometric needs for operations;
- defining (based on above) the working environment (room schedule, performances, potential for use, and utilization degrees) with the workplace planning procedure; and
- budgeting the outcome and, if needed, directing the costs back to activities for activity-based cost management.

Steering the Project Definition Group

Defining spatial resources is embedded in a complex system - a broad set of goals, a wide range of needs, different viewpoints of owner and user, and so forth. Even though in strategic workplace management there is no single right answer, there are certain limitations to decision making that cannot be exceeded. The most important points are economic and temporal limitations.

Figure 3 illustrates the control model for the workplace management system. Any system needs damped feedback to maintain it in balance. New viewpoints and creative ideas will be welcome if they produce new alternative solutions in an area possible to business. Therefore the structure of the steering model is based on the cybernetic closed-loop control. The controlled system is the organization defining its' spatial requirements. The outcome is their choice of the activities, rooms needed for activities, and consequences (economic and temporal) due to the activities and rooms.

The controller presents a representation of the controlled system. The representation is composed of two methods: the Workplace Planning Procedure created in this research and the Target Price Method (Haahtela & Kiiras, 2003). In the workplace planning process, the client does not inform spaces and areas needed, but describes activities he/she needs. The Workplace Planning Procedure adds dimension to the working environment and lists it as rooms needed. The Target Price Method calculates the budget based on the rooms and requirements set for the rooms. The budget can then be addressed back to the activities.

After the measurement the client knows the present or desired state of the system—what kind of working environment, how much the working environment demands resources, and from where the organizational activities originate. During the process participants

Figure 3. Closed-loop control as defined for workplace management (Pennanen, 2004)

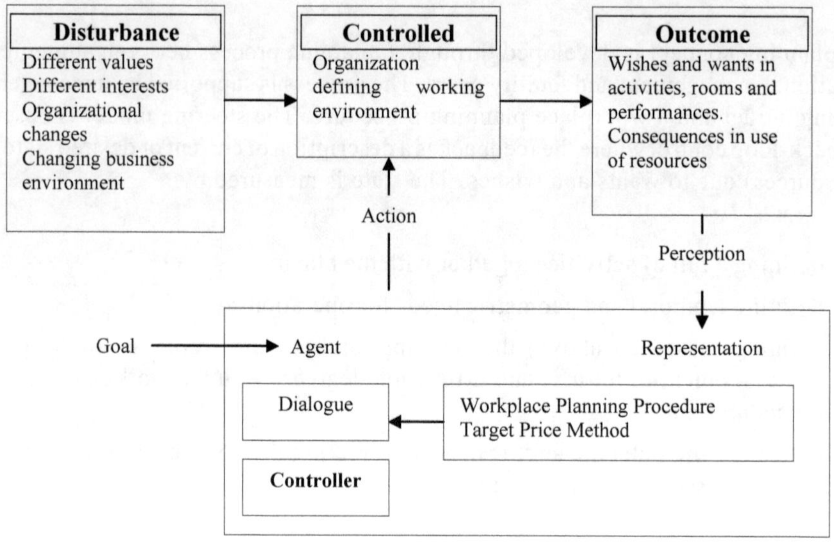

may find out new ideas that can reshape the organization's activities and working models. These new changes can be added to the workplace planning model for present and future use (new learning). The agent as controller is aware of the new goals (purposeful behavior).

Principal Process Phases

In order to deliver this service effectively, Haahtela developed a set of operational actions to manage the process. The programming service is mapped into nine stages as follows:

1. The client identifies potential facility need-based issues, and a directive is issued to the programmer to do workplace planning.
2. The workspace planner identifies the organization's primary level of needs, which are expressed as business drivers. A collective agreement is made on how best to measure the drivers between the client and the planner.
3. Resource use measurement: The planner engages in a set of group processes to develop the space quantification model:

   a) user functions, activities, and actions;

   b) space definitions;

c) performance results;

d) spatial costs and budget analysis.

4. Initial client strategy and operations dialogues: Once the initial state of the workspace is established, the client groups convene to validate the model.

   a) Client groups receive a summary of the state of system needs.

   b) Groups discuss accuracy and agree on state.

   c) Groups gain consensus on spatial metrics and their validity for use in dialogues.

   d) Groups seek to develop shared understanding of problem across system levels.

5. Shared group process with workspace planner:

   a) Dialogues initiated with strategy, operations, and planner.

   b) Planner offers opinions on state of system.

   c) Planner identifies cues or problem areas where poor performance is evident.

   d) Planner offers scenario advice on impacts on strategy scenarios, for example: if do A, then the value generated is X.

   e) Planner seeks joint agreement on what the problem may be.

Figure 4. Network diagram: Workspace Planning Process stages (Whelton, 2004)

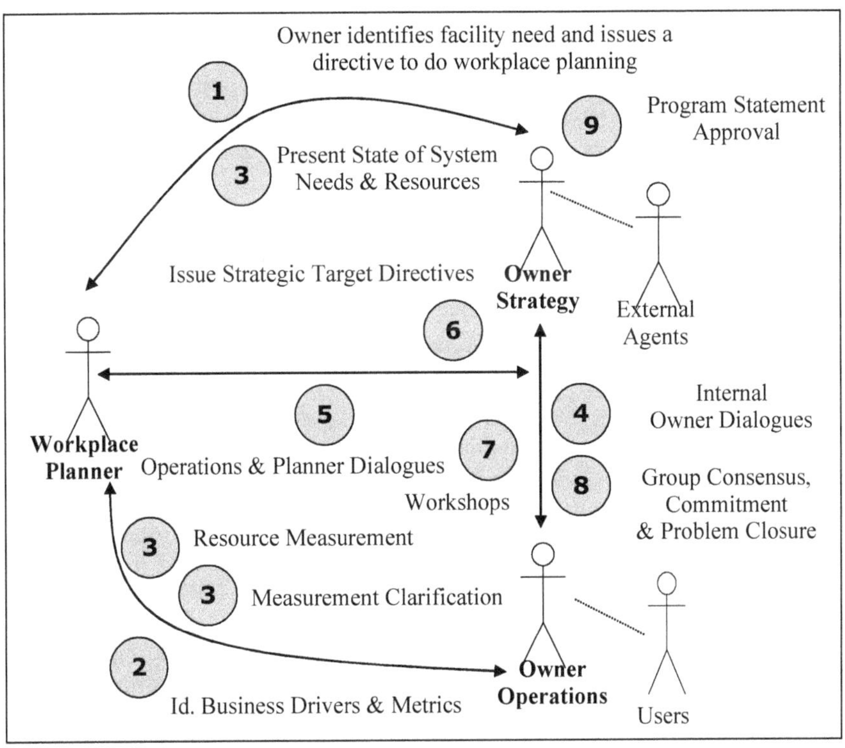

6. Client strategy offers a broad directive at this stage to operations:
   a) The directive may be broad or a specific target.
   b) Decision making to achieve the target may often be pushed to operations at this stage.
7. Operations and programmer workshop:
   a) development of scenario planning to establish targets;
   b) numerous cycles of measurement and planning iterations;
   c) numerous versions of program model are developed.
8. Agreement on problem definition, consensus and closure by group.
9. Strategy approves Program Statement.

Results from a High School Workplace Planning Project

Cygnaeus High School is a case study from Haahtela's current portfolio of programming projects. In this project the traditional architectural programming approach was not effective in consensus making, in that there was stalemate in the project development, primarily because of the project costs being too high. The traditional method was not successfully controlling the project needs within budget constraints.

The Haahtela process developed a set of stakeholder dialogues to establish means of working with these budget constraints. Subsequent changes in client operations and workplace designs were the results of these group dialogues. The case study reveals the learning involved about the state of needs and their ability to be fulfilled within the project constraints. Figure 5 illustrates the multiple visions that are emergent as the process evolves. Management requires a broad view of these visions of the problem as the process evolves. The problem framing may take a narrow focus depending on the dominant problem frame at a specific time period.

Pattern of Problem-Solving Steering

Figure 6 illustrates how the planner steered to the group dialogues towards a negotiated program target. The later targets in Versions 4, 4.1, and 4.2 reveal the results of the dynamics of the group dialogues. The dialogues centered around understanding changes in the workplace performance, and client functions and activities. Through operations design and workplace design, along with measurement of the decision impacts, the group finally converged on a solution that satisfactory to all stakeholders.

Once the workplace planner identifies constraints with the group, this triggers a subsequent search to work with the constraint. The search process typically extends across the boundaries of the organization horizontally, vertically, and external to the organizational boundary. The workplace planner facilitates this group process when discussing means of changing their organizational function and workplace strategy.

Figure 5. Managing shared project purpose (Whelton, 2004)

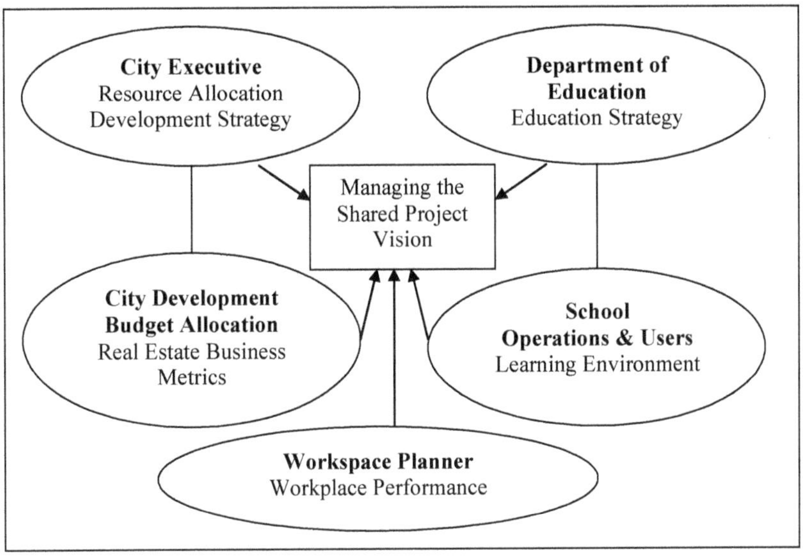

Figure 6. Program space targets and emergent values (Whelton, 2004)

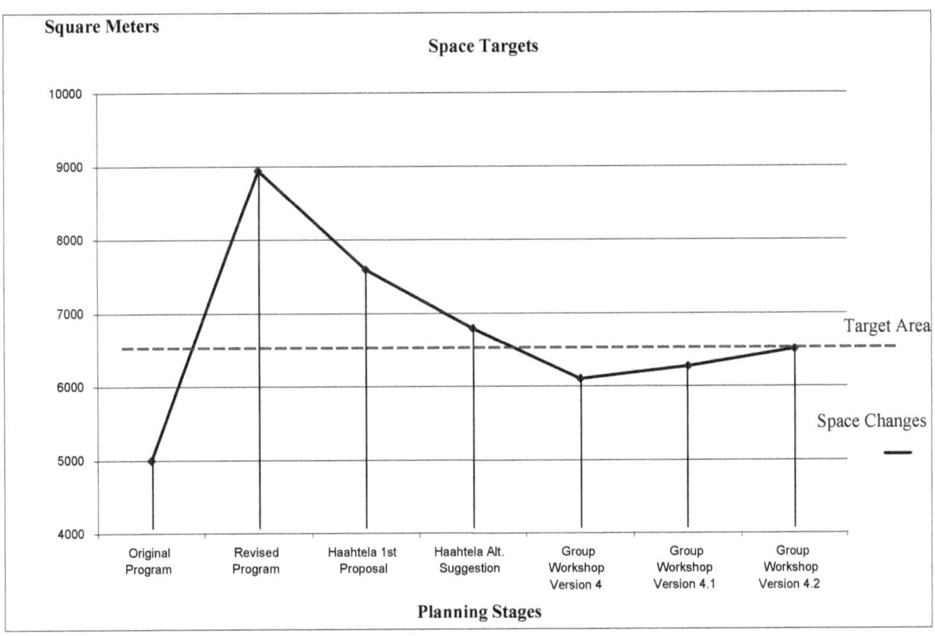

For example, in the Cygnaeus High School study, the workplace planner initiated a number of stakeholder discussions on ways of working within budget constraints. The workplace planner continued a line of inquiry to resolve the space issue at hand. This line of inquiry led to new innovations not considered before, and subsequently allowed valuable space to be reassigned to other purposes. The process often produced new ways of working for the owner groups. The high school case showed positive changes in teaching operations, for example, changes in computer education and in supporting functions, such as a new delivery method for food preparation. The groups realized how they could share spatial resources and decide on the value of their current and future operations and functions.

Conclusions and Future Research

The Haahtela management system reveals a set of capabilities to create collaborative conditions to create purpose statements for project stakeholders. After observing the management system in action on a number of projects, there is evidence of collaboration to generate client and stakeholder purpose. Future research is proposed to understand and develop the system characteristics that support this collaborative capacity. Such characteristics include the ability to:

- manage large amounts of selective information for large-scale and diverse client organizations;
- maintain group forums, information transparency, and common language to support shared understanding;
- facilitate the collaboration of self-organizing groups;
- facilitate judgment-making by diverse stakeholder groups;
- empathically understand user and operational functions;
- act as a collaborative mechanism for direct and indirect group dialogues;
- penetrate organizational boundaries to bring about new collaborations;
- parallel search within the workplace solution space and the client functions space;
- unearth real constraints and deal with them through group problem solving; and
- selectively check feasibility of system-wide constraints prior to design development.

Impact of Research

Projects, in theory, are construed as unique bodies of work that deliver value to the customer. Projects are inherently agents of innovation and change. Current management paradigms lose sight of the required innovative and creative dimensions of project definition, and of the opportunity this phase offers to create customer and stakeholder value. Achieving dynamic capability in this phase requires effective organizational sensemaking, understanding, and strategic and tactical action planning. The research seeks to promote the changing role of project managers, not only with traditional administrative roles, but one with a dynamic role that increases their capability to generate value.

The research contributes to knowledge by understanding the relationship of management action and the community of practice when generating robust program statements for capital facility projects. The analysis provides new insights into management action and the project definition phase. These include:

1. the explicit acknowledgement that the project phase operates in an environment that is dynamically complex;
2. the importance of management steering action and adaptive group learning;
3. the importance of facilitative leadership by the lead problem solver or manager; and
4. the collaborative capability of the project definition management system to consider the constraints and problem frames of diverse stakeholders.

Planning groups require dynamic capabilities and agility to respond with effective problem definitions of client needs. This research contributes towards understanding this group learning capability. At a practical level, it is hoped that the research will provide project managers with insights on how to manage planning groups in the early phase of strategic planning and project development.

From a knowledge management perspective, this research integrates the framework of a knowledge-based architecture with project management theory. It observes the project manager as a facilitator of the "shared problem space" and as a knowledge agent with steering ability. By understanding the context within which groups work collectively, only then can knowledge management systems be effectively created to support dynamic organizations.

Practical Tips and Lessons Learned

Project managers should consider the following when managing project definition meetings:

- Project definition is the act of the defining the problem to be solved.
- Acknowledge that the process of developing project purpose is complex, and changes in project purpose are inherent over the course of project definition.

- Stakeholders may or may not understand the implications of their needs and values.
- Focus on the creative process verses the administrative controlling aspects of project management.
- Ensure process quality through thorough stakeholder identification, sound meeting agendas, and meeting facilitation.
- Consider the impacts when choosing a particular problem-solving methodology.
- Organizational boundaries often limit the creative process. Search internally and externally for solutions to stakeholder needs.

References

Altschuld, J., & Witkin, B. (2000). *From needs assessment to action: Transforming needs into solution strategies.* Thousand Oaks, CA: Sage Publications.

Argyris, C. (1999). *On organizational learning.* Massachusetts: Blackwell Business.

Axelrod, R., & Cohen, M.D. (1999). *Harnessing complexity, organizational implications of a scientific frontier.* New York: The Free Press.

Barrett, P., & Stanley, C. (1999). *Better construction briefing.* UK: Blackwell Science.

Barton, R. (2000). Soft value management methodology for use in project initiation—a learning journey. *Journal of Construction Research, 1,* 109-122.

Brown, J.S., & Duguid, P. (1998). Organizing knowledge. *California Management Review, 40*(3).

Bruce, M., & Cooper, R. (2000). *Creative product design: A practical guide to requirements capture management.* Chichester, UK: John Wiley & Sons.

Carlile, P. (2002). A pragmatic view of knowledge and boundaries: Boundary objects in new product development. *Organization Science, 13*(4).

Checkland, P., & Scholes, J. (1999). *Soft systems methodology in action.* Chichester, UK: John Wiley & Sons.

Cherry, E. (1999). *Programming for design: From theory to practice.* New York: John Wiley & Sons.

CII. (1995). *Pre-project planning handbook.* Special publication 39-2. Construction Industry Institute, University of Texas at Austin: Pre-Project Planning Research Team.

CII. (1999). *PDRI: Project Definition Rating Index for building projects.* Implementation resource 155-2, Construction Industry Institute, University of Texas at Austin: Pre-Project Planning Research Team.

Corning, P. (2002). The re-emergence of "emergence": A venerable concept in search of a theory. *Complexity, 7*(8), 18-30.

Dove, R. (2003). *Knowledge management and agility: Relationships and roles. Handbook of knowledge management.*

Eodice, M. (2000). *A theory of requirements definition in engineering design.* PhD Thesis, Department of Mechanical Engineering, Stanford University, USA.

Goldratt, E.M. (1990). *Theory of constraints.* Croton-On-Hudson, NY: North River Press.

Graham, A., & Kruger, L. (2002). *Research in adaptive management: Working relations and the research process.* Research Paper, PNW-RP-538, United States Department of Agriculture, Forest Service, Pacific Northwest Research Station.

Green, S.D. (1994). Beyond value engineering: SMART value management for building projects. *International Journal of Project Management, 12*(1), 49-56.

Green, S.D., & Simister, S.J. (1999). Modeling client business processes as an aid to strategic briefing. *Construction Management and Economics, 17,* 63-76.

Haahtela, Y., & Kiiras, J. (2003). *Talonrakennuksen kustannustieto.* Helsinki: Haahtela-kehitys Oy (in Finnish only).

Hansen, K.L., & Vanegas, J.A. (2003). Improving design quality through briefing automation. *Building Research & Information, 31*(5), 379-386.

Hooks, I.F., & Farry, A.K. (2001). *Customer-centered products: Creating successful product through smart requirements management.* New York: AMACON.

Horgen, T., Joroff, M., Porter, W., & Schon, D. (1999). *Excellence by design: Transforming workplace and work practice.* New York: John Wiley & Sons.

Innes, J.E., & Booher, D.E. (1999). Consensus building and complex adaptive systems: A framework for evaluating collaborative planning. *Journal of the American Planning Association, 65*(4), 412-422.

Innes, J.E., & Booher, D.E. (2002). The impact of collaborative planning on governance capacity. Paper prepared for presentation at the *Annual Conference of the Association of Collegiate Schools of Planning,* Baltimore, November 21-24.

Kaglioglou, M., Cooper, R., & Aouad, G. (1999). The process protocol: Improving the front end of the design and construction process for the UK industry. *Harmony & Profit, CIB Working Commission W92, Procurement Systems Seminar,* Chiang Mai, Thailand, January 25-28. Retrieved from pp2.dct.salford.ac.uk/pdf/cibw92.pdf.

Kamara, J., Anumba, C., & Evbuomwan, N. (2000). Establishing and processing client requirements—a key aspect of concurrent engineering in construction. *Journal of Engineering, Construction and Architectural Management, 7(1),* 15-28.

Koskela, L. (2000). *An exploration towards a production theory and its application to construction.* Espoo: VTT Publications.

Lambert, S., Poteete, J., & Waltch, A. (2000). *Generating high-performance corporate real estate service.* Boston: MIT Press.

Liu, A., & Leung, M. (2002). Developing a soft value management model. *International Journal of Project Management, 20,* 341-349.

MacMillan, S., Steele, J., Austin, S., Kirby, P., & Spence, R. (2001). Development and verification of a generic framework for conceptual design. *Design studies, 22*(2), 169-191.

March, J.G. (1983). *The technology of foolishness, ambiguity and choice in organizations.* Norway: Universitetsforlaget.

Mayer, R.J., Painter, M.K., & Lingineni, M. (1995). *Information integration for concurrent engineering (IICE) towards a method for business constraint discovery (IDEF9).* Knowledge Based Systems, Inc. Texas: Human Resources Directorate Logistics Research Division. Retrieved from www/idef.com.

Melgrati, A., & Damiani, M. (2002). Rethinking the project management framework: New epistemology, new insights. *Proceedings of PMI Research Conference*, Seattle, Washington.

Murman, E., & Allen, T. (2002). *Lean enterprise value: Insights from MIT's Lean Aerospace Initiative.* New York: Palgrave.

Nutt, B. (1993). The strategic brief. *Facilities, 11*(9), 28-32.

Pena, W.M., & Parshall, S.A. (2001). *Problem seeking, an architectural programming primer* (4th Edition). New York: John Wiley & Sons.

Pennanen, A. (2004). *Workplace planning – User activity-based workspace definition as an instrument for workplace management in multi-user organizations.* Department of Architecture, University of Tampere, Finland. Available at http://www.haahtela.fi

Ruitenbeek, J., & Cartier, C. (2001). *The invisible wand: Adaptive co-management as an emergent strategy. Complex bio-economic systems.* Occasional Paper No. 34. Indonesia: Center for International Forestry Research.

Scharmer, initial. (2000). Self-transcending knowledge: Organizing around emerging realities. In I. Nonaka & D. Teece (Eds.), *Managing industrial knowledge: Creation, transfer and utilization* (pp. 68-90). London: Sage Publications.

Simon, H.A. (1996). *The sciences of the artificial* (3rd Edition). Cambridge, MA: MIT Press.

Smith, J., & Jackson, N. (2000). Strategic needs analysis: Its role in brief development. *Facilities, 18*(13/14).

Sterman, J.D. (2000). *Business dynamics, systems thinking and modeling for a complex world.* Boston, MA: Irwin McGraw-Hill.

Teece, D.J. (2002). Dynamic capabilities. In W. Lazonick (Ed.), *The IEBM handbook of economics.* Australia: Thomson.

Teicholz, E. (2001). *Facility design and management handbook.* New York: McGraw-Hill.

Thomke, S., & Nimgade, A. (2000). IDEO product development. *Harvard Business Review,* Ref No. 600-143. Boston, MA: Harvard Business School Publishing.

Ulrich, K., & Eppinger, S. (2000). *Product design and development* (2nd Edition). New York: McGraw-Hill.

Whelton, M., Ballard, G., & Tommelein, I.D. (2002). A knowledge management framework for project definition. *Electronic Journal of Information Technology in Construction (ITcon), Special Edition on Knowledge Management, 7*(13).

Whelton, M. (2004). *The development of purpose in the project definition phase of construction projects—implications for project management.* PhD Dissertation, Department of Civil and Environmental Engineering, University of California, Berkeley, USA.

Womack, J., & Jones, D. (1996). *Lean thinking: Banish waste and create wealth in your corporation.* New York: Simon & Schuster.

Woodhead, R.M. (1999). *The influence of paradigms and perspectives on the decision to build undertaken by large experienced clients of the UK construction industry.* PhD Dissertation, School of Civil Engineering, University of Leeds.

Chapter XV

Decision Support Systems and their Application in Construction

Alaa Abdou
United Arab Emirates University (UAEU), United Arab Emirates

Moh'd Radaideh
United Arab Emirates University (UAEU), United Arab Emirates

John Lewis
University of Liverpool, UK

Abstract

Decisions are activities that we face and deal with every day. Decision support systems are used to support and improve decision making. They help people make better and faster decisions than they could make themselves. The construction industry witnessed a growth in the application of knowledge-based expert systems in the eighties and early nineties, followed by the application of fuzzy, artificial neural networks and hybrid (integrated) systems. Potential applications of the Internet in the construction industry have generated many research projects recently. The purpose of this chapter is to understand decision support systems and their basic technologies, and to review their application in the construction industry. The construction industry is rapidly realising the need to integrate information technology and artificial intelligence into its processes in order to remain competitive.

Introduction

Decision support systems (DSSs) are designed, built, and used to assist in and support the decision-making process. They are used by managers in direct support of managerial decision making (Keen & Wagner, 1979). "The evolution of DSS from its conception in the 1970s to the present day has seen numerous extensions of the original notion. The modern-day study of DSSs must include a focus on conventional model-based systems, knowledge-based systems, artificial intelligence, expert systems, executive information systems, group support systems, data visualisation systems, and organisational decision support systems" (Marakas, 1999, p. 8). As we begin the 21st century, major changes can be observed in the way managers use computerised tools to support their decision making. Organisations can easily use the Internet and intranets to deliver high-value performance analysis applications to decision makers around the world. Today's decision support technologies can also create an interactive interface that allows users to view and process data models with standard Web browsers with great flexibility, efficiency, and ease (Turban & Aronson, 2001).

For the construction industry, one of the most urgent challenges to its fragmented nature is the need to improve business performance and client satisfaction. It is now recognised that the management of project and organisational knowledge is necessary if construction businesses are to remain competitive and effectively respond to the needs of their clients (Kamara, Anumba, & Carrillo, 2000).

This chapter focuses on decision support systems (DSSs) and their application in construction. It starts with an overview of the DSS concept and its fundamental principles. Following that, it presents a series of sub-sections, each of which focuses on a particular theoretical aspect of decision support technologies, starting with DSS basic components and ending with artificial intelligence (AI) and the Internet's impact on decision support systems. Finally, a number of case studies showing the application of DSSs in the construction industry are presented.

Learning Objectives

1. To understand the definition of decision support systems and their major components and categories.
2. To understand different decision support technologies and their background, components, theories, and limitations.
3. To highlight the relationship between DSSs and organisations and their people.
4. To highlight the impact of the Internet on decision support.
5. To present some examples of the applications of different decision support technologies in the construction industry.
6. To discuss the impact of decision support systems and their future.

Background

Management is the process needed to achieve organisational goals through its resources. It is the dynamic input that makes an organisation work. All managerial activities revolve around decision making at various levels. Turban and Aronson (2001) describe decision making as a process of choosing among alternative courses of action for the purpose of achieving a goal or goals. According to Simon (1977), managerial decision making is synonymous with the whole process of management. For a long time, managers have considered decision making as an innate talent based on (a manager's) creativity, intuition, judgment, and experience that gets better through a process of trial and error. However, with the complex business environment in which management operates today, decision making is becoming more complicated. Turban and Aronson (2001) seek to explain this increased complication. They recognised, firstly, the improved technology and communications and their impact in increasing the number of available alternatives; secondly, the continuous changes in the fluctuating environment within which organisations operate and the uncertainty and risk that arises from these changes; and finally, the high cost of errors due to the complexity and magnitude of operations, automation, and the chain reaction that an error can cause in many parts of the organisation. As a result of these trends and changes, it is dangerous to rely on a trial-and-error approach. They conclude that managers should take advantage of new computerised decision support systems that have been developed in their field. The following paragraphs give an overview of fundamental DSS concepts.

The Need for Decision Support Systems and their Benefits

Organisations, and the people in them, normally use DSSs to improve some aspect of their operations and their decision-making processes. According to Simon (1977), the human mind is limited in its abilities to process and store information. Computerised systems or DSSs enable people to quickly access and process vast amounts of stored information and knowledge. In addition to overcoming cognitive limits in processing and storage, computerised DSSs may be needed for various reasons, including: speedy computations, increased productivity, technical support and quality support for management staff, and finally for giving a competitive advantage over competing organisations (Turban & Aronson, 2001). On the other hand, the complex business environment in which management operates today, along with the increased demand for effective communication and collaboration, enable intranet/Web-based DSSs to play an essential role in facilitating communication and coordination, especially in group decision making. Other benefits of DSSs include promoting learning or training, which may be considered a positive impact, and increasing an organisation's control of its employees, which may provoke a more ambivalent reaction. However, DSSs cannot yet be designed to replicate distinctly human decision-making talents such as creativity, imagination, or intuition (Marakas, 1999).

Definitions

"A decision support system, as its name implies, is used to support decision making and is not intended to replace the decision maker. In fact, a DSS works under the assumption that the decision maker is familiar with the problem to be solved and the data required for its solution" (Juras, 1989, p. 1). In the early 1970s, Morton first articulated the concept of decision support system (Turban & Aronson, 2001). He defined a DSS as an interactive computer-based system that helps decision makers utilise data and models to solve unstructured problems (Gorry & Morton, 1971).

Reviewing the literature, several definitions can be found for decision support systems. The early definitions describe a DSS as a system intended to support managerial decision makers in semi-structured problems. For example, Little (1970) defined a DSS as a model-based set of procedures for processing data and judgments to assist the manager in his decision making. Another definition by Keen and Morton (1978) defined a DSS as a computer-based support system that couples the intellectual resources of individuals with the capabilities of the computer to improve the quality of decisions for managers who deal with semi-structured problems. Adopting a component-based definition, Bonczek, Holsapple, and Whinston (1980) defined a DSS as a computer-based system consisting of three interacting components: a language system, a knowledge system, and a problem processing system. Hicks (1993) defined a DSS as an integrated set of computer tools that allow a decision maker to interact directly with computers, creating information useful in making anticipated semi-structured and unstructured decisions.

In carrying out their own survey of definitions, Turban and Aronson (2001) conclude that the basis for defining DSS has been developed from the perceptions of what a DSS does (such as support decision making), and from ideas about how its objective can be accomplished.

Different researchers, including Mallach (1994), Marakas (1999), and Turban and Aronson (2001) discussed the common characteristics of DSSs, which can be summarised as follows:

- DSSs are used to support decision makers, not to replace them.
- Provide support in semi-structured and structured decision contexts.
- Support all phases of the decision-making process.
- Support various managerial levels from top executives to individual contributors.
- Interactive and user friendly.
- Usually, utilise models for analysing decision-making situations and incorporate a database of some sort.

Categories and Classifications of Decision Support Systems

When reviewing the literature, several ways of classifying DSSs and their applications are found. An early classification was provided by Alter (1980). This classification was based on the degree of action implication of system outputs and generated seven categories of DSSs. Turban and Aronson (2001) later clustered these seven categories into three groups: data-oriented systems, data/model-oriented systems, and model-oriented systems (see Figure 1).

Other proposed classifications include Donovan and Madnick (1977), who divided DSSs into 'institutional' DSSs, which are part of the fabric of an organisation and are used regularly; and 'ad-hoc' DSSs, which are typically developed for one-time use. Hackthorn and Keen (1981) grouped DSSs into three categories—personal, group, and organisational—depending on the level within an organisation at which they offered support.

A recent and comprehensive classification for DSSs was presented by Hollsopple and Whinston (1996). They had a six-category classification for DSSs, as follows:

- *Text-oriented DSS*: Where information is stored in a textual format which is accessed by the user. It allows documents to be electronically created, revised, and distributed. Recently, emerging technologies such as Web-based document imaging, hypertext, and intelligent agent technologies have been incorporated into this type of DSS.

Figure 1. Alter's classification of decision support systems

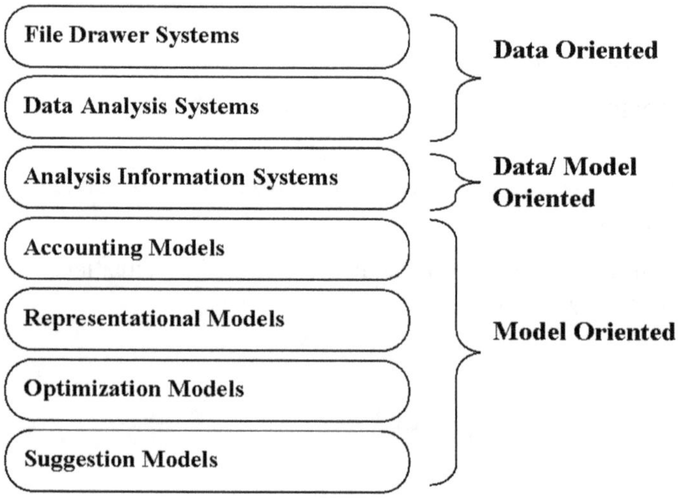

- *Database-oriented DSS*: Where the data organisation plays an important part in the DSS structure. This type of DSS often has strong report generation and query capabilities.
- *Spreadsheet-oriented DSS*: In this type of DSS, the model part component instructs the system to execute its self-contained instructions and allows the user to create, view, and modify procedural knowledge. "The most popular end user tool for developing DSS is Microsoft Excel. Excel includes dozens of statistical packages, a Linear Programming package (Solver), and many financial and management science models" (Turban & Aronson, 2001, p. 113).
- *Solver-oriented DSS*: This type contains an algorithm, procedure, or mathematical model in the form of a computer program, written in a programming language, such as C++ or Java, and intended to solve a particular problem type. These programs may be stand-alone, can be an add-in tool in a spreadsheet, or can be incorporated as problem-specific modules into a more broadly applicable DSS.
- *Rule-oriented DSS*: The knowledge components of this type make use of qualitative or quantitative rule-based reasoning, and are often implemented in the form of an expert system.
- *Compound DSS*: In this type, two or more DSS types previously described are integrated to solve a particular problem. It is also called a hybrid system.

Components of a Decision Support System

The components of a DSS can be divided into five main parts or subsystems as follows:

- The Data Management Subsystem
- The Model Management Subsystem
- The Knowledge-Based Management Subsystem
- The User Interface Subsystem
- The User

These five components are shown in Figure 2 and described below.

The Data Management Subsystem

"The data management component of a DSS is where the various activities associated with retrieval, storage, and organisation of the relevant data for the particular decision context are managed" (Marakas, 1999, p. 10). It is an essential aspect of any DSS. Most decision support features involve access to, or manipulation of, data (Ariav & Ginzberg, 1985). The data management subsystem includes: *a database, database management system, data directory,* and *query facility* (Turban & Aronson, 2001).

"*A database* is an integrated collection of data, organised and stored in a manner that facilitates its easy retrieval" (Marakas, 1999, p. 11). Several databases can be used in one DSS application, depending on the database type and source. An external database could be linked to the DSS internal database and be used in some problem application for extracting data from industry or marketing data for example. In many recent cases, external data are provided over the Internet. A database component is created, accessed, and updated by the "*database management system*" (DBMS). All data in the database with their definitions are sorted into a form of catalogue in the *data directory*, the main function of which is to answer questions regarding the availability and sources of data. A separate *query facility* performs the tasks related to access, manipulation, and querying of data, and may require users to formulate their requests in a special query language.

The Model Management Subsystem

Marakas (1999) stated that "the model management component performs the retrieval, storage, and organisational activities associated with the various quantitative models that provide the analytical capabilities for the DSS" (p. 10). Turban and Aronson (2001) describe it as: "...a software package that includes financial, statistical, management science, or other quantitative models that provide the system's analytical capabilities and appropriate software management" (p. 100). It has the following components: *model base, model base management system, model directory,* and finally *model execution, integration, and command processor*.

Figure 2. Basic components of the DSS

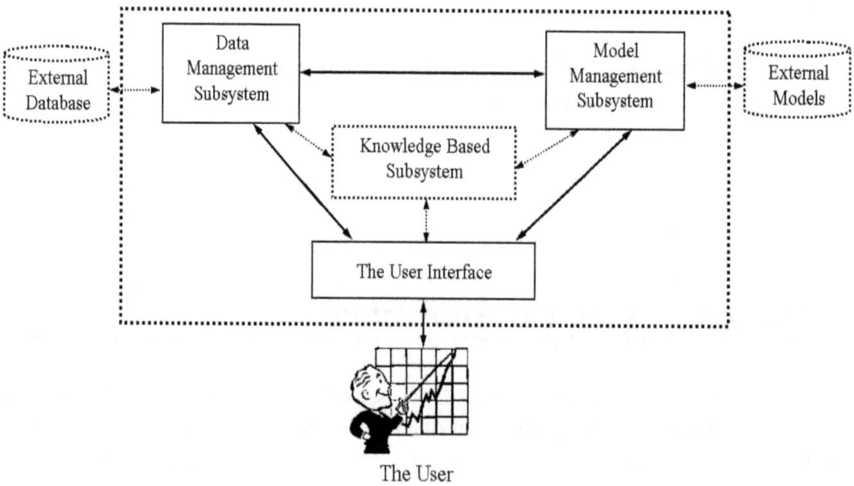

The *model base* contains various statistical, financial, mathematical, and other quantitative models, which provide the DSS with its analysis capabilities. These different models are created by the *model base management system* (MBMS) component using programming tools and languages such as C++ and Java. It is the component that is also responsible for storing and managing all models, generating new routines, updating and changing models, and manipulating model data. Similar to the database directory, the *model directory* is a catalogue of all models with their definitions and functions. A *model execution, integration, and command processor* is usually used to enable the model management subsystem to carry out certain activities such as controlling the actual running of the model, combining the operation of several models, accepting and interpreting modelling instructions from the user interface, and routing them to appropriate functions (Turban & Aronson, 2001).

Knowledge-Based Management Subsystem

The knowledge-based management subsystem is where the 'knowledge' of the DSS is stored. According to Marakas (1999), DSS's knowledge is defined as: "...the rules, heuristics, boundaries, constraints, previous outcomes, and any other knowledge that may have been programmed into the DSS by its designers or acquired by the DSS through repeated use" (p. 17). In situations where unstructured or even semi-structured problems are complex or require expertise for their solutions, the DSS is equipped with a knowledge-based management component in order to provide the required expertise for solving the problem or provide helpful knowledge for the decision-making process. The knowledge component consists of one or more intelligent systems. "A decision support system that includes such a component is called an intelligent DSS, a DSS/ES, an expert support system, active DSS, or knowledge-based DSS" (Turban & Aronson, 2001, p. 107).

User Interface Subsystem

The basic function of the interface is to allow the user to access and interact with the system's components in an efficient manner. In addition to the hardware and software, the user interface subsystem also includes factors that deal with ease of use, accessibility, and human-machine interactions, and it is managed by software called the user interface management system which provides the capabilities listed in the DSS (Turban & Aronson, 2001). According to Hix and Hartson (1993), user interface development includes interaction development and interface software development. Interaction development is concerned with how the user interface works in response to what a user sees, hears, and does during interacting with the system, while the interface software development is the means for implementing the code that initiates the interaction components. The interface is an important element for a successful decision support system. Marakas (1999) states that "in the past, lack of support for many DSS implementation[s] has been linked to a poor user interface that served only to discourage use rather than facilitate it" (p. 21). Recently, Web browsers have been used effectively as a user interface for many DSSs. This is illustrated in the case studies later in this chapter.

The User

The DSS user is considered one of the most important elements in the entire life cycle of a DSS. Besides their interaction with the system in use, users are in many cases involved in conceptual design and construction of the DSS, and in its testing and implementation. According to Alter (1980), terms like manager and decision maker fail to reflect the heterogeneity that exists among users and usage patterns. He defined the user as the person who communicates directly with the decision support system, regardless of method or intention. Furthermore, the decision maker is the person who ultimately makes or takes the decision, based on the outcome of the DSS. In general, managers expect the DSS interface to be 'user friendly', as they have limited time to learn or deal with the DSS. On the other hand, a user who is a technical specialist tends to be more willing to use a more detailed and complex interface.

Decision Support System Technologies

Decision support can be provided by one or more decision support technologies/tools. When reviewing the literature, the term decision support system is sometimes used as an umbrella to describe any/all computerised systems that are used in supporting decisions, such as information systems, expert systems, neural networks, and neurofuzzy expert systems. Turban and Aronson (2001) introduced a broader term for different management support technologies, namely "management support systems" (MSSs). They defined these as "collections of computerised technologies used to support managerial work and decision making" (p. 1). This refers to the application of any technology, either independent from or in combination with other information technologies, that supports management tasks in general and decision making in particular.

Decision support system technologies, which could be viewed as another class of DSSs, include:

- Executive and Enterprise Information Systems
- Group Decision Support Systems and Groupware Technologies
- Knowledge Management Systems
- Intelligent Decision Support Systems (Expert Systems, ANNs, Fuzzy Logic, and Intelligent Agent)
- Internet-Based Management Systems
  - Internet-Based Decision Support Systems
  - Internet-Based Intelligent Decision Support Systems

A brief technical description of these different technologies/tools is presented in the following paragraphs.

Executive and Enterprise Information Systems (EISs)

Executives are the top level of management in any organisation, and they include the chairman, the president, or the chief executive officer (Mallach, 1994). Executive information systems (EISs), also known as "executive support systems" (ESSs), originally evolved from traditional DSSs to serve the needs of top executives in an organisation. They normally feature additional capabilities in terms of simplicity of user interface and speed of processing to suit top executives. Turban and Aronson (2001) describe an EIS as:

> *"... a computer-based system that serves the information needs of top executives. It provides rapid access to timely information and direct access to management reports. EIS is very user friendly, is supported by graphics, and provides exceptions reporting and drilldown capabilities. It is also connected to the Internet, intranets, and extranets." (p. 308)*

EISs in some cases include modelling and/or intelligent features. Recently EISs have evolved to serve many people in the organisation in addition to its executives. Different users across the enterprise can use the system for different purposes. Such systems are called "enterprise information systems." Enterprise systems have an important role in the enterprise resource-planning (ERP) concept, which has received a great deal of attention recently. The main objective of ERP is to integrate all functions and departments across the organisation into a single computer system that serves the needs of the entire enterprise. According to Turban and Aronson (2001), the promised benefits for ERP include: improved quality, increased efficiency, productivity, and profitability for the organisation.

Special classes of enterprise systems are those that deal with the supply chain and its management. The term "supply chain" relates to the procurement of materials and components from their sources or suppliers into organisations where they are used in their operations. Supply chain management tools are available for a variety of decision activities. These include optimisation of manufacturing processes, scheduling, inventory management, and procurement (Turban & Aronson, 2001). The concept of supply chain management has received a great deal of attention in the construction industry in recent years.

Group Decision Support Systems and Groupware Technologies

When people work together as groups making important decisions in an organisation, communication and coordination are essential. Communication is the first step towards effective coordination, and without coordination there will not be collaboration. In many cases, where team members are working in different locations or even in different time zones, the communication and computer technologies used become important factors in successful group works (of an organisation). Teamwork situation is the norm in construction projects. Thus, with the increased demand for effective communication and collaboration, project management has evolved and has become a collaborative process, improving overall team motivation through a deeper involvement by the entire project team. Improving the overall project team has become the main objective of 'collaborative project management' (Abdou, 2000).

A group decision support system (GDSS) is a decision support system whose design reflects the way in which the group decision is taken. Group support systems enable the emerging concept of collaboration management/support. They support communication facilities such as electronic meetings (e-meetings), for example. According to Turban and Aronson (2001), such collaborative computing is known by a number of terms, including groupware, group support systems (GSSs), and computer-supported-cooperative work (CSCW). Furthermore, the term "groupware" refers to software products that provide collaborative support to groups by enabling teams to share opinions, data, information, knowledge, and other resources. Groupware typically contains capabilities such as electronic audio/or video conferencing, electronic document sharing, voting, and diary synchronisation. The Internet, intranets, and extranets are the vehicles and enablers of today's groupware.

Knowledge Management Systems (KMSs)

The concept of knowledge management (KM) has received a great deal of attention recently. The objectives of KM in an organisation are to promote knowledge growth, knowledge communication, and knowledge preservation (Steels, 1993). Effective KM presents many complex problems and can be tackled from several viewpoints: socio-organisational, financial and economic, technical, human, and legal (Barthes, 1996). Knowledge management might also involve a range of activities such as: financing, classifying, storing, and maintaining an organisation's knowledge, and determining how to benefit from it in future projects. According to Turban and Aronson (2001), knowledge management systems (KMSs) facilitate knowledge management in/across an organisation by insuring knowledge flow from the person(s) who know, to person(s) who need to know, while knowledge evolves and grows during the process. KMSs can be implemented using a variety of technologies. Among the most important technologies, Dyer (2000) identified e-mail and messaging, document management, search engines, enterprise information portals, data warehouses, Web-based training, groupware, and workflow management tools.

Intelligent Decision Support Systems

Different definitions can be found in the literature for the term "artificial intelligence" (AI). Rich and Knight (1991) described it as the study of how to make computers do things, which at the moment people do better. Mallach (1994) described AI as an inclusive term for several areas of computing that attempt to mimic processes that humans carry out without much conscious thought. He also reports on Massimo Negrotti, who collected more than 100 definitions of AI during the 8th International Joint Conference for Artificial Intelligence (IJCAI) held in Karlsruhe, Germany, in 1983. He highlights the variation in the definitions, ranging from the technologically oriented to the pure research oriented, and from the intellectual to the commercial.

An "intelligent decision support system" is a type of decision support system that includes one or more artificial intelligence components such as expert systems, intelligent agents, or artificial neural networks, and are thus designed to employ reasoning based on expertise or case-based knowledge in order to solve complex problems.

Under the AI umbrella, "expert systems" (ESs) have evolved. An expert system attempts to capture and mimic experts. It is a computer-based application that employs a set of rules based on human knowledge to solve problems that require human expertise (Marakas, 1999). According to Mallach (1994), the function of an ES is to replicate the reasoning process that a human decision maker would go through in reaching a decision, diagnosing a problem, or suggesting a course of action. An ES is ideal for modelling reasoning in relation to a problem, and reproducing expertise that could be expressed and justified by 'what if' analysis. However, in the case of a complex problem and/or one that is resistant to solution through the 'what if' approach, the use of neural networks could be suitable.

Another type of intelligent DSS that has arisen out of developments in AI technology is the "neural network" or "artificial neural network" (ANN). An ANN attempts to mimic the way the human brain functions. It operates by determining a set of relationships between the values (quantitative or qualitative) of the user-specified input and output variables in a problem. For example, in a building cost estimation problem, the inputs may be floor area, number of floors, or building type, and the outputs could be total cost or a probability distribution of costs. The ANN determines the relationships in a so-called "training" phase, in which the user provides it with a set of problems with known inputs and outputs. The user may then use the ANN to predict the outputs from new cases where only input values are known. An ANN is useful for tackling problems that are not easy to couch in a rule-based form. It also has a number of further benefits, for example, in being relatively tolerant of gaps in the user's data, both in training and when being used for prediction.

The applicability and utility of both expert systems and artificial neural networks may be enhanced by adopting a fuzzy approach, first introduced by Zaden in 1965, to handling data. Fuzzy data has an inherent uncertainty. For example, suppose a decision requires entities to be classified into groups (Is this project "big," "medium," or "small"?), but the boundaries between the groups are ill-defined. Human decision makers often handle this complexity unconsciously, but conventional ES and ANN systems cannot. The application of "Fuzzy Set Theory" and related techniques allow these systems to cope with this type of uncertainty.

Some of the above systems, especially expert systems and artificial neural networks, can be used in the construction of a class of software programs known as "intelligent agents." An intelligent agent (IA) is a piece of software that, once put into operation, carries out its task with some degree of autonomy (Turban & Aronson, 2001). The tasks performed by individual agents may be routine, for example, the search for and/or retrieval and filtering of information on a topic from sources on the Internet. However, agents with different functions can be designed to interact with each other and hence address complex problems "cooperatively.. This technology is relatively new and has received a great deal of attention as a result of the rapid growth of the Internet. According to Turban and Aronson (2001), intelligent agent technology has the potential to become one of the most important tools in the 21st century. They stated that agents "can overcome the most critical limitation of the Internet-information overflow—and make electronic commerce a viable organisational tool" (p. 690). For further discussion of agent systems and examples of their application, see for example Wooldridge (2002).

When two or more types of management support systems are integrated in order to solve a particular problem, this is normally referred to as a "hybrid/integrated systems."

Internet-Based Decision Support Systems

With developments in computer technology and the introduction of the Internet with its collaborative features, communication and collaboration have become very important for successful group decision making and for the development of groupware to support this process. This is especially true where the groups involved are geographically distributed. The widespread availability and use of the Internet, intranets, and extranets provides good opportunities to share expertise and knowledge among mass audiences around the globe. This new infrastructure for communication creates a unique platform for decision support, the Web-based DSS, or Internet-based DSS. Molenaar and Songer (2001) summarise the benefits of this platform as: greater accessibility, more efficient distribution, more effective administration, and a greater degree of flexibility across a user's individual operator platform. By implementing expert systems (and other intelligent systems) as knowledge servers, it becomes economically feasible for the organisation to publish expertise on the Net (Turban & Aronson, 2001). Such a system is called an Internet/Web-based intelligent decision support system.

Decision Support Systems and People in the Organisation

In any organisation, regardless of the level (individual, team, group, or organisation) at which the decision is made, decisions are made by individuals and are often affected by several human factors, including power, politics, and culture, that exist in the organisation. Organisational culture determines the way that people who work in the organisation think, respond, and behave in the decision-making process. According to Marakas (1999), an organisation's culture is a system of shared meaning and values for practices that have evolved over time under the influence of several factors, including beliefs, symbols, rituals, and myths. It often determines the firm's ability to change or adapt to changes in surrounding conditions. All the above factors serve to shape the decision-making context and, as a result, affect the kind of decision support required. In order to have a successful DSS, the problem needs to be fully identified and stated along with the decision-making processes and context in order to be successfully modelled into the desired DSS.

Another issue affecting the people working in an organisation is business process reengineering (BPR). With the level of competitive pressures in the current market and the increased demand for effective communication and collaboration, organisations are seeking innovative ways to conduct business, as these may be necessary for maintaining profitability or even ensuring survival. BPR is normally employed when major IT projects, such as enterprise resource-planning (ERP) or electronic commerce (EC), are undertaken (Turban & Aronson, 2001). It involves changes in structure, organisational culture, and processes (Hammer & Stanton, 1995). In the construction industry, electronic commerce is currently affecting a lot of procurement processes. A number of Internet-based electronic marketplaces that connect architects, engineers, and contractors are already in existence, and this trend is expected to accelerate in the near future.

The Application of Decision Support Systems in Construction: Case Studies of Recent Work

The construction industry witnessed a growth in the application of knowledge-based expert systems in the '80s and early '90s, followed by the application of artificial neural networks and fuzzy systems. Potential applications of the Internet in the construction industry have recently generated many research projects. This new infrastructure for communication creates a unique platform for decision support systems. In this section, samples of recent case studies are presented to demonstrate the application of DSSs in the construction industry, while considering the relationship between the trends in these applications and the trends in emerging information technology.

According to Wanous (2000), the application of knowledge-based/expert systems has affected different areas in the construction process such as planning, contractual disputes, site investigations, equipment selection, monitoring, and risk management. He identified several references to systems developed for use in each of the areas mentioned above and noted that most of these systems were developed during the late '80s and early '90s (see for example the work of Alkass & Harris, 1988; Alshawi & Hope, 1989; Baldwin & Oteifa, 1993; Boussabaine, 1991; Thorpe, Tah, & McCaffer, 1992).

Following this period, there was a wave of application of ANNs and fuzzy systems tackling topics such as planning, duration and cost forecasting, project selection, and contractor pre-qualification (see for example the work of Adeli, 2001; Boussabaine & Kaka, 1998; Boussabaine, Thomas, & Elhag, 1999; Emsley, Lowe, Duff, Harding, & Hickson, 2002; Provenzano, 2003; Tah & Carr, 2000). The last decade has also witnessed a move towards "hybrid (integrated) systems" and "Web-based systems." The following paragraphs give more details of some related examples.

The research work of Salford and Stanford universities provides good examples of hybrid/integrated systems. The University of Salford conducted a research project, titled "Gallicon," during 1999 and 2000. The aim of Gallicon was to demonstrate the benefits of applying an integrated project database for improved project information exchange to the water and housing sectors of the construction industry. Two integrated systems were developed using the OSCON (Open Systems for Construction) integration technology. The first system was a distributed environment designed to improve the conceptual design of wastewater treatment plants, and minimise development costs by integrating the process of design, cost estimation, and project planning. The second system was a "back office" application that created a Web site for customer relationship management in housing projects, by integrating the process of design, cost estimation, and project planning. The project utilised a central database that was accessed by the various applications (CAD, cost estimating, project planning, and client management applications) to share data. The study concluded that the use of a knowledge management system would allow information from past projects to be used more effectively in new projects. This may help address issues of captured knowledge that may be lost when continuity of personnel is lost (Sun, Aouad, Bakis, Birchall, & Swan, 2000).

The Center for Integrated Facility for Engineering (1999) at Stanford, conducted a research project, performed by Sheryl Staub, investigating the integration of design, costing, and scheduling in the construction of a pharmaceutical plant. The goal of the project was to use a shared 3D model throughout the entire design and construction process, and to link the 3D model to a cost-estimating program for automated quantity take-off and to the construction schedule for day-to-day planning of construction work with a 4D model (with time as the fourth dimension). This approach was found to allow the project team to track revisions of the design, estimate, and schedule in a more integrated way than was previously available.

Potential applications of the Internet in the construction industry have produced many recent research projects. The first wave of Web-based systems facilitated document management and collaboration activities among project participants in order to save time, communication cost, and paperwork.

Tam (1999) investigated the potential of the Internet to facilitate information transfer between construction project participants in Hong Kong. He developed a system called the "Total Information Transfer System." It comprised document management and collaboration functions such as data exchange and remote login, white boarding chat, video cam, e-mail system, and a search engine.

In Abdou (2000), more than 20 new, Web-based commercial systems for project and construction management were reviewed. The study concluded that, by providing better document management, collaborative activities, and online business process models, along with e-commerce facilities, Web-based tools have the ability to reduce time and cost for the construction industry.

Alshawi and Ingirige (2003) conducted a research project to address and examine the emerging paradigm of performing project management over the Internet. The research described the pitfalls in current project management practices amidst the challenges they have to face due to increasing project complexities. The authors pointed out that the government, construction industry, and clients are all seeking to bring about change in the construction industry to improve quality, competitiveness, and profitability, and to increase value to clients. The research findings highlight how Web-enabled project management and its associated features are strongly linked to fulfilling the above endeavours. However, the authors concluded that in order to successfully embrace Web-enabled project management, equal consideration should be given to technology, processes, and people.

Moving construction procurement and project control processes into the online Internet environment was the subject of a recent wave of research work. Abudayyeh, Temel, Al-Tabtabai, and Hurley (2001) suggest the use of an Intranet-based project management system, focusing on cost control as a mechanism for improving the quality and timeliness of information. They developed a Web-based cost control system, where data and cost information were centrally stored and processed by a database management system. The system architecture is shown in Figure 3. It is a visual representation of how information is transferred within the system. A password is needed to access the data entry and retrieval section of the Web site. Once in the Web site, one can submit data, which is then stored in the database tables. Queries are then used to develop reports. These reports could be displayed on demand. The authors highlight the potential benefits of using Web technology for project control. They mentioned the instant, automated online reports that could be produced, up-to-the-minute and on-demand, which will lead to improvements in decision-making processes.

Tserng and Lin (2002) propose Internet-based online subcontract bidding and negotiation architecture. The Internet-based DSS, titled "Accelerated Subcontracting and Procuring" (ASAP), aims to support the processes that general contractors use for selecting sub-contractors. The main objective of this system was to speed up the subcontracting process of construction projects by taking advantage of (IT initiatives of) e-commerce and combining it with the use of electronic information exchange standards, that is, eXtensible Markup Language (XML). It intends to take the subcon-

Figure 3. Cost control system architecture (based on Abudayyeh, Temel, Al-Tabtabai, & Hurley, 2001)

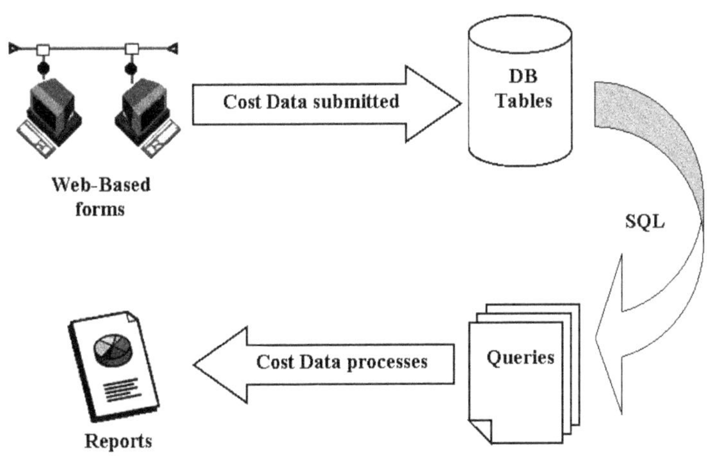

Figure 4. Accelerated subcontracting and procuring system architecture (based on Tserng & Lin, 2002)

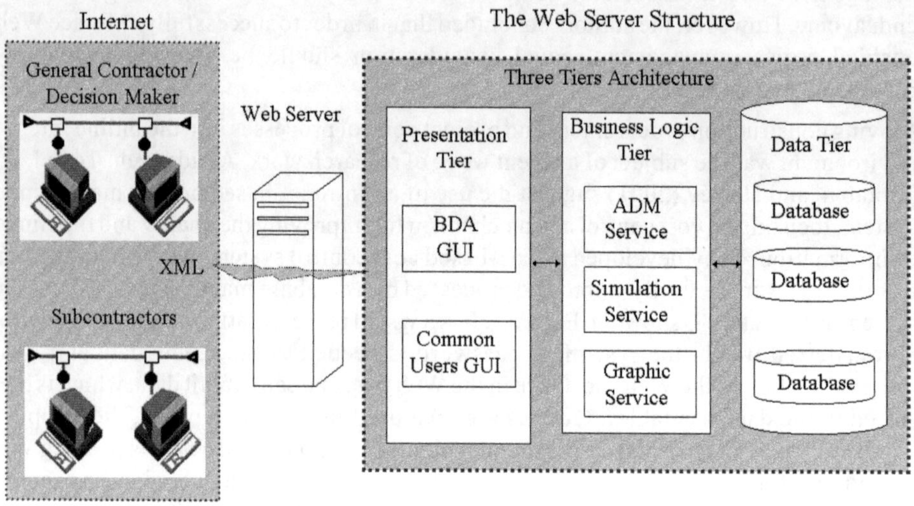

tracting and procurement process into reengineering through the Internet. The ASAP model adopts the three-tier architecture that is widely applied nowadays. The three tiers are the presentation, business logic, and data tiers (see Figure 4 for more details). The system performs scenario simulation so that the general contractors can better appreciate the trade-off between risk and revenue for different subcontracting combinations.

Chau, Cao, Anson, and Zhang (2003) have developed a DSS prototype called "the Construction Management Decision Support System" (CMDSS) for project control and inventory management. A data warehouse was integrated with On-Line Analysis Processing (OLAP) in order to change the data into multidimensional cubes that could be observed from all perspectives. The system enables the construction manager to view data from various perspectives and offers significantly reduced query time. According to the authors, the *data warehouse* technology has not yet been applied in construction management practice. The integration of data warehouse and On-Line Analysis Processing into a DSS seems to have considerable potential. It enables the right data to be tracked down and provides the required information in a direct, rapid, and meaningful way (Chau et al., 2003).

Abdou, Lewis, and Radaideh (2003) have proposed an Internet-based decision support system for the appraisal stage of healthcare construction projects. The underlying philosophy of the approach was to integrate architectural space programming and cost-estimating processes with a risk assessment approach, in order to obtain a more accurate inception cost estimate along with a preliminary space-program outline. The system's main objectives focus on assisting decision makers in examining different space program alternatives with their associated capital budgets. In addition, the system helps to reflect risk factors associated with healthcare projects and affecting budget approval decisions. The Internet is utilised as a mechanism for communicating and updating project data and cost information. The system also proposes a mechanism for using the recurring bidding

Figure 5. Healthcare inception decision support system architecture (based on Abdou, Lewis, & Radaideh, 2003)

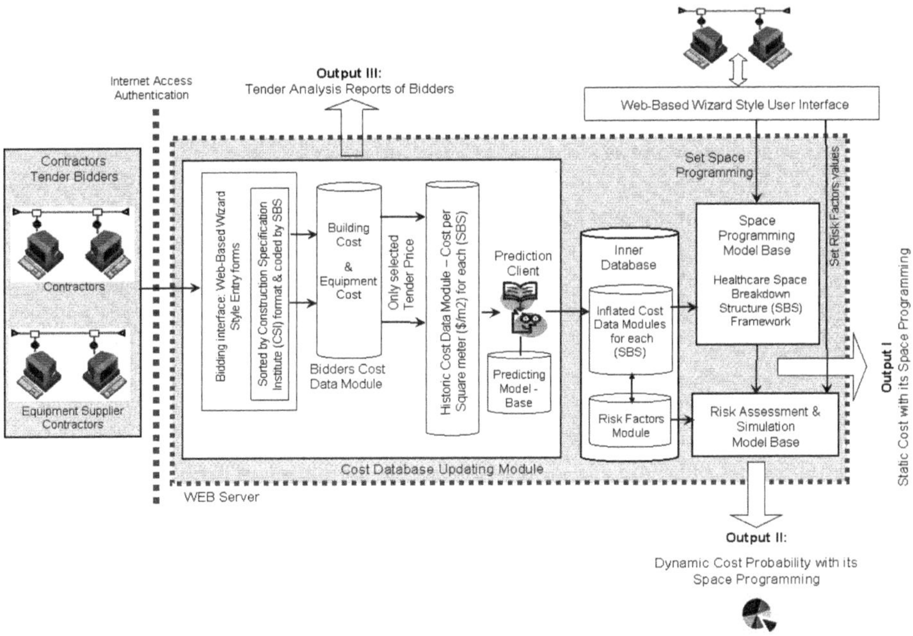

process in the tender stage of healthcare projects as a source of a reliable, in-house cost index for the system. Furthermore, reports can be generated in the bidding stage to assist the clients' decision makers in tender selection. This will also help in reducing paperwork and will speed up the bidding process. The layout of the proposed system architecture and its data flow are illustrated in Figure 5.

Another recent example of an Internet-based project control system has been developed by Moselhi, Li, and Alkass (2004). The integrated time and cost control (IT/CC) system has been developed using Microsoft Access 2000, ASP, HTML, JavaScript, Java Applet, and VB script. It is a three-tier client/sever system, coded using Java Applets and Active Server Pages (ASP). It involves a presentation tier, an application logic tier, and a data tier.

According to the authors, the IT/CC system has a number of interesting features: (1) it maps project work breakdown structure into an object-based model; (2) it generates earned value-based project status reports at project, control object, and resource levels; (3) it uses a set of resource performance indicators to analyse the reported project variances and to detect possible factors behind the deviations from the project baseline; and (4) it takes advantage of the Internet to provide an efficient data-sharing environment, and accordingly provides timely generation and dissemination of site progress reports.

The Implementation of Decision Support Systems and their Future

In the era of information technology in which we are living, rapid changes are taking place in our way of our life and behaviour. Turban and Aronson (2001) described this as an "information and knowledge revolution." Furthermore, decision support systems are important enablers of this revolution and transformation. "Inherent to this rapid transformation is a host of managerial and social problems: impact on organisational structure, resistance to change, rapidly increasing unemployment levels, and so on" (Turban & Aronson, 2001, p. 778). The impact of DSSs can be noticed at different organisational, individual, and societal levels. Huber (1990) states that the use of MSS and other computer-assisted communication technologies leads to significant impacts at different organisational levels. Among the impacts he mentions are an increase in the number of people participating as decision sources and a decrease in the number of people participating in traditional face-to-face communication, more frequent development and use of databases, more rapid and more accurate identification of problems and opportunities, and less time required to authorise actions and make decisions. Other impacts could occur in the degree of centralisation of power and authority. In the case of a distributed MSS, decentralisation may be decreased. The impact of a DSS on people could be positive or negative. According to Turban and Aronson (2001), DSSs could be viewed, according to one opinion, as having the potential to cause massive unemployment due to increased productivity and a reduction in the required skill levels. According to another view, DSSs could increase employment levels, because automation makes products and services more affordable, and as a result, demand increases. On a social level, the impact could be noticed in the emergence of electronic/virtual communities.

The future of DSSs can be seen as laying in the direction of more integration with artificial intelligence and the Internet technologies. DSSs will be available for dissemination, or rented via the Internet for a reasonable fee to large numbers of users. DSSs that facilitate electronic commerce are expected to evolve to provide more and better applications. The construction industry, with its huge materials marketplace, is expected to witness more application of DSSs in the area of electronic commerce, supply chain, and inventory management, with more intelligent agents' applications. It is also expected that the DSS will be integrated with enterprise resource-planning (ERP) to provide better supply chain management for construction projects.

Barriers to the take-up of DSS solutions in the construction industry are a result of several factors, including lack of skills, the limited IT/network/Internet infrastructure in the organisation, cost issues, culture issues, and to some extent, the rigid processes used in the industry that may be difficult to model in many cases. To overcome these barriers, several factors need to be taken into consideration. The IT/network/Internet infrastructure of construction organisations must be reliable. The culture of an organisation should be capable of adapting from conventional habitual ways of thinking and working, to those required when using a DSS and computerised processes. Prior to considering a DSS, a team composed of technical and executive management from the organisation should be formed to review the processes. Any process that cannot cope or is difficult to model must be redesigned or even replaced by another, more appropriate process. In such a context, business process reengineering may need to be considered. Finally, construction organisations are advised to provide more training for their staff.

Conclusion

The concept of knowledge management (KM) has received a great deal of attention in recent years. The construction industry is rapidly realising the potential benefits of information technology to keep up with the future growth of project knowledge and technologies.

The implementation of decision support systems or information systems alone will not lead to effective knowledge management of construction projects. Organisations and their decision-making processes, along with people issues, need to be taken into account. Business process reengineering needs to be considered, especially in the project procurement and control areas. This could enable the industry to benefit from the new revolution of information technology to reduce cost and increase value.

The evolving Web technologies, with their potential for communication and information sharing, provide a unique platform for DSSs. The future will witness more integration between construction processes and the Internet technologies, as well as intelligent agents.

Practical Tips and Lessons Learned

- A DSS system is field specific. It is developed and used to support the process of decision making in a specific field such as enterprise business, healthcare, or construction.
- Processes must be carefully reviewed and evaluated prior to considering a DSS.
- The IT, network, and Internet infrastructure of the organisation are essential factors for successful implementation of a DSS.
- DSSs support decision makers, but do not replace them. Talents such as creativity, imagination, or intuition shouldn't be ignored in the course of making a decision.

References

Abdou, A. (2000). Collaborative management tools in the information technology era. *Proceedings of the 7th Annual Conference of the Project Management Institute, Project Management and IT* (pp. F1-F16), Bahrain, February.

Abdou, A., Lewis, J., & Radaideh, M. (2003). An Internet-based decision support system for healthcare project appraisal: A conceptual proposal. *Construction Innovation, 3*(3), 145-155.

Abudayyeh, O., Temel, B., Al-Tabtabai, H., & Hurley, B. (2001). An Intranet-based cost control system. *Advances in Engineering Software, 32*(2), 87-94.

Adeli, H. (2001). Neural networks in civil engineering: 1989–2000. *Computer-Aided Civil and Infrastructure Engineering, 16*(2), 126-142.

Alkass, S., & Harris, F. (1988). Expert system for earthmoving equipment selection in road construction. *Construction Engineering and Management, 114*(3), 425-440.

Alshawi, M., & Hope, A. (1989). Expert systems and contractual disputes: Extensions of time under JCT 80. *Construction Management and Economics, 7*(1), 65-74.

Alshawi, M., & Ingirige, B. (2003). Web-enabled project management: An emerging paradigm in construction. *Automation in Construction, 12*(4), 381-393.

Alter, S.L. (1980). *Decision support systems: Current practices and continuing challenges.* Reading, MA: Addison-Wesley.

Ariav, G., & Ginzberg, M. J. (1985). DSS design: A systematic view of decision support. *Communications of the ACM, 28*(10), 1045-1052.

Baldwin, A.N., & Oteifa, S.A. (1993). Development of knowledge-based systems for contractors' cost estimators: A role for protocol analysis. *Computer Systems in Engineering, 4*(2-3), 193-200.

Barthes, J.P. (1996). ISMICK and knowledge management. *Proceedings of ISMICK'96, knowledge management: Organisation, competence and methodology* (pp. 9-13), The Netherlands, October.

Bonczek, R.H., Holsapple, C.W., & Whinston, A.B. (1980). The evolving roles of models in decision support systems. *Decision Sciences, 11*(2), 339-356.

Boussabaine, A.H. (1991). *An expert system prototype for construction planning and productivity analysis.* Unpublished PhD, University of Manchester, Institute of Science and Technology, Manchester, UK.

Boussabaine, A.H., & Kaka, A.P. (1998). A neural networks approach for cost flow forecasting. *Construction Management and Economics, 16*(4), 471-479.

Boussabaine, A.H., Thomas, R., & Elhag, T.M.S. (1999). Modelling cost flow forecasting for water pipeline projects using neural networks. *Engineering, Construction and Architectural Management, 6*(3), 213-224.

Center for Integrated Facility for Engineering, Stanford University. (1999). *Design-cost-schedule integration for the Sequus Pharmaceuticals' pilot plant.* Retrieved January 2004 from www.stanford.edu/group/4D/projects/projects.shtml.

Chau, K.W., Cao, Y., Anson, M., & Zhang, J. (2003). Application of data warehouse and decision support system in construction management. *Automation in Construction, 12*(2), 213-224.

Donovan, J.J., & Madnick, S.E. (1977). Institutional and ad hoc DSS and their effective use. *DATA BASE, 8*(3), 79-88.

Dyer, G. (2000). *KM Crosses The Chasm, Knowledge Management Magazine.* Retrieved January 2004 from www.kmmag.com/articles/default.asp?ArticleID=170.

Emsley, M.W., Lowe, D.J., Duff, A.R., Harding, A., & Hickson, A. (2002). Data modelling and the application of a neural network approach to the prediction of total construction costs. *Construction Management and Economics, 20*(6), 465-472.

Gorry, G.A., & Morton, M.S.S. (1971). A framework for management information systems. *Sloan Management Review, 13*(1), 55-70.

Hackathorn, R.D., & Keen, P.G. (1981). Organisational strategies for personal computing in decision support systems. *MIS Quarterly, 5*(3), 21-27.

Hicks, J.O. (1993). *Management information systems: A user perspective.* West Information Publishing Group.

Hix, D., & Hartson, H.R. (1993). *Developing user interfaces: Ensuring usability through product and process.* New York: John Wiley & Sons.

Holsapple, C.W., & Whinston, A.B. (1996). *Decision support systems: A knowledge-based approach.* Minneapolis, MN: West Publishing Company.

Huber, G.P.A. (1990). A theory of the effects of advanced information technologies on organisational design, intelligence and decision making. *Academy of Management Review, 15*(1), 47-71.

Juras, P.E. (1989). The next generation of decision support (decision support computer systems). *CPA Journal.* Retrieved January 2003 from www.nysscpa.org/cpajournal/old/07505724.htm.

Kamara, J.M., Anumba, C.J., & Carrillo, P.M. (2000). Integration of knowledge management within construction business processes. *Proceedings of the UK National Conference on Objects & Integration for Architecture, Engineering & Construction* (pp. 95-105), March.

Keen, P.G., & Morton, M.S.S. (1978). *Decision support systems: An organisational perspective.* Reading MA, Addison-Wesley.

Keen, P.G., & Wagner, G.R. (1979). DSS: An executive mind support system. *Datamation, 25*(12), 117-122.

Little, J.D.C. (1970). Models and managers: The concept of a decision calculus. *Management Science, 16*(8), B466-B485.

Mallach, E.G. (1994). *Understanding decision support systems and expert systems.* McGraw-Hill.

Marakas, G.M. (1999). *Decision support systems in the twenty-first century* (International edition). Englewood Cliffs, NJ: Prentice-Hall.

Molenaar, K.R., & Songer, A.D. (2001). Web-based decision support systems: A case study in project delivery. *Journal of Computing in Civil Engineering, 15*(4), 259-267.

Moselhi, O., Li, J., & Alkass, S. (2004). Web-based integrated project control system. *Construction Management & Economics, 22*(1), 35-46.

Provenzano, initial. (2003). A fuzzy-neural network method for modeling uncertainties in soil-structure interaction problems. *Computer-Aided Civil and Infrastructure Engineering, 18*(6), 391-411.

Rich, E., & Knight, K. (1991). *Artificial intelligence* (2nd edition). New York: McGraw-Hill.

Simon, H.A. (1977). *The new science of management decision*. Englewood Cliffs, NJ: Prentice-Hall.

Steels, L. (1993). Corporate knowledge management. *Proceedings of ISMICK '93* (pp. 9-30), Compiegne, France.

Sun, M., Aouad, G., Bakis, N., Birchall, S., & Swan, W. (2000). GALLICON—a prototype for the design of water treatment plants using an integrated project database. *International Journal of Computer-Integrated Design and Construction, 2*(3), 157-165.

Tah, J.H.M., & Carr, V. (2000). A proposal for construction project risk assessment using fuzzy logic. *Construction Management and Economics, 18*(4), 491-500.

Tam, C.M. (1999). Use of the Internet to enhance construction communication: Total Information Transfer System. *International Journal of Project Management, 17*(2), 107-111.

Thorpe, A., Tah, J.M.H., & McCaffer, R. (1992). A knowledge-based decision support system for drainage cost estimating. *International Journal of Construction Information Technology, 1*(1), 11-28.

Tserng, H.P., & Lin, P.H. (2002). An accelerated subcontracting and procuring model for construction projects. *Automation in Construction, 11*(1), 105-125.

Turban, E., & Aronson, J.E. (2001). *Decision support systems and intelligent systems* (6th edition). Englewood Cliffs, NJ: Prentice-Hall.

Wanous, M. (2000). *A neurofuzzy expert system for competitive tendering in civil engineering*. Unpublished PhD, Liverpool University, UK.

Wooldridge, M. (2002). *An introduction to multi-agent systems*. Chichester, England: John Wiley & Sons.

Chapter XVI

A Knowledge Management Portal System for Construction Projects Using Knowledge Map

H. Ping (Jack) Tserng
National Taiwan University, Taiwan

Yu-Cheng (Vincent) Lin
National Taiwan University, Taiwan

Abstract

Knowledge management (KM) has become an important term in the construction industry. Knowledge management involves creating, securing, capturing, coordinating, combining, retrieving, and distributing knowledge. Most know-what, know-how, and experience exists only in the minds of individual participants during the construction phase of construction projects. The knowledge can be reused and shared among the involved engineers and experts to improve the construction process and reduce the time and cost of solving problems. Sharing and reusing knowledge depends on acquiring and preserving of both tacit knowledge and explicit knowledge as the property of a corporation. This study addresses application of knowledge management in the construction phase of construction projects and proposes a construction activity-based knowledge management (ABKM) concept and system for general contractors. This chapter proposes a practical methodology to capture and represent construction project knowledge by using knowledge maps. Using knowledge maps, users can get an overview of available and missing knowledge in core project areas and take appropriate management in tacit and explicit knowledge. Also, a Web-based system is developed to assist and present project-related knowledge by providing keyword and map search on the Internet environment. The ABKM system is then applied in a case study of a construction precast building project in Taiwan to verify our proposed methodology and demonstrate the effectiveness of sharing knowledge special in the construction phase. By effectively using information and Web technologies during the construction phase of a project, knowledge can be captured and managed to benefit future projects. The combined results demonstrate that an ABKM-like system can be an effective tool for all experts and engineers participating in construction projects by utilising the knowledge management concept and Web technology.

Introduction

Knowledge is the true asset of a marketing-oriented organisation, and its integration across departments and disciplines should be emphasised (Carneiro, 2001). Many organisations are now engaged in knowledge management (KM) efforts in order to leverage knowledge both within their organisations, and externally to their stakeholders and customers (Malhotra, 2000, 2001). These assets, or knowledge, can be classified as either tacit or explicit. Explicit knowledge is that which has been codified and expressed in formal language; it can be represented, stored, shared, and effectively applied (Nonaka & Takeuchi, 1995). Tacit knowledge is knowledge that is difficult to express, represent, and communicate (Nonaka & Takeuchi, 1995). The distinction between these two types of knowledge is relevant because each must be managed differently. Knowledge management in the construction phase mainly deals with the process of creating value from construction operation, organisation, to company knowledge. Valuable knowledge can be available in different forms and media, such as in the minds of experts, in operation procedures, and in documents, databases, intranets, and so forth; however, knowledge management in the construction phase of projects aims at effectively and systematically collecting and sharing the experience and knowledge of the project by Web-based and intranet technologies.

The reuse of information and knowledge minimises the need to refer explicitly to past projects, reduces the time and cost of solving problems, and improves the quality of solutions during the construction phase of a construction project. If experience and knowledge are shared, then the same problems in construction projects do not need to be repeatedly solved. Reduced problem solving has the following advantages: (1) the cost of problem solving is reduced; and (2) the probability of repeated problems is decreased. Several enabling activities should be considered to help to achieve the ultimate goal of efficient experience and knowledge reuse; experience and knowledge should be preserved and managed—that is, they should be captured, modelled, stored, retrieved, adapted, evaluated, and maintained (Bergmann, 2002).

Knowledge reuse and update improves the performance of future activities and projects. Most of the data and information for construction projects are stored in paper-based documents; these consist of contracts, specifications, notes, discussion, and field reports. In order to facilitate the information management and enable knowledge reuse, it is important to convert the paper document into electronic versions to be shared and applied in other and future projects. Information and knowledge of a project can then be identified as project components in the project management and preserved in a Web-based system that provides the platform for the exchange and storage of information and knowledge.

Learning Objectives

1. Understand the application of knowledge management for construction projects.
2. Understand each phase of knowledge management specific for construction projects.
3. Form a case study for knowledge management using knowledge map in the construction phase.
4. Study procedures of implementing knowledge management for construction projects.
5. Study procedures of applying knowledge mapping in knowledge management.

Problem Statement

Construction projects are complex and time consuming. Construction projects are characterised by their complexity, diversity, and the non-standard nature of the production (Clough, Sears, & Sears, 2000). Professional competency in project management is attained by combining knowledge acquired during training and skills developed through experience, as well as the application of the acquired knowledge (Edum-Fotwe & McCaffer, 2000). Whatever successful and unsuccessful projects have been executed by general contractors, a valuable record of each one should be kept to identify best and worst company practices. During the construction phase of projects, an effective means of improving construction management is to share experiences among engineers, which helps to prevent mistakes that have already been encountered in past projects. Drawing on experience avoids the need to solve problems from scratch: Problems that have already been solved do not need to be solved again. According to a survey conducted for this study, most engineers and experts agree that KM is necessary and expect that knowledge management may benefit a construction project. However, no suitable platforms exist to assist senior engineers or experts in sharing and collecting their know-how and experiences when general contractors execute a project.

This situation represents a major loss for general contractors who do not preserve the know-how and experiences of senior engineers and experts. When these engineers and experts complete projects or leave the company, they normally take domain knowledge with them and leave little or nothing that will benefit subsequent projects or the company. From the perspective of knowledge management, this know-how and these experiences of construction engineers and experts are the most valuable, because their accumulation depends not only on manpower, but also on the spending of much money and time.

Figure 1. Knowledge management applied in construction projects

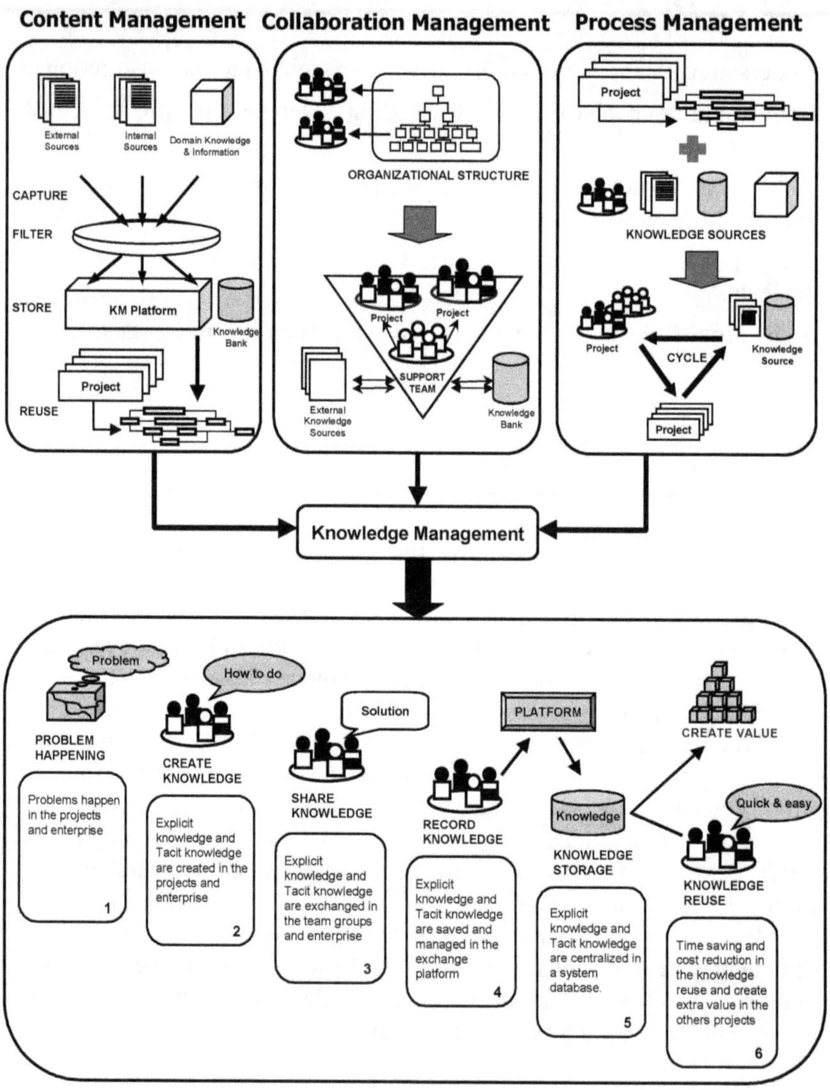

Research Objectives

The main purpose for conducting this research is to develop an activity-based knowledge management (ABKM) system for construction projects, to provide knowledge exchange and management service in the construction phase of a project for the reuse of domain knowledge and experience in future and other related projects. Usually, the management of knowledge in a general contractor revolves around projects. Thus the capture, transfer, reuse, and maintenance of the construction project knowledge are critical (Kamara, Augenbroe, Anumba, & Carrillo, 2002). Knowledge management in the construction phase of a project is a knowledge-intensive organisational environment where knowledge creation has critical importance for general contractors. To be competitive, a general contractor has to make innovative use of knowledge created and accumulated through project activities, and share it across other relative projects. Engineers and experts participating in projects act as knowledge workers, facilitating the collection and management of knowledge between current and past projects. In order to apply knowledge management to construction projects, the process and content of working construction knowledge need to be modified because of construction project characteristics. In order to improve the knowledge capture function, a knowledge map as a tool allows the user to quickly note key concepts, identify important processes and tools, and gain insights into associated behaviours. Figure 1 illustrates the main concepts of knowledge management applied in construction projects. With appropriate modification, the ABKM system can be utilised in construction companies to support knowledge management functionality for any construction project.

Knowledge Management in Construction Phase

In construction projects, knowledge management is a discipline that promotes an integrated approach to the creation, capture, access, and use of a profession's domain knowledge on products, services, and processes. During the construction phase of a project, most project-related problems, solutions, experiences, and know-how are in the heads of individual engineers and experts. Implicit knowledge usually is not documented or stored in a system database. To reuse the knowledge in other projects and also preserve it as corporation property, how to capture the implicit knowledge and make it become available as explicit knowledge is important in the execution of knowledge management in the construction phase. Experience, problem solving, know-how, know-what, and innovation are created in the construction phase of any project. By practicing knowledge management, tacit knowledge can be reused for other projects and speed the improvement of operations in the construction phase.

Figure 2. Five phases of construction knowledge management

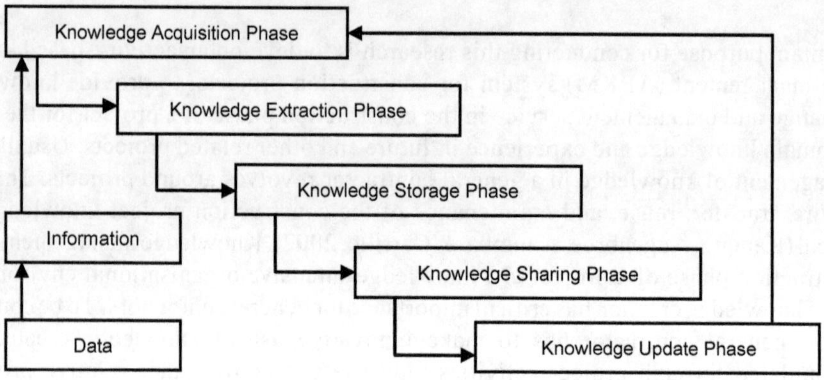

Life Cycle of Constuction Knowledge Management

There are five phases in the construction knowledge management life cycle as shown in Figure 2. They are knowledge acquisition, knowledge extraction, knowledge storage, knowledge sharing, and knowledge update. Each phase is briefly outlined in the following descriptions.

Knowledge Acquisition Phase

The knowledge to be shared must be acquired. In the construction phase, most information and knowledge is acquired on the job site. Hence, the collection of knowledge on a job site plays an important role in the knowledge acquisition phase. Most work of knowledge acquisition is done in the office, because all information and tacit knowledge sent back from the job site can be transferred to explicit knowledge.

Knowledge Extraction Phase

In this study, we define knowledge extraction as a process by which the expert's thoughts and experience are captured. Some knowledge that must be extracted for reuse and storage may only be available from the memories of experts and engineers. In a broader view, knowledge extraction may also include capturing knowledge from other sources such as problem-solution descriptions, construction operation process digital record, virtual communication, and collaboration. However, knowledge is usually captured when the knowledge worker interviews the experts or engineers who answer the questions. In the system, all the descriptions of experience and know-how knowledge are necessary to be edited directly by experts or engineers using unstructured text ways because the knowledge is not necessary to convert into rules or a decision tree. Furthermore, knowledge work will assist with experts and engineers, and deal with the digital process recording work if they are important or valuable enough to be kept as the company asset.

Knowledge Storage Phase

The collected knowledge can be stored for future use. During knowledge storage, all information and knowledge are centralised and stored in the knowledge bank (central database) to prevent the collection of redundant information and knowledge. All information and knowledge can be stored in the system by ensuring that data are all electronic and in standard format for each type of file, such as a specific document format or a drawing format.

Knowledge Sharing Phase

Knowledge sharing is the ultimate goal of knowledge management. After the development of knowledge management, people who need to apply knowledge on a particular project can access relevant knowledge for reuse. If necessary, they can adapt it to a new project and solve the new problem.

Knowledge Update Phase

Available knowledge and experience should be continually updated. Reused experience can be evaluated in the context of a new problem to be solved. It can be evaluated in terms of the appropriateness of the selected experience, or in terms of the accuracy of the retrieved experience. Such evaluation is important to ensure the continued improvement of the process by which experience is reused. Invalid knowledge must be identified and be removed or updated. The updating of knowledge can be triggered by a negative experience evaluation or can be conducted as a precaution.

Knowledge Content in Construction Projects

Most knowledge content in the construction phase of a project can be classified into two broad categories—tacit knowledge and explicit knowledge. Tacit knowledge is personal, context-specific knowledge that is difficult to formalise, record, or articulate; it is stored in the heads of people (Hart, 1992). The tacit component is primarily developed through a process of trial and error in practice. Tacit knowledge is personal knowledge embedded in individual experience, and shared and exchanged through direct, face-to-face contact (Tiwana, 2000). Tacit knowledge can be communicated directly and effectively. In contrast, the acquisition of explicit knowledge is indirect: It must be decoded and re-coded into one's mental models, and is then internalised as tacit knowledge. Explicit knowledge can be codified and transmitted in a systematic and formal language. Explicit knowledge is formal knowledge that can be packaged as information. It can be found in the documents of organisations, including reports, articles, manuals, patents, pictures, images, video, audio, software, and other forms. It can also be found in organisational documents, such as organisational charts, process maps, mission statements, domains of experience, and others.

Explicit knowledge is easier to collect and manage during the construction phase of a project because the information and knowledge are available in document form. Document management can preserve and manage information and knowledge without extracting explicit knowledge. Therefore, explicit knowledge is easier than tacit knowledge to manage. However, the main problem with tacit knowledge concerns its effective extraction into structured information and unstructured information. After tacit knowledge is extracted, the structured information and unstructured information can be maintained and managed with the assistance of document management tools.

Concepts of Activity-Based Knowledge Management

Similar to project scheduling management, knowledge management is based on the concept of undertaking activities of project planning and control. Figure 7 presents an overview and conceptual framework of activity-based knowledge management used in construction projects. Knowledge and information associated with activities in previous projects may be reused and applied in future projects. Information and domain knowledge from all projects are divided and saved as "activity" units in categories related to the projects for collection and management. The main advantage of activity-based knowledge management is the ease with which the information and knowledge can be understood and reapplied.

When knowledge is saved in project units, the knowledge includes both tacit and explicit knowledge. In terms of explicit knowledge, project-related information or knowledge usually include specification/contract, reports, drawing, change order, and data. Actually, each project does not contain one-to-one information or knowledge because some of them belong to activity-based information. In contrast, tacit knowledge may include process records, problems faced, problems solved, expert suggestions, know-how, innovation, and experience notes. The information and knowledge is better saved as activity-based units because the result makes it easier to be classified and searched by

Figure 3. Activity-based knowledge management

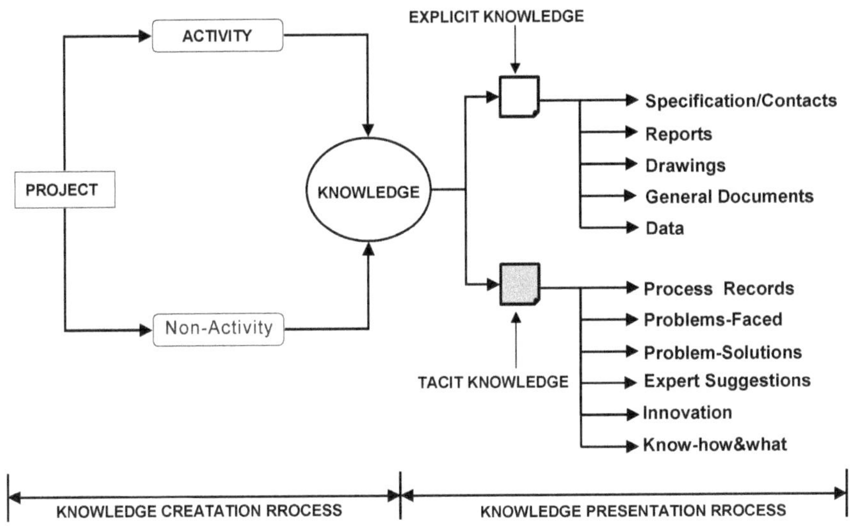

users. In addition, users may search and refer to related information and knowledge from related activities in past projects. The tacit and explicit knowledge of activity-based knowledge management is the same as the duration and relationship of activity-based project management.

The relationship of current and past activities is important for users to link related information and knowledge together. According to Figure 3, not only the information and knowledge of the current project can be applied, but also same or similar activities of past projects can be referred to as experiences are recorded. When experts or engineers enter related information and knowledge into the system, they need to add the relationships for the activities in the project. Of course, the system is designed to link same or similar activities together based on high similarity automatically or manually.

Some of the information and knowledge belongs to the whole project without clear classification of project units. To let users access the information by the Web-based portal, the basic electronic information concerning the project (i.e., specifications, contracts, reports, and drawings) may be saved as explicit knowledge. In terms of tacit knowledge, advanced electronic information and knowledge for the project (i.e., meeting records and e-courses) are saved in the system. The information and knowledge will be saved under the catalogue of current projects. The benefit of non-active classification is the collection of related electronic information in the centralised system.

Knowledge Bank

To enrich the knowledge bank in the system, the system is designed to encourage all engineers and experts to submit their domain knowledge and valuable experience to the knowledge bank. Traditionally, companies included mostly numeric, structured data in their data warehouse. From this point of view, decision support systems are divided into two components: data warehouses deal with structured data; knowledge management involves unstructured data (Ponniah, 2002). It is a need to integrate both structured (such as data and text) and unstructured information (such as image, video, audio, image, and drawing) in the knowledge bank for further reference and decision making. The main purpose of a knowledge bank is to provide the rich source of content concerning all projects and gather project-related explicit information and tacit knowledge together for involved engineers and experts.

Knowledge Map

A knowledge map is used to categorise the content within a particular discipline area. A knowledge map (Wexler, 2001) is a consciously designed communication medium using graphical presentation of text, models, numbers, or symbols between makers and users. Knowledge mapping helps one understand the relationships between knowledge stores and dynamics. Davenport and Prusak (1998) note that developing a knowledge map involves locating important knowledge in the organisation and then publishing some sort of list or picture that shows where to find it. The knowledge map plays important roles in implementing knowledge management. All captured knowledge can be summarised and abstracted through the knowledge map. The knowledge map also gives a useful blueprint for implementing a knowledge management system.

The knowledge map in the system mainly deals with the assistance to find the needed knowledge easily and effectively. Furthermore, applying a knowledge mapping technique has the following two advantages. First, the knowledge map is represented in a simple, clear, visual presentation in the knowledge management system. Second, the mapping methodology helps users to identify key critical and available knowledge areas to the project.

A knowledge map can be defined as a diagrammatic and graphic presentation of knowledge linking the relationships between knowledge and knowledge attribute. As shown in Figure 4, a knowledge map consists of two components:

1. Knowledge diagram: Graphical representation of knowledge, having node, sub-node, and linkage:

Figure 4. A concept map of the activity-based knowledge management

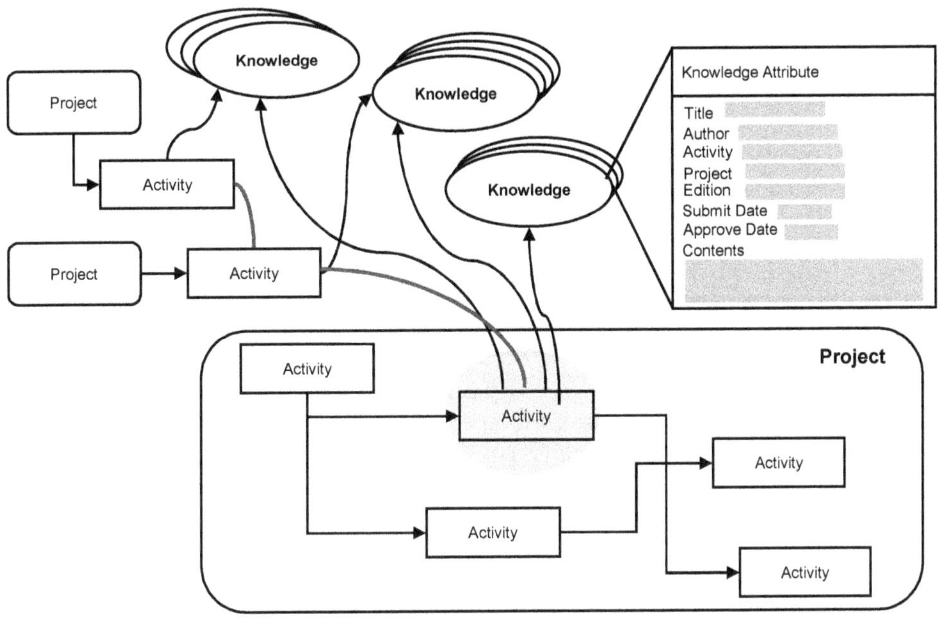

- *Node*: Rectangular object (denoting project or activity representation)
- *Sub-node*: Ellipse object (denoting captured knowledge)
- *Linkage*: Arrow between nodes implying relationships among knowledge

2. Knowledge attribute: Descriptive representation of knowledge feature

A knowledge map approach is employed in this chapter to represent explicit and tacit knowledge location within a construction project. A knowledge map in the system provides users with a robust foundation to capture, share, and use project-related or activity-related knowledge. Advantages gained from building the knowledge map in the system can be summarised as follows:

- efficient capture of knowledge;
- understanding of relationships between knowledge; and
- formalisation of all knowledge inventories within a construction project.

Procedures for Building the Knowledge Map

According to the knowledge management framework, procedures are proposed for building the knowledge map using the ABKM system. The procedure consists of the following five phases: knowledge determination phase, knowledge extraction phase, knowledge attribute phase, knowledge linking phase, and knowledge validation phase.

Knowledge Determination Phase

The purpose of this phase is to provide a uniform, text-based intermediate representation of the knowledge types specific to construction projects. This phase covers defining knowledge and baseline taxonomy within a project. The scope of the knowledge map decides whether the knowledge map is constructed throughout a specific project. After deciding the scope, we determine the detail level of knowledge analysis. It is necessary and important to determine the proper level of detail to meet project-based knowledge demand effectively. When analysing the source of knowledge within a project, activity is suggested as a unit to analyse the construction project knowledge. We analyse related knowledge based on an activity-unit of the construction project. In other words, we consider which experience and know-how should and can be captured according to the activity of the selected project. After the analysis process of this phase, the types of tacit and explicit knowledge specific to the activity will be considered. Furthermore, all capturing and documenting knowledge with regards to the activity are saved in this activity category.

Knowledge Extraction Phase

Knowledge is extracted through the project execution. There are two types of knowledge extracted from the projects. They are tacit knowledge and explicit knowledge. Tacit knowledge and explicit knowledge may exist in any project. After identifying knowledge through those activities of the project, we decide which knowledge needs to be extracted from the activities of the projects. Suggested knowledge extraction techniques include interviewing with experts, making system analysis, and digital process record.

Knowledge Attribute Phase

Knowledge attributes illustrate the basic description of extracted knowledge and derive relationships with project and similarity activity (see Figure 4). The main purpose of knowledge attribute is to provide the relationship and available knowledge information

Figure 5. An example of a knowledge attribute of a activity unit with the help of a knowledge map

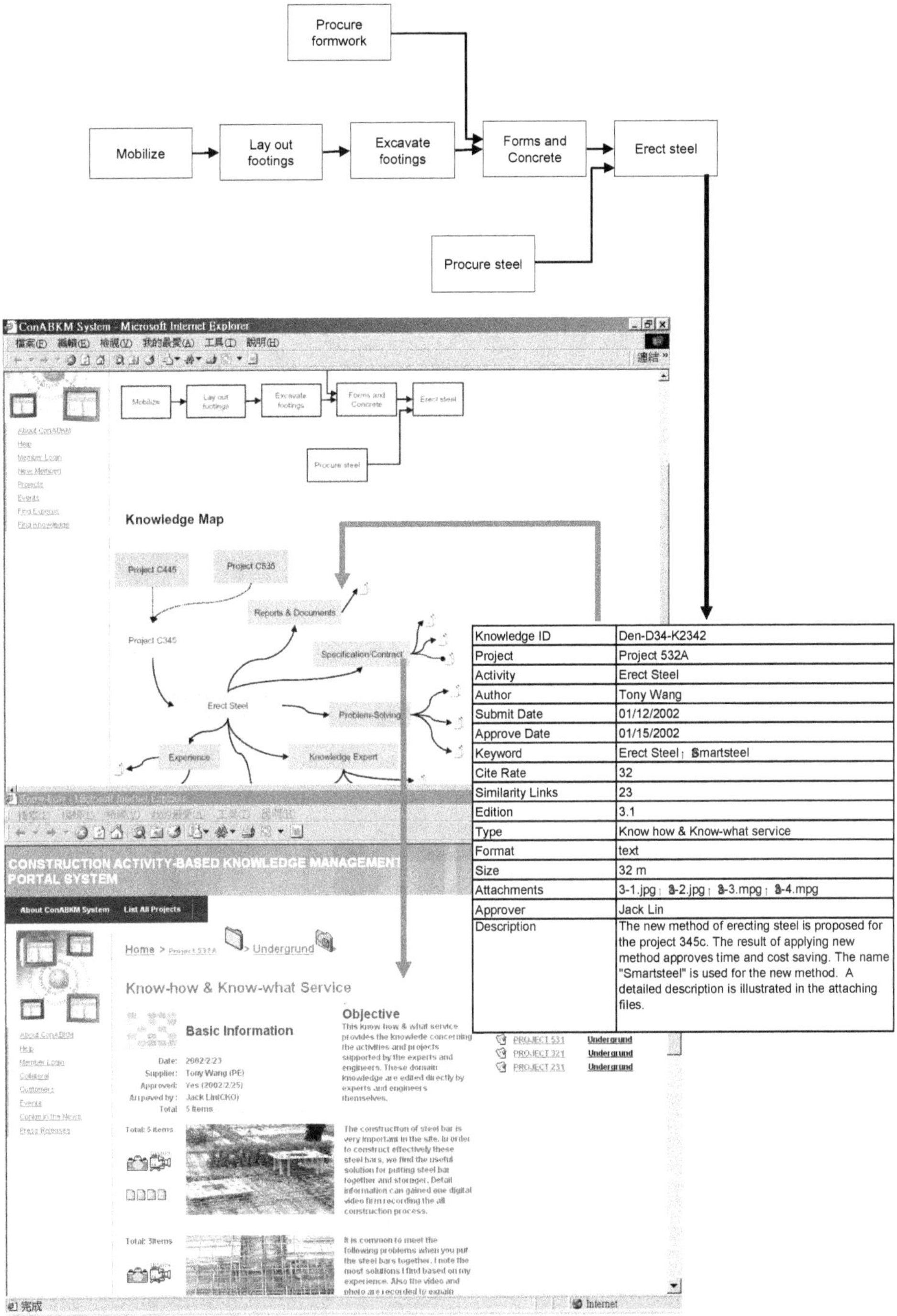

Knowledge ID	Den-D34-K2342
Project	Project 532A
Activity	Erect Steel
Author	Tony Wang
Submit Date	01/12/2002
Approve Date	01/15/2002
Keyword	Erect Steel ; Smartsteel
Cite Rate	32
Similarity Links	23
Edition	3.1
Type	Know how & Know-what service
Format	text
Size	32 m
Attachments	3-1.jpg ; 3-2.jpg ; 3-3.mpg ; 3-4.mpg
Approver	Jack Lin
Description	The new method of erecting steel is proposed for the project 345c. The result of applying new method approves time and cost saving. The name "Smartsteel" is used for the new method. A detailed description is illustrated in the attaching files.

for knowledge workers and general users. Knowledge attributes include the keywords, description, project name, activity name, contributor, and attached files. Figure 5 illustrates an example using the knowledge attribute with the help of a knowledge map.

Knowledge Linking Phase

The knowledge link is identified after completing the knowledge attribute. The knowledge link is first indicated when the tacit or explicit knowledge is available and documented, and is later confirmed. Three types of knowledge linking are proposed in the chapter. One is *activity link activity* based on high similarity. Second is *activity link knowledge* based on the relationship between activity and knowledge. The third is *knowledge link knowledge* based on high knowledge similarity. When the contributor creates a new link, the link needs to be examined and confirmed before the knowledge map is published.

Knowledge Validation Phase

All knowledge maps need to be validated before a map is published. All the validation processes must be communicated with domain experts, knowledge workers, and knowledge map producers in the enterprise knowledge management division.

The System

This section describes in detail the ABKM system. There are three search functions supported in the system. They are project category search, keyword search, and expert category search. With the function of keyword search and project category, users can find the knowledge they need directly form the activities of selected projects. Also, the system provides another service for users to find the related knowledge of the domain experts. The information of each domain expert is provided to the people who want to find the domain knowledge-related experts (Figure 6).

Figure 7 illustrates the system architecture. The server of the ABKM system supports four distinct layers— interface, access, application, and database layers; each has its own responsibilities.

The interface layer defines administrative and end-user interfaces. Users can access information through Web browsers such as Microsoft Internet Explorer or Netscape Navigator. Administrators can control and manage information via the Web browser or using a separate server interface. The access layer provides system security and restricted access, firewall services, and system administration functions. The application layer defines various applications for collecting and managing information. These applications offer indexing, full text search, collaborative work, and document manage-

Figure 6. Knowledge searching process

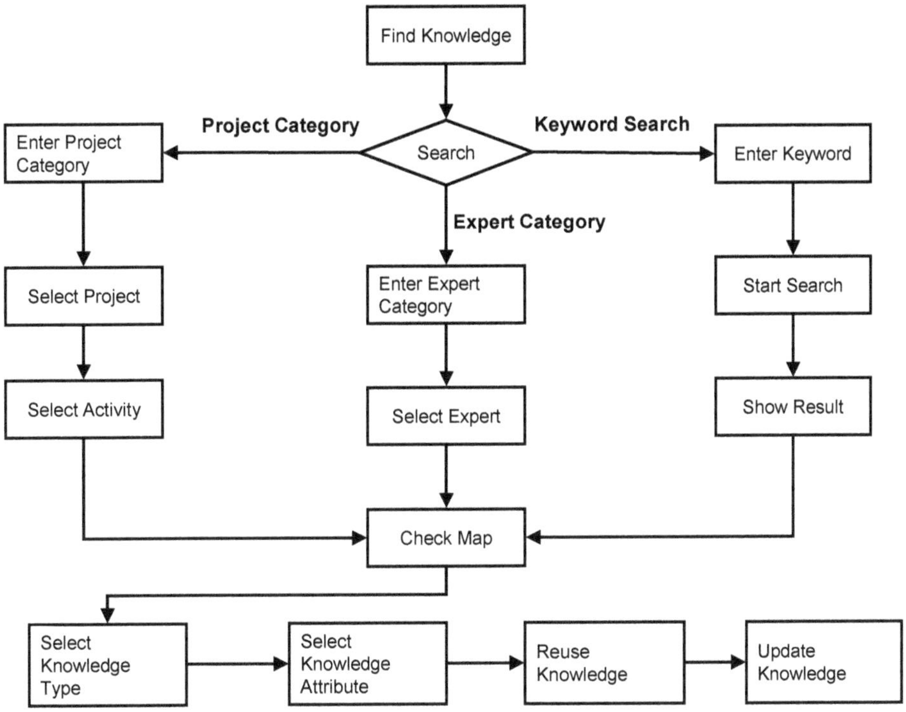

ment functions. The database layer consists of a primary SQL Server 2000 database and a backup database (also based on SQL Server 2000).

All project information and knowledge in the ABKM system is centralised in a system database. Project participants may have access to all or some of these documents through the Internet, as determined by their levels of access authorisation. Any information/knowledge about the project can be obtained from and deposited into the system database only through a secure interface. The Web and database servers are distributed on different computers, between which a firewall can be built to protect the system database against intrusion.

The ABKM portal services described in this study are made available to all participants of the corporation through a specially designed portal, which also serves as a messaging (mail) server for the company (organisation). The portal is a key element of the proposed system and consists of three content areas - public, member, and knowledge manager areas. The public area is open to anyone who is interested in the project. The project member and knowledge manager are accessible only by members with passwords.

Anyone can access information in the public area, so project managers can reasonably use this space for public relations, for example, by posting project descriptions, news, and announcements.

Figure 7. System architecture

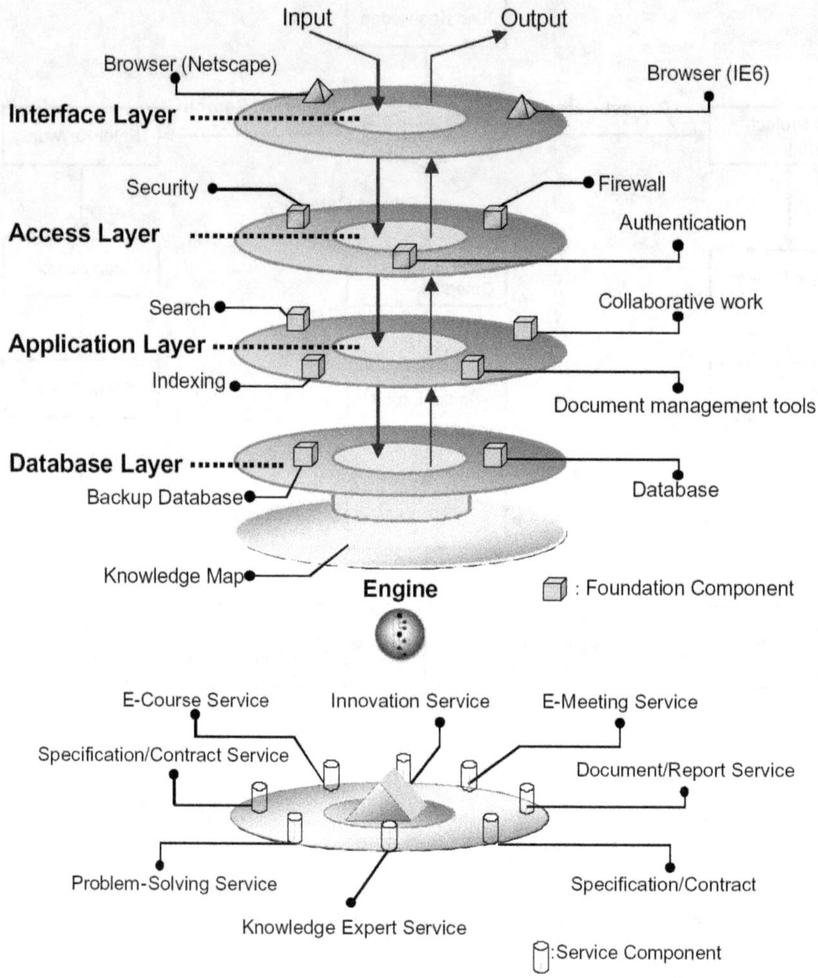

In the project member area, project members can use ABKM services, such as messaging, accessing data, and accessing files. Only project members can access information and knowledge concerning the projects.

Only knowledge managers can log into the manager area, where they can access all information in the project server. Knowledge managers have the right to use the administration service to: (1) post and edit news and announcements, and (2) add new user accounts.

Table 1. The descriptions of function provided by ABKM services

Category	*Function description*
Specification\Contract Managers	Each project's specifications and contracts are stored in the active and non-active categories of a project category. One of the best benefits of this is it is easy to find all specifications and contracts for all projects without wasting time and mass storage space. Paper-based specifications and contracts from past projects need to be scanned by knowledge workers into electronic format (PDF files) then subsequently saved into the system database. All specifications and contracts of past and current projects can be found and downloaded in PDF format from the ABKM system.
Document/Report Managers	All project-related documents and reports are centralized in the server of the ABKM system. The main purpose of these services is to be the e-information center for the company. All users who need to find project-related information or documents can just access the system and get what they need. Therefore, project-related documents must be transferred to electronic format so they can be preserved in the system and be accessible. These documents and reports will be saved under the categories of active and non-active for related-project categories. In the ABKM system, documents and reports files must be uploaded in PDF format, the standard file format. It is convenient for users to reuse these documents and reports without the problems of different software versions or file format.
Video/Photo Managers	In order to reduce the cost of training for problem-solving, it is necessary to record videos and photos by the on-site engineers during the construction phase of a project. With the use of digital cameras and camcorders, all operational processes and problem-solving of a project can be recorded and preserved as company knowledge asset. For future related-projects, digital records can be reused and shared for experts and managers to reduce the time to solve problems.
E-Courses Services	The purpose of e-course services is to provide e-learning to all participants in the construction phase of a project. Without time and location limitations, it is helpful to provide e-courses for needed people via the internet or intranet. Especially for junior engineers, it is convenient for them to learn and understand how a project operates in the construction phase.
E-Meeting Services	The purpose of e-meeting services is to record all project-related meetings and discussions. All meetings or discussions are recorded and saved as digital files in the ABKM server. Also, video conferencing is one of the e-meeting services. The contents of meetings or discussions can be recorded and preserved in the ABKM system.
Expert Yellow Page Services	Experts Yellow Page is a service that assists with finding related experts in the enterprise. It is just like a yellow page. All related-area experts will be mapped in the system based on their specific and professional trade. Users can access this service to find and contact the experts they need in a project.
Problem-solving Services	During the construction phase of a project, lots of problems may occur in the construction phase. Some problems may be solved or improved by engineers or experts, and some are unsolved. To avoid making similar mistakes in other projects, the valued experience and know-how can be saved using problem-solving services. During or before the same problems happen, engineers may find existing solutions concerning the problem.
Collaboration Services	During the construction phase of a project, collaboration services allow experts and engineers to perform tasks and work on projects together by the collaborative communication platform. The communications support video conferencing, voice, multimedia, and conversations.
Knowledge Communities	Knowledge communities provides the platform to let experts and engineers to discuss the project-related topic and share their idea and concept with others or participants. The benefit of knowledge communities in the system is to store the all unstructured project-related information.
Message Services	The message manager is a web-based project-oriented mail system that allows project participants to receive notice mail when the knowledge bank is updated.

All project-related information and knowledge in the ABKM system are classified and stored in the system database. Different kinds of project information/knowledge are stored in different tables in the system database. The ABKM system backs up all databases in the system to ensure the reliability of data storage. The backup database is in another building to protect it against disasters such as earthquakes or fire. All functions are briefly outlined in Table 1, and Figure 8 illustrates some applications of knowledge sharing in the ABKM system.

Figure 8. The application of knowledge sharing in the ABKM system

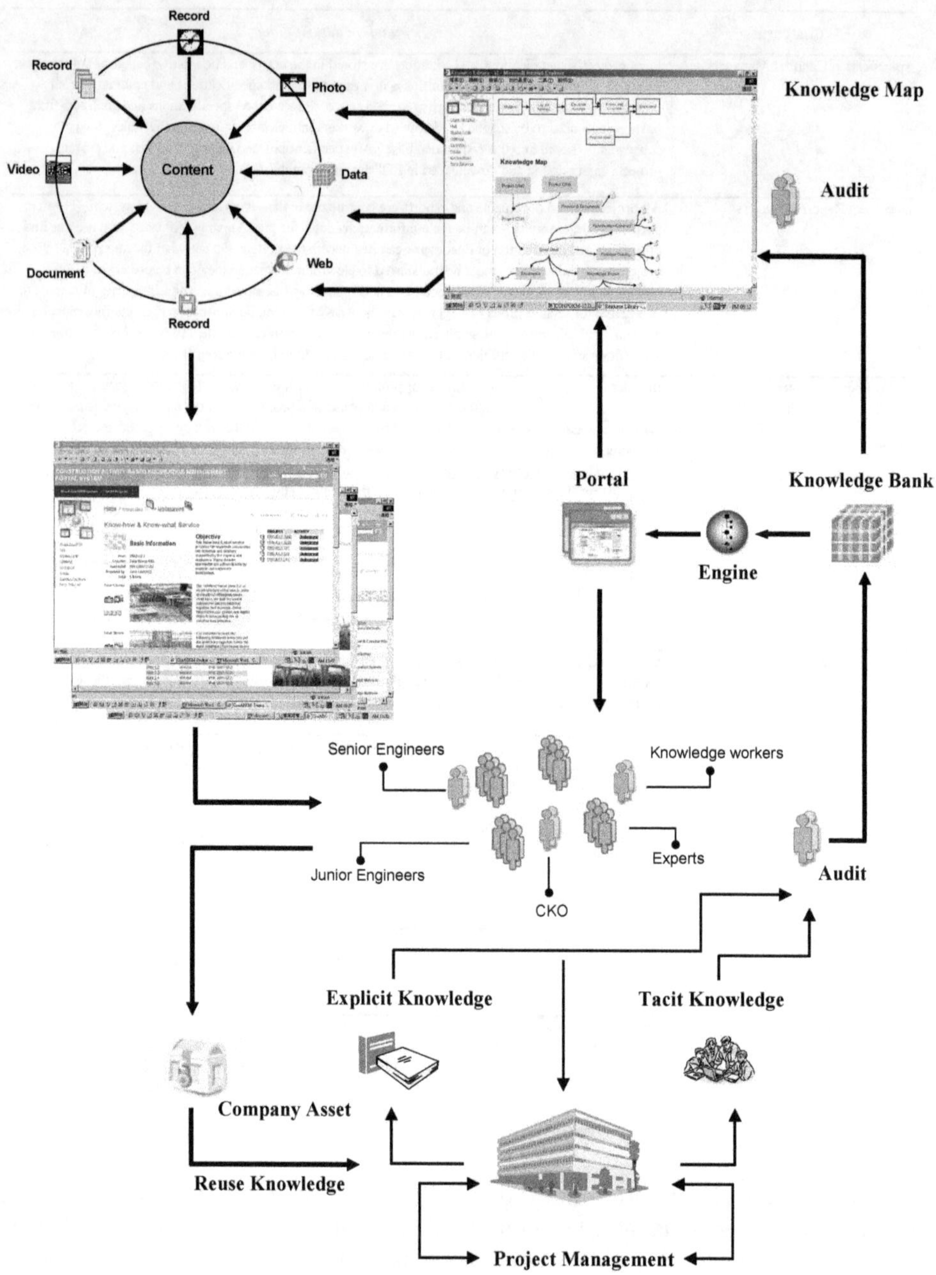

Case Study

In the following case, the general contractor won the bid of a precast project, to construct a high-tech office building within seven months in Taipei. However, the contractor has no experience of precast construction, and the contractor estimates that the construction period might need to extend to nine months at least to complete the project. Therefore, the contractor decided to hire two construction specialists, particularly in precast construction, to assist the senior engineers in order to finish the project on time as required. For the future, precast projects could process more easily; the company decided to take a good advantage of knowledge management to pass on the valuable know-how to the engineers and manage it well to keep the knowledge inside the company. In the following case study, the contractor applied and practiced knowledge management by using the knowledge management concept and map. In addition, Table 2 presents the details of various stages for construction knowledge management implementations.

Knowledge Acquisition Phase (Step 1-Step 5)

A senior engineer collects related information/documentation and coverts it into digital format. After the related information and documentation have been digitised, the senior engineer writes descriptions/notes concerning that digital information and packages them as a knowledge set for submission.

Knowledge Extraction Phase (Step 6-Step 10)

A senior engineer and a knowledge worker record all the operating procedures by taking digital video and photographs. The senior engineer discussed progress with two experts every two days to accelerate the solving of the problem in question. All discussions were recorded and summarised as suggested by experts. Discussions with experts continued for five months, until the problem was solved. The senior engineer recorded and summarised his experience and domain knowledge in the system to enable the problem's solution to be reused in other and future projects. The domain knowledge included the problem description (including documents, photographs, drawings, and specifications), the problem's solution (including related documents, photographs, and video of processes), and expert suggestions (such as notes, discussions, and meeting records). Knowledge was extracted according to each process defined as related to the activity of a project. Domain knowledge and experience was organised according to the attribute of the activity concerned. Most senior engineers are required to provide their own knowledge concerning the tasks for which they are responsible.

Table 2. Scenario description for using ABKM system

Step		Phase	Scenario
Knowledge Acquisition	1	Collect Information	Senior engineer collects all activities-related documents/information/data
	2	Digital Information	Senior engineer transfers paper-based information/documents into the digital information/data set
	3	Edit Information	Senior engineer edits the description and comment/note for digital document/information
	4	Package Information	Senior engineer packages the description and comment/note with attaching related files that can illustrate the explanation or example
	5	Submit Information	Senior engineer submmits package that includes the description and comment/note with attaching related files for approving
Knowledge Extraction	6	Record Operation & Event	Senior engineer or a knowledge worker records digital related process information for the operation of successful and failure events
	7	Edit Knowledge	Senior engineer or a knowledge worker edits the description and note/comment for the records of video and photo
	8	Manage Knowledge	Knowledge workers collect and manage the coordinated information (includes grouping the meeting records and knowledge communities)
	9	Package Knowledge	Senior engineer or a knowledge worker packages the description and comment/note with attaching related files
	10	Submit Knowledge	Senior engineer or a knowledge worker submmits the package that includes the description and comment/note with attaching related files
Knowledge Storage	11	Approve Knowledge	knowledge worker audits; knowledge worker/expert checks and approves the submission of knowledge package before the classification and storage
	12	Classify Knowledge	Knowledge worker attributes knowledge and classifies this knowledge by placing it in an appropriate position (the activities of project map) in the system.
	13	Store Knowledge	Knowledge worker stores knowledge package into knowledge bank based on the classification.
	14	Backup Knowledge	Knowledge Package is automatically backuped from the knowledge bank to another database for the safety purposes.
	15	Publish Knowledge	Knowledge package is published after knowledge map is validated and announced for the re-use and application
Knowledge Sharing	16	Search Knowledge	Junior engineer found past related knowledge/experience by using knowledge map and keywords or domain expert search
	17	Refer Knowledge	Junior engineer refers and studies past existing knowledge/experience that is stored in the activities of knowledge map
	18	Modify Knowledge	Junior engineer modifies the original knowledge package based on new projects or others existing projects
	19	Apply Knowledge	Junior engineer applies the modified existing knowledge package in other existing projects or future projects
	20	Collect Feedback	Junior engineer collects feedback from the applied-original or modified knowledge package
Knowledge Update	21	Collect Information	Junior engineer collects all paper-based and electronic format of document/information/data
	22	Renote Knowledge	Junior engineer edits the digital document/information by adding detail description and comment/note
	23	Repackage Knowledge	Junior engineer packages the description and comment/note with attaching related files that can illustrate the explanation or example
	24	Approve Knowledge	Knowledge package would be approved to be processed under an accurate procedure before saving in the knowledge bank
	25	Republish Knowledge	Knowledge package is republished in the knowledge map and announced for the reuse or application, based on the original knowledge

Knowledge Storage Phase (Step 11-Step 15)

When the submitted knowledge set is approved, a knowledge worker attributes knowledge and classifies this knowledge by placing it in an appropriate position (the activities of project map) in the system. In other words, users can find and read related domain knowledge directly, just by clicking these activities of the project (see Figure 5). All knowledge maps need to be validated to perform well before the map is published. All the validations must be conducted with domain experts, knowledge workers, and knowledge map producers in the enterprise knowledge implantation term. Finally, the knowledge set is automatically backed up from the knowledge bank to another database. After approving and storing knowledge, the system sends a message regarding the updating of the knowledge to the appropriate users automatically.

Knowledge Sharing Phase (Step 16-Step 20)

In a new project that began just after the precast building was constructed three month ago, a junior engineer with no prior experience meets a similar problem and tries to locate past knowledge/information to help him solve the problem. The junior engineer uses the keyword search to find the expert who has domain knowledge concerning the precast building. The junior engineer finds the experts and retrieves, refers to, and studies the knowledge set (including digital video and documentation) supported by these senior engineers. He starts to apply and reuse the knowledge in his own project. Also, the junior engineer gives some feedback and offers knowledge that can be reused when others face new problems. Additionally, some senior engineers in other projects reuse the same knowledge to solve the same problem at reduced cost.

Knowledge Update Phase (Step 21-Step 25)

After applying knowledge and information to the other similar projects, the junior engineer solves his problem and finds a new solution in collaboration with some senior engineers. Finally, the junior engineer notes and submits the new suggestion and experience to the project map, associated with the original knowledge. Furthermore, the knowledge is updated later because of further feedback, and another solution is provided regarding the same problem. After the approval process has been completed, the updated knowledge set is republished in the activities of the project map, and a notice message will be sent to the authorised members.

Conclusion

Web technologies and knowledge management concepts can be effectively used during the construction phase of a project, to enable knowledge to be captured and reused in similar projects in the future. This study presents the application of knowledge management in the construction phase of construction projects by using a Web-based portal. The concept of activity-based knowledge management (ABKM) is presented, and a system for use as a knowledge-sharing platform in construction projects is presented. The construction activity-based knowledge management system maps valuable information and knowledge into activity units during the construction phase of a project. The

ABKM system is advanced, at least in the following respect: The ABKM system provides insight into the factors that have an impact on construction management activities, and so helps engineers to share knowledge and improve the results of the entire construction project. Junior engineers can interact with the computer to gain domain knowledge, and thus prepare for and participate in a construction project. Briefly, the ABKM system can assist engineers by providing structured and unstructured information to support reference to, and the reuse of, explicit and tacit knowledge. The integration of knowledge management and Web-based technologies appears to be a promising means of improving construction operation and management, especially in the construction phases of projects.

The content of the knowledge bank in the system not only provides specific problem-solutions, but also includes all domain knowledge and experiences gained on projects. Although effort is required to update the explicit/tacit knowledge, the developed system will support construction management by: (1) providing an effective and efficient computerised environment to support knowledge management tasks, and (2) facilitating the implementation of a Web-based knowledge management system pertinent to these activities in the projects.

The demonstration of the system in a case study of the new precast office building located in Taipei indicates that the ABKM system effectively promotes the sharing and reuse of knowledge for new construction projects. The case study also highlighted the need for improving knowledge management and exchange platforms. However, the received feedbacks based on the use of the system are as follows: (1) the content of knowledge bank in the system is not enough to support and provide the junior engineers to get past experience and knowledge in the beginning; (2) most of senior engineers and experts have a low willingness to share their knowledge and experience without the proper reword policy and strategy; (3) it takes time and is very inconvenient for senior engineers to edit and record the knowledge without any assistance form knowledge workers, and (4) most senior engineers agree the ABKM system is a useful platform for them to edit and manage their knowledge and experience.

The knowledge map in the system mainly deals with the assistance to find the needed knowledge easily and effectively. The primary purposes of using a knowledge map in the system are as follows: The knowledge map (1) is represented in a clear and visual way to identify key knowledge areas that are most strategic and critical to the project, and (2) deals with the assistance for users to find the needed knowledge easily and effectively.

The major contributions of the study are as follows: (1) elucidating the content and procedure of knowledge management in construction projects; (2) proposing procedures for building knowledge maps for construction projects, and (3) developing a construction knowledge management system specific for managing construction projects with reference to activities by using a knowledge map.

Practical Tips and Lessons Learned

- Knowledge management is necessary in any life cycle phase of construction projects.
- The content and method of knowledge management in a construction project will be different in each life cycle phase.
- Without the Internet, it is impossible to run knowledge management successfully for construction projects.
- A knowledge map is one of best ways to illustrate whether the knowledge is available and where the knowledge is located in construction projects.
- The way and method of application in construction knowledge management will differ from the size and culture of various construction corporations.

References

Bergmann, R. (2002). *Experience management: Foundations, development methodology, and Internet-based applications.* Germany: Springer-Verlag.

Carneiro, A. (2001). The role of intelligent resources in knowledge management. *Journal of Knowledge Management, 5*(4), 358-367.

Clough, R.H., Sears, C.A., & Sears, S.K. (2000). *Construction project management* (4th edition). New York: John Wiley & Sons.

Davenport, T.H., & Prusak, L. (1998). *Working knowledge.* Boston, MA: Harvard Business School Press.

Edum-Fotwe, F.T., & McCaffer, R. (2000). Developing project management competency: Perspectives from the construction industry. *International Journal of Project Management, 18,* 111-124.

Hart, A. (1992). Knowledge acquisition for expert systems (2nd edition). New York: McGraw-Hill.

Kamara, J.M., Augenbroe, G., Anumba, C.J., & Carrillo, P.M. (2002). Knowledge management in the architecture, engineering and construction industry. *Construction Innovation, 2,* 53-67.

Malhotra, Y. (2000). *Knowledge management and virtual organizations.* Hershey, PA: Idea Group Publishing.

Malhotra, Y. (2001). *Knowledge management and business model innovation.* Hershey, PA: Idea Group Publishing.

Nonaka, I., & Takeuchi, H. (1995). *The knowledge-creating company.* Oxford: Oxford University Press.

Ponniah, P. (2002). Data warehousing fundamentals: A comprehensive guide for IT professionals. New York: John Wiley & Sons.

Tiwana, A. (2000). *The knowledge management toolkit—Practical techniques for building a knowledge management system.* Englewood Cliffs, NJ: Prentice-Hall.

Wexler, M.N. (2001). The who, what, and why of knowledge mapping. *Journal of Knowledge Management, 5*(3), 249-263.

Chapter XVII

An Integrative Knowledge Management System for Environmental-Conscious Construction

Zhen Chen
Massey University, New Zealand

Stephen C.W. Kong
Loughborough University, UK

Heng Li
Hong Kong Polytechnic University, Hong Kong

Qian Xu
Hong Kong Polytechnic University, Hong Kong

Abstract

This chapter introduces an integrative knowledge management prototype named E+ for environmental-conscious construction based on a comprehensive integration of current environmental management (EM) techniques and tools in construction. The overall objective is to apply the theory of knowledge management (KM) in EM in construction, and the authors will achieve it though developing the E+ model and its tools for dynamic EM in construction. The approaches applied in this chapter include system analysis and development, literature review, questionnaire survey and interview, and case study. The results from this chapter include the E+, a comprehensive review of EM tools adopted in construction, and a demonstration of the implementation of the E+. Furthermore, the authors hope that the adoption and implementation of the E+ can effectively improve contractors' performance in EM and reduce adverse environmental impacts in construction.

Learning Objectives

Due to the purpose of the integrative methodology of KMS for EM in construction, this chapter mainly contributes to existing theory for EM in construction in the area of quantitative analytical approaches and their integrative implementation. According to the literature review and questionnaire survey for this research, the lack of effective, efficient, and economical (E3) quantitative analytical approach is one of obstacles to implementing EM in construction. Therefore, there are four points of contribution of this research to the existing theory or practice for EM in construction:

1. This research has developed an integrative methodology (E+) to implementing EMS and KM in construction, with a rigorous dynamic Environmental Impacts Assessment (EIA) model based on various functional-different approaches to EM in a construction cycle. The E+ prototype was originally created in both the theory and practice for EM in construction, and it is open to further integration of various functional-different approaches for EM in construction other than the three EM tools presented in this chapter. Because the E+ is both EMS oriented and process oriented in construction, it can thus help contractors to implement EM from a messy situation to a normalised system, and to effectively share EM knowledge and information internally and externally.
2. The CPI method integrated in the E+ model is a quantitative approach to predicting and levelling complex adverse environmental impacts potentially generated from construction and transportation due to the implementation of a construction plan. As a result, the CPI method has been integrated into E+ EM Toolkit A, one functional section of the E+ system, to carry out the task in environmental-conscious construction planning.
3. The IRP method is a quantitative approach to reducing the waste of construction materials on a construction site, and it is designed to effectively be implemented by using a bar-code system. The IRP is then integrated in the E+ EM Toolkit B, another functional section of E+, as a basic component.
4. The Webfill method is an e-commerce model designed for the trip-ticket system to effectively reduce, reuse, and recycle C&D waste. Although there is a lack of data to prove the efficiency in reality, the computer simulation results and a questionnaire survey from another research (Chen, 2003) have proven that the Webfill system can effectively realise the design function. As a result, the Webfill is also integrated in the E+ model as an important component of the E+ EM Toolkit C.

Readers can obtain the socio-technical perspectives from the introduction of the E+ prototype and its toolkits, and know how E+ can work for a dynamic EIA process in construction with integrated supports from E3 quantitative analytical approaches in the toolkits.

Introduction

The adverse environmental impacts of construction—such as construction and demolition waste, noise and vibration, dust, hazards emissions and odours, soil and ground contamination, water pollution, wildlife and natural features demolition, and archaeological destruction —have been of concern since the early 1970s, and received more and more academic and professional attention in the construction industry especially after the ISO 14000 series of EM standards was enacted. In this regard, quantitative analytical approaches to EM in construction are currently not as available as qualitative approaches such as regulations and practical guides due to the difficulties in the transformation of practical data to abstraction data that are needed in the calculation for EM. However, it is hard to accept an environmental management system (EMS) without adequate support from EM tools and knowledge warehouse (KW), which is the essential component of an enterprise's knowledge management system (KMS) where knowledge is developed, stored, organised, processed, and disseminated (SAP INFO, 2004). From this point of view, the overall objective of this chapter is to put forward a methodology of an integrative KMS based on the developed holistic approach for environmental-conscious construction project management, which can facilitate knowledge management (KM) initiatives for improved competitiveness of construction enterprises in EM. This objective will be achieved step by step though four sub-objectives. They are: (1) to illustrate an integrative KM prototype to capture and reuse data, information, and knowledge for dynamic EM in construction project management; (2) to describe quantitative EM tools that can be integrated into the integrative KM model for dynamic environmental-conscious construction project management; (3) to describe the interaction among the quantitative EM tools within the integrative KM model and key information techniques to develop the integrative KMS for dynamic EM in construction project management; and (4) to demonstrate the implementation of the integrative KMS through a case study.

Background

EM in construction has received more and more attention over the past 30 years. For example, studies on noise pollution (USEPA, 1971), air pollution (Jones, 1973), and solid waste pollution (Skoyles & Hussey, 1974; Spivey, 1974) from construction sites were all individually conducted in the early 1970s. Although the expression of EM in construction came out in the early 1970s, after the U.S. National Environmental Policy Act of 1969 was enacted (Warren, 1973), the concept of EM in construction was introduced later in the '70s, when the role of environmental inspector was defined in the design and construction phases of projects to provide advice to construction engineers on all matters in EM (Spivey, 1974; Henningson, 1978). However, there has been little enthusiasm for establishing an EMS in construction organisations until two main important standards, BS 7750 (released by the BSI Group in 1992) and the ISO 14000 series (released by the International Organisation for Standardisation (ISO) in 1996) were promulgated to guide the construction industry from passive CM on pollution reduction to active EMS for pollution prevention.

Implementation of EM

In the 1990s, the Construction Industry Research and Information Association (CIRIA) conducted a series of socio-technical reviews on environmental issues and have undertaken initiatives relevant to the construction industry after the introduction of BS 7750 (Shorrock et al., 1993; CIRIA, 1993-95; Guthrie & Mallett, 1995; Petts, 1996). Thereafter, research works on EM have also been led to the implementation of EMS and the registration of ISO 14001 EMS by authoritative institutions in the construction industry, such as the CIOB (Clough & Antonio, 1996), the FIDIC (1998), the Construction Policy Steering Committee (CPSC) (1998), and the CIRIA (Uren & Griffiths, 2000).

To assess the implementation of EMS in construction, several socio-technical investigations have been conducted independently by researchers in different countries since 1999. For example, Kein, Ofori, and Briffett (1999) conducted a field study in Singapore to assess the level of commitment of ISO-9000-certified construction enterprises to EM. They found that contractors in Singapore were aware of the merits of EM, but were not instituting systems towards achieving it. Ofori, Briffett, Gang, and Ranasinghe (2000) then conducted a survey to ascertain the perceptions of construction enterprises on the impact of the implementation of the ISO 14000 series on their operations. Major problems were identified, such as the shortage of qualified personnel, lack of knowledge of the ISO 14000 series, indistinct cost-benefit ratio, disruption and high expenses on changing traditional practices, and resistance from employees. The CIRIA (1999) led a self-completion questionnaire survey of the state of environmental initiatives within the construction industry and of sustainability indicators for the civil engineering industry in the United Kingdom.

Tse (2001) conducted an independent questionnaire survey in the Hong Kong construction industry to gain a further understanding of the difficulties in implementing the ISO 14000 series. Lo (2001), also in Hong Kong, made an effort to identify nine critical factors for the implementation of ISO 14001 EMS in the construction industry based on critical factors drawn from an investigation in another industry. The CPSC (2001) in Australia conducted a questionnaire survey of the New South Wales construction industry on EM with industry leaders. Chen, Li, and Wong (2000) conducted a questionnaire survey of main contractors in five main cities in mainland China and found that there are five classes of factors influencing the acceptability of the ISO 14000 series including governmental regulations, technology conditions, competitive pressures, cooperative attitude, and cost-benefit efficiency. In addition, Zeng, Tam, Deng, and Tam (2003) conducted a questionnaire survey of the mainland China construction industry to discover the conditions of implementation of the ISO 14000 series. All these questionnaire surveys clarified real situations in adoption and implementation of the ISO 14000 series in the local construction industries, and provided relative perspectives on how to conduct EM of the construction industry.

In addition to the questionnaire survey, case studies are further applied to investigate the acceptability of the ISO 14000 series in the construction industry. For example, Valdez and Chini (2002) conducted a literature review and case study of a construction contracting firm certified for the ISO 14001 EMS in the United States. They concluded that the positive aspects of certification outweigh the negative aspects, and recommended adding government support and the combined use of the ISO 14000 series with other EM and matrices.

Conflicts with EIA

The Environmental Impacts Assessment (EIA) is a process to identify, predict, evaluate, and mitigate the biophysical, social, and other relevant environmental effects of development proposals or projects prior to major decisions and commitments being made (IAIA, 1997). Although the EIA has been accepted by the construction industry in different countries under governmental regulations to evaluate environmental impacts of construction projects, the implementation rate of ISO 14001 EMS accreditation is normally much lower than the implementation rate of EIA in the construction industry. For example, a remarkable disagreement/deviation between the rate of ISO 14001 EMS accreditation and EIA implementation in some countries indicates that contractors there have not really implemented EM and accepted the ISO 14000 series (Chen, 2003). According to the *Official Report on the State of the Environment in China 2001* (China EPB, 2002), the annual implementation rate of EIA for construction projects was 97% in 2001 in mainland China. In addition, a further investigation on the implementation rate of EIA in mainland China indicates that the average EIA rate from 1995 to 2001 is 88%, with an increasing rate of 23% (China EPB, 2002). By contrast, the percentage of the construction enterprises that have been awarded environmental certificates versus total governmental registered construction enterprises in mainland China is as low as 0.083%, and statistical figures also indicate that most construction enterprises have not yet adopted or accepted the ISO 14000 series in mainland China (Chen, 2003). Because of the disagreement between the implementation rates of EIA and EMS, there may be little coordination between the EIA process and EMS implementation in construction projects, and thus the EIA may not really serve as a tool to promote EM in the construction industry in those countries. For that reason, adverse environmental impacts such as noise, dust, waste, and hazardous emissions still occur frequently in construction projects in spite of their EIA approvals prior to construction. As a result, the failure to implement mitigation measures adequately or monitor environmental impacts following the approval of projects is often cited as a major shortcoming of the EIA process, and the contents and recommendations of EIA are often disregarded when EMS are implemented (Sánchez & Hacking, 2000).

KM Will

The growing consciousness, requirements, and initiatives of KM are existent in order to manage the intellectual capital and get benefits from previous construction processes and projects (Zyngier, 2002; & Zarli, Rezgui, & Kazi, 2003). For example, the C-Sand project (c-sand.org.uk) has been conducted in the United Kingdom to foster organisational practices that enable knowledge creation for subsequent sharing and reuse, and to promote sustainable construction (Khalfan, Bouchlaghem, Anumba, & Carrillo, 2003). As one of the largest contracting companies in the United States, Centex Construction Group (centex-construction.com) faces some knowledge-related business challenges that are not always associated with the construction industry. For instance, they have a technology infrastructure in place where all professionals in the company have computing power, that is, laptops and/or desktops. All offices and job sites are connected to a nationwide WAN via dial, ISDN, and Frame. Remote access is Web based and available from anywhere to lead some initiatives to increase knowledge sharing and provide better information access across the company's diverse landscape (Velker, 1999). Beyond the development of knowledge warehouse (KW) in the construction industry, socio-technical research also reflected that there was a majority of agreement to the statement that KM is an extension of IT (Zyngier, 2002). The progress of KM in construction also reflects the trend of construction enterprises away from traditional blue-collar operations towards a more knowledge-based CM.

According to the survey results, EMS implementation requires practical approaches for contractors to conduct EM. That is, although the governmental regulations have been identified as a major factor influencing the implementation of the EMS and the EIA in the construction industry, the construction industry is still a negative receiver if there are not enough technology conditions to support EMS implementation, especially the techniques or tools that can help contractors to conduct EM in construction projects where most of the negative environmental impacts are generated. Even the positive bodies in the construction industry have high willingness to implement EM; the E3 EM tools are essential to them (Chen, 2003). Based on this consideration, the authors of this chapter will integrate several EM tools and an EMS-based dynamic EIA process developed previously into an environment of KM, entitled E+, for E3 EM in construction project management.

The E+

Methodology

The E+ is an integrative methodology for E3 EM in construction projects in which an EMS-based dynamic EIA process is applied within a knowledge support system for active knowledge capture and reuse about environmental-conscious CM during construction. The successful implementation of an EMS in a construction project requires far more than just the apparent prevention and reduction of adverse or negative environmental impacts in a new project and its construction process development cycles during pre-construction stage, continuous improvement of the EM function based on institutionalisation of change throughout an onsite organisation to reduce pollution during construction stage, or efficient synergisms of pollution prevention and reduction such as waste recycle and regeneration during construction and post-construction stages. It necessitates a complete transformation of the CM in an environmentally conscious enterprise, such as changes in management philosophy and leadership style, creation of an adaptive organisational structure, adoption of a more progressive organisational culture, revitalisation of the relationship between the organisation and its customers, rejuvenation of other organisational functions (i.e., human resources engineering, research and development, finance, and marketing) (Azani, 1999). In addition to the transformation for the EM in construction enterprises, the E+ for the E3 implementation of the EM in all phases of construction cycle—including the pre-construction stage, the construction stage, and the post-construction stage—is necessarily activated, together with other rejuvenated CM functions such as human resources, expert knowledge, and synergetic effect.

There are already some approaches to effectively implementing the EM onsite at different construction stages. For example, the CPI approach, which is a method to quantitatively measure the amount of pollution and hazards generated by a construction process and construction project during construction, can be utilised by indicating the potential level of accumulated pollution and hazards generated from a construction site at the pre-construction stage (Chen et al., 2000), and by reducing or mitigating pollution level during construction planning stage (Li, Chen, & Wong, 2002). In addition to the CPI approach, an analytic network process (ANP) approach to construction plan selection (Chen, Li, & Wong, 2003), a life-cycle assessment (LCA) approach to material selection (Lippiatt, 1999), and a decision programming language (DPL) approach to environmental liability estimation (Jeljeli & Russell, 1995) also provide quantitative methods for making decisions on EM at the pre-construction stage. For the construction stage, a crew-based incentive reward program (IRP) approach, which is realised by using a bar-code system, can be utilised as an onsite material management system to control and reduce construction waste (Chen et al., 2002a). For the post-construction stage, an online waste exchange (Webfill) approach, which is further developed into an e-commerce system based on the trip-ticket system for waste disposal in Hong Kong, can be utilised to reduce the final amount of construction and demolition (C&D) waste to be landfilled (Chen et al., 2002b). Although these approaches to EM in a construction project have been proven effectively, efficiently, and economically applicable in a corresponding construction stage, it has also been noticed that these EM tools can be further integrated for a total EM purpose in construction project based on the interrelationships among them. The integration can bring about not only a definite utilisation of current EM tools, but also an improved environment for contractors to maximise the advantages of utilizing current EM tools due to sharing EM-related data, information, and knowledge in construction project management.

As mentioned above, the EMS is not as acceptable as the EIA at present in some countries and areas such as mainland China partly due to the lack of E3 EM approaches in construction, besides the governmental regulations; the tendency of practical EM in construction is to adopt and implement the EMS when the EIA report/form of a construction project has been approved. As a result, the E+ for contractors to enhance their environmental performance, which integrates all necessary KW and EM tools, simply appropriates to the occasion.

The proposed E+ aims to provide high levels of insight and understanding regarding the EM issues related to the project management in a construction cycle. In fact, current EIA process applied in construction projects is mainly conducted prior to the pre-construction stage when a contractor is required to submit an EIA report/form based on the size and significance of the project, and the EIA process for the construction stage is seldom conducted in normalised forms. Due to the strong alterability of the environmental impacts in the construction cycle, commonly encountered static EIA process prior to construction cannot accommodate the implementation of the EMS in project construction, and a dynamic EIA process is thus designed for the E+. In addition, several EM tools are to be combined with the ISO 14000 Model (ISO, 2002), which is a frame for implementation of the ISO 14000 series, according to their interrelationships with which various EM-related data, information, and knowledge in construction can be captured, organised, and reused (see Figure 1). Because the main task of the EM in the construction cycle is to reduce adverse environmental impacts, the dynamic data transference in the framework is the prime focus of the E+ methodology. Thus, a prototype of the E+ is put forward in Figure 1.

Operation

Operation of the E+ model needs an integrative software environment in which various EM tools can work together with the EMS process. There are three main steps to develop the E+ system (software environment) as follows:

1. **Feasibility study.** The feasibility study is conducted not only prior to the establishment of the E+ model, but also before the system analysis and realisation of the E+ software environment. First of all, it is important to analyse whether such an E+ system is necessary for the EMS-based dynamic EIA process in project construction, and this is to be done prior to the establishment of the E+ model. Next, if the E+ system is necessary, it is required to search for enough quantitative EM tools to support the E+ system, and this is to be done before the system analysis and development of the E+ system. The feasibility study is essential for both a practicable E+ model and an E3 E+ system.
2. **System analysis and realisation.** The system analysis and realisation is to be conducted after the E+ model has been established. The aim of this step is to realise the E+ system from a model to a software environment with computer programming. Being limited by the length of this chapter, no further discussion is presented here to illustrate the development of the E+ system.
3. **Application test.** The application test is a trial process for the developed E+ system. There is also no further discussion related to this step due to the under-construc-

Figure 1. The prototype of the E+

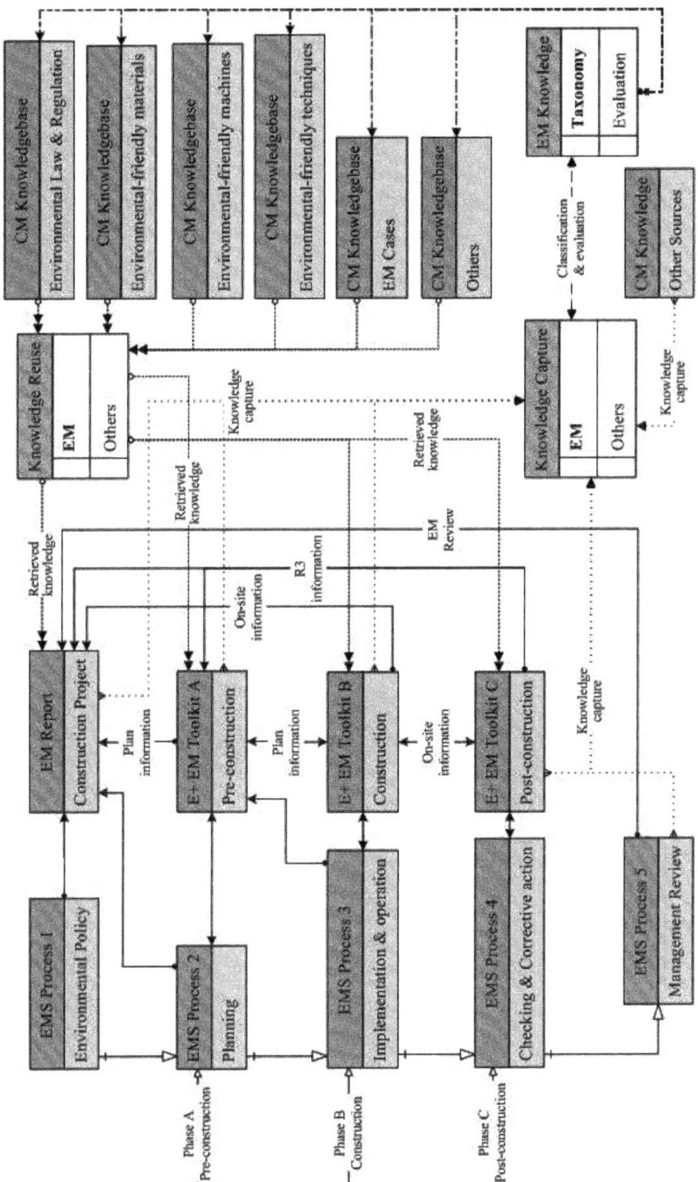

tion of the E+ system. However, an experimental case study is conducted to demonstrate the E3 EM function of the E+ system.

The following discussions focussed on several core EM tools adopted in the E+ model, and interrelationships among these EM tools as to working for the EMS-based dynamic EIA process. The EM tools selected for the E+ model in this chapter include: the CIP approach for adverse environmental impacts indicated at the pre-construction stage, the IRP approach for material management onsite at construction stage, and Webfill approach for residual and waste material and equipment exchange at post-construction stage; no further EM tools will be discussed here.

EM Tools for the E+

CPI

The Construction Pollution Index (CPI) (Chen et al., 2000) is an approach to quantitatively measure the amount of pollution and hazards generated by a construction project or a construction process during construction. Because the value of CPI reflects the accumulated amount of adverse environmental impacts generated by a construction project within its project duration, its utilisation in construction planning is easily realised though a CPI histogram, similar to the resource histogram in a Gantt chart used in construction scheduling. By integrating the concept of CPI into a commonly used tool for construction project management such as Microsoft Project, a system to neatly combine EM with project management is then formed, and project managers can use the CPI histogram to identify the periods in which the project will generate the highest amount of pollution and hazards, and reschedule the whole project to level extremely high CPI (Li et al., 2002).

In the E+ prototype, the CPI approach is set to the E+ EM Toolkit A for construction planning at the pre-construction stage (see Figure 1). The E+ EM Toolkit A captures data from three kinds of sources:

- **Source one:** EMS Process, including EMS Processes 2 and 3.
- **Source two:** E+ EM Toolkits, including Toolkits B and C.
- **Source three:** KW of EM, including knowledgebase of environmental law and regulation, environmental-friendly construction materials, environmental-friendly construction machines, environmental-friendly construction techniques, and EM cases.

Meanwhile, the E+ EM Toolkit A transfers data to these three kinds of data sources and the EM report, such as the EIA report of a construction project.

IRP

The Incentive Reward Program (IRP) approach (Chen et al., 2002b) is an approach to quantitatively measure the amount of material waste generated by a construction project or a process during construction.

In the E+ prototype, the IRP approach is set to the E+ EM Toolkit B for construction at the construction stage (see Figure 1). The E+ EM Toolkit B captures data from three kinds of sources:

- **Source one:** EMS Process 3.
- **Source two:** E+ EM Toolkits, including Toolkits A and C.
- **Source three:** KW of EM, including knowledgebase of environmental law and regulation, environmental-friendly construction materials, environmental-friendly construction machines, environmental-friendly construction techniques, and EM cases.

Meanwhile, the E+ EM Toolkit B transfers data to these three kinds of data sources and the EM report, such as the EIA report of a construction project.

Webfill

The Webfill approach (Chen et al., 2002b) is an e-commerce method to increase the amount of construction and demolition waste exchanged for reuse and recycle among different construction sites and material regeneration manufacturers. As a matter of fact, disposal of C&D waste to landfills usually comes with a monetary charge in many countries (Chen, 2003). For example, to dispose of C&D waste in an orderly manner to disposal facilities by trucks, a trip-ticket system (TTS) was conducted of the Hong Kong construction industry in 1999, which requires contractors to pay for the disposal of their C&D waste in terms of waste disposal receipts issued to them. The Webfill approach sets a TTS-based e-commerce model conforming to the external requirement, and simulation results indicate that the Webfill-enhanced TTS can apparently reduce the total amount of the C&D waste through encouraging the increase of waste reuse and recycle.

In the E+ prototype, the Webfill approach is set to the E+ EM Toolkit C for post-construction work at the post-construction stage (see Figure 1). The E+ EM Toolkit C captures data from three kinds of sources:

- **Source one:** EMS Process 4.
- **Source two:** E+ EM Toolkit B.
- **Source three:** KW of EM, including knowledgebase of environmental law and regulation, environmental-friendly construction materials, environmental-friendly

construction machines, environmental-friendly construction techniques, and EM cases.

Meanwhile, the E+ EM Toolkit C transfers data to these three kinds of data sources and the EM report, such as EIA report of a construction project.

Interrelationships

The interrelationships among the EMS process, the EIA process, the EM Toolkit, and the knowledge capture and knowledge reuse processes can be put up in agreement with EM-related data transfers. There are six kinds of EM-related data transfers in the E+ system. The first kind occurs between the EMS process and the EIA process; the second kind occurs among the EM toolkits and the EIA process; the third kind occurs among the various EM toolkits; the fourth kind occurs from the knowledge reuse entity to the EM toolkits and the EIA process; the fifth kind occurs from the EM toolkits, the EIA process, and the EMS process to the knowledge capture entity; and the sixth kind occurs from the knowledge capture entity to the knowledge reuse entity through several essential KWs such as knowledgebase of environmental law and regulation, environmental-friendly construction materials, machines, techniques, and EM cases. Because all these data are generated from different construction stages, integrative data transfer in the E+ system can thus provide up-to-date information to the EIA process, and the dynamic EIA process is realised accordingly. In order to completely clarify the interrelationship potentially existing in the E+ system, some of the EM-related data and their transfers are summarised in Table 1.

An Experimental Case Study

The experimental case study conducted here combines data such as wastage at different construction stages (Vaid & Tanna, 1997) from several separated cases (Chen et al., 2000, 2002) and authors' experiences with a virtual construction project because there are no data available at present to demonstrate the utilisation of the E+ system from only one construction project. In this case, the aim of this experimental case study focuses mainly on the utilisation of the E+ model. Data adopted are for reference only, although there is practical background to support them. Case studies for real construction projects can be further conducted in the future when the E+ software environment has been realised.

The experimental case study conducted in Table 2 demonstrates the process of the E+ model. The process of the EMS-based dynamic EIA in the experimental case study is divided into three stages corresponding with the construction cycle, including pre-construction stage, construction stage, and post-construction stage (see Table 2). The EM-related data for the EMS-based dynamic EIA provided by the E+ system are different from stage to stage. At the pre-construction stage, there are two kinds of data for the EIA

Table 1. Interrelationship among EM-related data in the E+ system

Data Host	Data Name	Transfer to	Captured from	Usefulness
CPI Host (Toolkit A)	CPI and CPI_i	EIA Host and KW	-	Data update for EIA report
	CPI_i	EMS Process 1 and KW	-	Construction planning
	CPI_{waste}	IRP Host and KW	-	Quantity survey of waste
	h_i	KW	KW	Hazard magnitudes
	D_i	KW	KW	Construction duration
	$\Delta Q^i(j)$	-	IRP Host and KW	Wastage rate survey
	Q_{sold} and Q_{bought}	-	Webfill Host and KW	Wastage rate survey
	undefined	-	EMS Process 3&5	Pollution & hazard survey
IRP Host (Toolkit B)	$\Delta Q^i(j)$	CPI Host and KW	-	Wastage rate
	$\Delta Q^i(j)$ and $C^i(j)$	EMS Process 3 and KW	-	Reward
	$\Delta Q^i(j)$	Webfill Host and KW	-	Quantity survey of waste
	CPI_{waste}	-	CPI Host and KW	Wastage rate
	$Q^i_{de}(j)$ and $Q^i_{re}(j)$	-	EMS Process 3&4 and KW	Quantity suvey of waste
	Q_{sold} and Q_{bought}	-	Webfill Host and KW	Quantity survey of waste
Webfill Host (Toolkit C)	Q_{sold} and Q_{bought}	IRP Host and KW	-	Quantity survey of waste
	Q_{sold}	EMS Process 4 and KW	-	Quantity suvey of waste
	Q_{bought}	EMS Process 3 and KW	-	Deliver to crews
	Q_{sold} and Q_{bought}	CPI Host and KW	-	CPI_{waste}
	undefined	-	IRP Host and KW	Waste for exchange
	undefined	-	EMS Process 4 and KW	Price of waste
EIA Host	CPI and CPI_i	-	CPI Host and KW	Data update for EIA report
	$\Delta Q^i(j)$	-	IPR Host and KW	Data update for EIA report
	Q_{sold} and Q_{bought}	-	Webfill Host and KW	Data update for EIA report
	undefined	-	EMS Process 1 and KW	Data update for EIA report
	undefined	-	EMS Process 2 and KW	Data update for EIA report

Note: 1. CPI_{waste} represents the CPI_i that involves waste impact only;
2. Q_{sold} is the quantity of C&D waste a contractor sold; and
3. Q_{bought} is the quantity of regenerated materials or reusable material a contractor bought.

Table 2. Case study of the E+ implementation for a dynamic EIA in a construction cycle

		EIA data at different construction stages								
		Pre-construction		Construction				Post-construction		
Construction Tasks	Duration (day)	Original CPI_{waste}	Q_{bought} (ton)	Relay CPI_{waste}	$\Delta Q^i(j)$ (ton)	Q_{sold} (ton)	Q_{bought} (ton)	Final CPI_{waste}	Total Q_{sold} (ton)	Total Q_{bought} (ton)
Caisson Pile	31	0.07	0.0	0.05	0.1	0.1	0.0	0.05	0.1	0.0
Braced Excavation	57	0.13	0.0	0.15	0.5	0.3	0.0	0.10	0.5	0.0
Transportation	435	0.00	0.0	0.05	0.0	0.0	0.0	0.05	0.0	0.0
Support System-Building	42	0.10	22.6	0.12	1.2	0.2	5.5	0.10	0.6	28.1
Support System-Demolition	28	0.60	0.0	0.50	0.0	2.1	10.0	0.45	20.5	10.0
Foundation Construction	43	0.10	0.1	0.20	3.3	3.2	1.1	0.18	5.6	1.2
Structural RC Work-Rebar	155	0.06	3.2	0.05	1.5	5.5	1.8	0.04	6.0	5.0
Structural Work-Form	155	0.36	5.7	0.30	1.2	0.2	1.1	0.28	0.5	6.8
Structural Work-Concrete	156	0.05	0.0	0.03	3.1	5.1	0.0	0.02	5.1	0.0
Mansion Work	176	0.20	0.5	0.10	5.6	3.3	2.3	0.09	3.5	2.8
Structural Steel	94	0.10	0.0	0.01	0.5	0.0	0.0	0.01	0.0	0.0
Finish Work-Wall	211	0.29	0.0	0.18	2.5	3.6	0.0	0.16	3.6	0.0
Finish Work-Ceiling	211	0.39	0.0	0.26	3.2	5.5	0.0	0.25	5.5	0.0
Finish Work-Floor	181	0.12	0.0	0.08	1.7	2.1	3.0	0.07	2.1	3.0
Total		2.57	32.1	2.08		31.2	24.8	1.85	53.6	56.9

including the original set of CPI_i and the Q_{bought} requested by each crew; at the construction stage, there are four kinds of data for the EIA including the relay set of CPI_i, the $Q^i(j)$, the Q_{sold}, and the Q_{bought}; and at the post-construction stage, there are three kinds of data for the EIA including a final set of CPI_i, a total Q_{sold}, and a total Q_{bought}. Because the functions of current EM tools integrated in the E+ model are different, for example, the CPI approach deals with total adverse environmental impacts of construction processes, while the IRP approach and the Webfill approach deal with the C&D waste only; the CPI_i in this case study is thus represented by a CPI_{waste}, which represents the CPI_i that involves waste impact only (see Table 1).

Moreover, the case study puts forward and utilises the concepts of original CPI_{waste}, relay CPI_{waste}, and final CPI_{waste} to demonstrate the process of the EMS-based dynamic EIA, and looks at these three kinds of CPI as essential data in an EIA report. The original CPI_{waste} means the CPI_{waste} that is valued before a construction process, the relay CPI_{waste} means the CPI_{waste} that is devalued during a construction process, and the final CPI_{waste} means the CPI_{waste} that is finally valued after a construction process. Because the value of the CPI_{waste} is regarded as one important data in an EMS-based EIA process, the changing process of the three kinds of CPI_{waste} appropriately incarnates or reflects the process of a dynamic EIA. Thus an EMS-based dynamic EIA process is realised.

It is necessary to notice that in order to value each CPI_{waste}, expert experiences have to be used corresponding to changed amounts of Q_{sold} and Q_{bought}. The expert experiences required are stored in the knowledge reuse entity, which is distilled from crude data in KW including knowledgebase of environmental law and regulation, environmental-friendly construction materials, machines, techniques, and EM cases. Therefore the EM-related construction knowledge effectively supports the process of the EMS-based dynamic EIA.

The result of this experimental case study indicates that the implementation of the E+ system can finally reduce the adverse environmental impacts of project construction. For example, the total value of the original CPI_{waste} of all construction tasks in the experimental case study is 2.57, the total value of the relay CPI_{waste} of all construction tasks in the experimental case study is 2.08, and the total value of the final CPI_{waste} of all construction tasks in the experimental case study is 1.85. That is, the E+ system draws support from several EM tools such as the CPI approach, the IRP approach, and the Webfill approach, and realises an EMS-based dynamic EIA process, while the benefits of various EM tools can be shared within the E+ environment through EM-related data transfer and integrative data, information, and knowledge utilisation.

Future Trends

Recommendations on the integrative KMS for environmental-conscious construction come from the usefulness, efficiency, and benefit of the E+ prototype and EM tools, which have been demonstrated in this chapter. However, due to the limitations of current research, it is thus recommended that further research be conducted on both the development of the E+ software environment and the development of more E3 EM tools for the E+ system.

First of all, the E+ model can be further developed to Web-based environmental information and KMS for a contractor to implement EM in construction project management. According to the essential theory and practice of EM in construction, the environmental information is required in construction planning, construction material management, and C&D waste exchange, whilst the EM knowledge in construction is essential to support decision making by using various EM tools. Because both environmental information and EM knowledge are needed in the E+ system, and the Internet is particularly suitable to implement effective and flexible CM by mobile site management units, the key issues of the development of such an E+ system are: how to establish a Web-based software system and enable managers in different construction sites to use and share environmental information and knowledge on the same platform of E+; and how to capture, transfer, and reuse necessary data between the E+ system and current CM system. Moreover, additional functional components such as E+ EIA—besides the E+ toolkits—are also under consideration.

Besides the development of the E+ system, further research is also needed in the development of fully user-oriented EM tools and their integration in the E+ system. The fully user-oriented EM tools can enable a contractor or construction manager to use the EM toolkits easily, without the help of the tool developers. For example, the fully user-oriented tool of CPI can enable them to define the CPI of each construction process in a construction plan and then level the extremely high CPI, whilst the fully user-oriented tool of EnvironalPlanning (Chen, 2003) can enable them to transfer necessary data from construction plan alternatives to an ANP environment and thus select the most environmental-friendly construction plan. In addition to the development of the fully user-oriented EM tools, improvements in the functional-different approaches for EM in construction are also necessary, focusing on the innovation of these approaches. For example, the CPI of a construction process is currently defined by experts' experience, and this treatment is definitely practicable; however, in order to receive a wide recognition and minimise the arbitrary decision or subjective error on the definition of the CPI of each construction process, we suggest development of an objective calculation method to define the CPI of each possible construction process a contractor may use in construction planning. So both the development of the fully user-oriented EM tools and the consummation of current functional-different approaches for EM in construction are required in further research.

Beyond the consummation of the E+ system and its components, additional functional-different approaches for EM in construction are also necessarily to be developed in order to improve the performance of the E+ system. Currently, the potential functional-different approaches to implementation of EM in construction include the life cycle analytical (LCA) approach and risk analytical approach for E+ EM Toolkit A, and the EIA template for the new E+ EIA Toolkit.

Although this research project has been accomplished with satisfying results, there are some limitations not only within the research but also in the duplicate implementation of the EM tools developed in the research. The limitations of the research exist in the following areas:

- This research has not accomplished an E+ software environment to further demonstrate its usefulness, efficiency, and benefit for EM in construction.
- The CPI method has not been developed into a fully user-oriented tool to help contractors deal with any CPI problem in construction planning.

As a conclusion, it is recommended that further research and development for the E+ system focus on the development of a Web-based E+ system, the consummation and innovation of current EM tools in construction, and the development of new EM approaches for the E+ system.

Conclusion

This chapter presents an integrative prototype to knowledge-based dynamic EM in construction, which integrates various EM approaches and knowledge bases with a general EMS process throughout all construction stages in a construction project. The EM approaches integrated in the E+ are divided into three categories, including EM Toolkit A for pre-construction, EM Toolkit B for construction, and EM Toolkit C for post-construction. These EM toolkits are further integrated with an ISO 14001 EMS and EM knowledge capture and reuse entities for an integrative KMS for environmental-conscious construction. In addition to the proposed E+ prototype, an experimental case study has also been conducted to demonstrate the usefulness, efficiency, and benefit of the E+ system. The E+ is expected to effectively, efficiently, and economically assist contractors in enhancing both their EM techniques and environmental performances in construction project management, and to overcome the weakness of static EIA formally applied in the construction industry in some countries by the dynamic EIA process where the necessary data for an EIA report can be updated in the construction cycle.

Practical Tips and Lessons Learned

For the efficient management of environment-related knowledge, it is essential to integrate data and information from various dispersed tools and applications. In order to promote the implementation of the E+ model, further research is needed to transfer the E+ model to a computer software environment, improve current EM tools, and develop more EM approaches as subsidiary components of the E+ system to deal with all adverse environmental impacts of construction on the purpose of total EM in construction project management. Although the software environment of the E+ has not been presented in this chapter, the demonstration of the E+ model in the experimental case study enabled a closer understanding of how the E+ system can be effectively applied for EM in construction, and it also unveiled that the E+ methodology is flexible in the integrative implementation of functional-different quantitative approaches to EM in construction.

References

Azani, C.H. (1999). An integrative methodology for the strategic management of advanced integrated manufacturing systems. In J. Sarkis & H.R. Parsaei (Eds.), *Advanced manufacturing systems: Strategic management and implementation* (pp. 21-41). Australia: Gordon and Breach Science.

Chen, Z. (2003). *An integrated analytical approach to environmental management in construction.* PhD Thesis, Department of Building and Real Estate, Hong Kong Polytechnic University.

Chen, Z., & Li, H. (2003). An analytic network process model for demolition planning. *Proceedings of the CIB Student Chapters International Symposium on Innovation in Construction and Real Estate*, Hong Kong Polytechnic University, September.

Chen, Z., Li, H., & Wong, C.T.C. (2000). Environmental management of urban construction projects in China. *ASCE Journal of Construction Engineering and Management, 126*(4), 320-324.

Chen, Z., Li, H., & Wong, C.T.C. (2002a). An application of bar-code system for reducing construction wastes. *Automation in Construction, 11*(5), 521-533.

Chen, Z., Li, H., Shen, G.Q.P., & Xu, W. (2002c). An empirical model for decision-making on ISO 14000. *Construction Management and Economics, 22*(1), 55-75.

Chen, Z., Li, H., & Wong, C.T.C. (2002d). Webfill before landfill: An e-commerce model for waste exchange in Hong Kong. *Journal of Construction Innovation, 3*(1), 27-43.

Chen, Z., Li, H., & Wong, C.T.C. (2002e). EnvironalPlanning: An analytic network process model for environmentally conscious construction planning. Accepted for publication in the *ASCE Journal of Construction Engineering and Management.*

Chen, Z., Li, H., & Wong, C.T.C. (2003a). An integrative analytical system for environmental management in construction projects. *Automation in Construction, 13*(5), 621-628.

Chen, Z., Li, H., & Wong, C.T.C. (2003b). E+: An integrative methodology for dynamic EIA in construction. *Proceedings of the 3rd International Post-Graduate Research Conference*, Salford University, Lisbon, Portugal, April.

Chen, Z., Li, H., & Wong, C.T.C. (2003c). Environmental priority evaluation for construction planning. *Proceedings of the 2nd International Conference on Innovation in Architecture, Engineering and Construction (AEC)*, Loughborough University, UK, June.

Chen, Z., Li, H., & Wong, C.T.C. (2003d). An integrative methodology for dynamic EM in construction. *Proceedings of the ARCOM Doctoral Workshop*, Glasgow Caledonian University, Glasgow, UK, 18 June.

Chen, Z., Li, H., Wong, C.T.C., & Love, P.E.D. (2002b). Integrating construction pollution control with construction schedule: An experimental approach. *Environmental Management and Health, 13*(2), 142-151.

China EPB (Environmental Protection Bureau). (2002). *Official report on the state of the environment in China.* Retrieved 26 December from www.zhb.gov.cn/bulletin/country_intro.php3.

CIRIA. (1993). *Environmental issues in construction—a review of issues and initiatives relevant to the building, construction and relevant industries. Volume 2—technical review.* Publication Code: SP94. London: Thomas Telford.

CIRIA. (1994). *Environmental assessment.* Publication SP96. London: Thomas Telford.

CIRIA. (1994). *Environmental handbook for building and civil engineering projects: Checklists, obligations, good practice and sources of information.* Publication Code: SP97&98. London: Thomas Telford.

CIRIA. (1995). *A client's guide to greener construction: A guide to help clients address the environmental issues to be faced on building and civil engineering projects.* Publication Code: 120. London: Thomas Telford.

CIRIA. (1999). *Environmental issues in construction—research campaign. Executive summary.* Publication Code: PR74. London: Thomas Telford.

Clough, R.H., & Antonio, N. (1996). *Environmental management in construction: Model forms to assist implementation.* UK: The Chartered Institute of Building (CIOB).

CPSC. (1998). *Environmental management systems—guidelines.* Construction Policy Steering Committee, NSW Government, Australia. Retrieved 26 March from www.cpsc.nsw.gov.au/environment/.

CPSC. (2001). Construction industry CEO survey: NSW construction industry survey of industry leaders. Retrieved 26 March from www.cpsc.nsw.gov.au/docs/strategic-info/CEO-Survey-Fax-2001.pdf.

FIDIC. (1998). *Guide to ISO 14001 certification/registration* (test edition). Switzerland: International Federation of Consulting Engineers.

Guthrie, P., & Mallett, H. (1995). *Waste minimisation and recycling in construction: A review.* Publication Code: 122. London: CIRIA.

Henningson, J.C. (1978). Environmental management during construction. *ASCE Journal of the Construction Division, 104*(4), 479-485.

IAIA (International Association for Impact Assessment). (1997). *Principles of environmental impact assessment best practice.* Retrieved 26 March from www.iaia.org/principles/.

Jeljeli, M.N., & Russell, J.S. (1995). Coping with uncertainty in environmental construction: Decision-analysis approach. *ASCE Journal of Construction Engineering and Management, 121*(4), 370-380.

Jones, K.H. (1973). Synthesis approach to determining research needs in civil engineering; air pollution a test case. *ASCE Journal of the Environmental Engineering Division, 99*(4), 461-467.

Kein, A.T.T., Ofori, G., & Briffett, C. (1999). ISO 14000 series: Its relevance to the construction industry of Singapore and its potential as the next industry milestone. *Construction Management and Economics, 17*(4), 449-461.

Khalfan, M.M.A., Bouchlaghem, N.M., Anumba, C.J., & Carrillo, P.M. (2003). Knowledge management for sustainable construction: The C-SanD project. In K.R. Molenaar & P.S. Chinowsky (Eds.), *Winds of change: Integration and innovation in construction. Proceedings of the 2003 Construction Research Congress,* Honolulu, Hawaii, 19-21 March. Reston, VA: American Society of Civil Engineers.

Li, H., Chen, Z., & Wong, C.T.C. (2001). Application of barcode technology for an incentive reward program to reduce construction wastes in Hong Kong. *Computer-Aided Civil and Infrastructure Engineering, 18,* 313-324.

Li, H., Chen, Z., Wong, C.T.C., & Love, P.E.D. (2002). A quantitative approach to construction pollution control based on resource leveling. *International Journal of Construction Innovation, 2*(2), 71-81.

Lippiatt, B.C. (1999). Selecting cost-effective green building products: BEES approach. *ASCE Journal of Construction Engineering and Management, 125*(6), 448-455.

Lo, C.H. (2001). *Critical factors for the implementation of ISO 14001 environmental management system in Hong Kong construction industry.* MSc Thesis, Hong Kong Polytechnic University.

McMullan, R. (1998). *Environmental science in building* (4th edition). Basingstoke, England: Palgrave Macmillan.

Ofori, G., Briffett, C., Gang, G., & Ranasinghe, M. (2000). Impact of ISO 14000 series on construction enterprises in Singapore. *Construction Management and Economics, 18*(8), 935-947.

Petts, J. (1996). Environmental assessment: Good practice. *Proceedings of the Construction Industry Environmental Forum Conference on Good Practice in Environmental Assessment.* Publication Code: SP126. London: CIRIA.

SAP INFO. (2004). *SAP INFO glossary.* Retrieved on 22 July from www.sap.info/public/en/glossary.php4/displayglossarystart.

Shorrock, L. et al. (1993). *Environmental issues in construction—a review of issues and initiatives relevant to the building, construction and related industries. Volume 1—overview including executive and technical summaries.* Publication SP93. London: CIRIA.

Skoyles, E.R., & Hussey, H.J. (1974). Wastage of materials. *Building,* (February), 95-100.

Spivey, D.A. (1974). Construction solid waste. *ASCE Journal of the Construction Division, 100*(4), 501-506.

Tse, R. (2001). The implementation of EMS in construction firms: Case study in Hong Kong. *Journal of Environmental Assessment Policy and Management (JEAPM), 3*(2), 177-194.

Uren, S., & Griffiths, E. (2000). *Environmental management in construction.* Publication Code: C533. London: CIRIA.

USEPA. (1971). Noise from construction equipment and operations, building equipment, and home appliances. Washington, DC: Environmental Protection Agency.

Vaid, K.N., & Tanna, A. (1997). *Wastage control of building materials in construction of mass housing projects* (pp. 19-35). Mumbai, India: NICMAR (National Institute of Construction Management and Research).

Valdez, H.E., & Chini, A.R. (2002). ISO 14000 standards and the US construction industry. *Environmental Practice, 4*(4), 210-219.

Velker, L. (1999) Caution: KM under construction—challenges and solutions from a geographically diverse contractor. *KMWorld Magazine, 8*(5).

Warren, F.H. (1973). Environmental impact on project schedules. *ASCE Journal of Professional Activities, 99*(3), 299-306.

Zarli, A., Rezgui, Y., & Kazi, A.S. (2003). Present and future of European research on information technologies in construction. In K.R. Molenaar & P.S. Chinowsky (Eds.), *Winds of change: Integration and innovation in construction. Proceedings of the 2003 Construction Research Congress,* Honolulu, Hawaii, 19-21 March. Reston, VA: American Society of Civil Engineers.

Zeng, S.X., Tam, C.M., Deng, Z.M., & Tam, V.W.Y. (2003). ISO 14000 and the construction industry: Survey in China. *ASCE Journal of Management in Engineering, 19*(3), 107-115.

Zyngier, S.M. (2002). *Knowledge management strategies in Australia: Preliminary results of the survey of the knowledge management uptake in Australian companies.* School of Information Management and Systems, Monash University, Caulfield East, Victoria, Australia.

Chapter XVIII

Configurable Environments: A Vision for Future Project Information Technologies

William J. O'Brien
University of Texas at Austin, USA

Raja R.A. Issa
University of Florida, USA

Zhigang Shen
University of Florida, USA

Haiyan Xie
University of Florida, USA

Abstract

Configurable environments are presented as a specific vision for information technologies to support design and construction professionals. Configurable environments combine data integration capabilities with visualization and analysis tools in a format customized to the needs of specific users and projects. A key component of the approach is the deployment of formal user models to direct configuration of generalized tools and distributed data. This chapter details the need for and technical aspects of the configurable environments approach in the context of an illustrative case study of construction space management. Issues with current practice and implications for future practice are discussed in the context of project professionals' information processing and evaluative capabilities and needs. Also discussed is the relationship of the configurable environments approach to current technologies and to broader visions for information technology development in design and construction.

Introduction

The intention of this chapter is to outline a novel approach—*configurable environments*—to make information more accessible and useful to design and construction practitioners. The configurable environments approach combines personalization with complex collection and processing of data. "Personalization" remains a current buzzword encompassing various approaches, including custom filters, data visualization, user modeling, intelligent user interfaces, and so forth. Collection and processing of data is a broad concept that includes interoperability, autonomous agents, and mediation. At least by inspiration and example, the configurable environments (CE) approach combines elements of all the above.

Novel aspects of the CE approach are the complexity of the problem addressed (e.g., real problems in design and construction) and the ability to support users who, while professionals, are not narrow technicians or computer scientists. For design and construction users, the authors envision a toolkit that: (1) provides complex analysis and visualization tools to provide what-if type analysis capabilities, (2) draws data from multiple sources, and (3) is configurable to the needs of the user (and hence is not a static toolkit). CEs differ from current approaches that provide users a set of (rigid) applications that make use of specific sets of information. These current approaches have been characterized as "information tunnels" (Tannenbaum, 2002). Further, the authors argue that such tunnels persist in an environment with data standards designed for interoperability.

The chapter presents the case for, vision, and pathways to configurable environments as follows: After some brief background, a motivating case study concerning space management on construction sites is used to highlight issues with current practice and limitations of existing technologies. This supports generation of the configurable environments vision for practice. Supporting research in problem-solving environments, user modeling, and information integration are reviewed, enabling detailed description of the CE information architecture. The chapter concludes by considering future trends, including the potential scope of CEs in practice and pathways between existing technologies and development of commercial CEs.

Learning Objectives

Upon completion of this chapter, the reader shall be able to:

1. Understand the limitations of current information technology applications to support personalized views and analysis of project information, with particular reference to the challenges facing project superintendents.
2. Understand the concept of and proposed implementation architecture for the configurable environments (CE) approach.

3. Identify methods supporting the CE approach, including user modeling and problem-solving environments, and relate these approaches to knowledge management needs in design and construction.
4. Describe the potential scope of CEs in relation to broader visions for development of information technology applications in design and construction.
5. Identify the implications of CEs for knowledge management, particularly with regard to organizational development and individual work experience.

Background

There have long been calls for information technologies to support design and construction practice. While the past decade has seen increasing penetration of technology in the design office and on the construction job site, it is reasonable to say that few information technologies are making a dramatic, large-scale difference across the industry. A recent case study of a Finish project (Kam & Fischer, 2002) that tested the ability of information standards to support interoperability provides hints as to why radical innovations in practice are difficult to deploy: Information technologies are not yet able to bridge domains of practice with relative ease. The setup costs are high in both time and technical expertise, making wide-scale deployment difficult. Limitations of existing technologies are also reflected in a recent survey of 88 construction practitioners in the Southeast U.S., a majority of whom agreed with statements that the technology is immature with respect to the needs of the industry, and that the industry is still learning how to best make use of existing technologies (El-Mashaleh, 2003).

A specific opportunity and challenge of information technology is to leverage electronically stored information to support decision making. An ever-increasing amount of project information is generated and stored in computer format. Yet even with advanced applications based on shared data standards, project professionals still face considerable difficulty accessing needed information. In the authors' viewpoint, a primary reason for these difficulties is that most data and tools are designed and operated in the context of specific applications. This creates the problem of information tunnels: rich data and application areas with knowledge that is not easily transferable to other application areas. Information tunnels per se would not be a difficulty if projects had identical problems; walls between tunnels could be torn down or professionals could develop heuristic integration mechanisms that functioned with a high degree of reliability, speed, and accuracy. But each project faces unique challenges, making integration across tunnels difficult for both machines and humans. Hence, we envision information technologies that provide project professionals much greater configuration control in how they access project information, regardless of its source or tunnel.

The CE approach draws from current construction information technology research in visualization, data integration, and application development. Broadly, the CE approach is a specific instance of broad visions supported by information technology research and development organizations such CIB W78 (*w78.civil.auc.dk*) and FIATECH (*www.fiatech.org*). In a review of W78 research, Amor, Betts, Coetzee, and Sexton (2002) note that a major theme is the "modeling of processes and products integrated with visualization and simulation of information life cycles," with particular emphasis on flexible approaches to information representation and management. Similarly, the FIATECH technology roadmap calls for development of integrated systems, information on demand, and scenario-based planning systems to explore alternatives (FIATECH, 2003). The authors' CE approach is designed to be flexible in response to user needs while supporting complex analysis.

Space Management: A Sample Application for Configurable Environments

The configurable environments approach is broadly applicable to a wide range of problems in design and construction. For the purposes of this chapter, we discuss an application in construction space management. Space management is long understood to be an information-intensive problem that relates problems of scheduling, trade coordination, and procurement. As such, a stream of research has been performed to classify types of space use and/or propose algorithms and tools to aid practitioners (e.g., Akinci, Fischer, Levitt, & Carlson, 2002; Riley, 2000; Riley & Sanvido, 1997; Thabet & Beliveau, 1997; Tommelein, Levitt, Hayes-Roth, & Confrey, 1991; Tommelein & Zouein, 1993). This research provides a rich research base from which a configurable environment implementation can be built. However, all this research is directed at solving specific types of space management problems. No one application individually resolves all potential problems or is suitable to all projects. The research must be organized in a framework that allows its collective capabilities to bear on the space management problem.

Sample Project

A motivating case, shown in Figure 1, is the Tomorrow Square in Shanghai, China.

The building includes a six-story base and a 60-story tower, which has office space in the lower part and a hotel in the upper part. The hotel floors consist of flat concrete slabs, and the office floors are concrete beam and slab construction. The building is divided into three construction areas: A, B, and core areas for fast track construction, which allows at any moment for many parallel activities. As with most urban areas, site storage space is limited. Local rules and traffic conditions mean shipments are allowed only at night, exacerbating the need for timely and correctly sized and placed materials deliveries.

Consider the specific challenge faced by a superintendent on the Tomorrow Square project. The schedule is tight, with many precedence relationships. (Elements of the schedule for the tower construction are shown in Figure 2.) The drywall, electrical, plumbing, and mechanical subcontractors are working on Floors 4-7. The masonry and

Figure 1. Tomorrow Square in Shanghai, Architect: John Portman & Associates

Figure 2. Partial construction schedule for Tomorrow Square tower (dark bars indicate critical path activities)

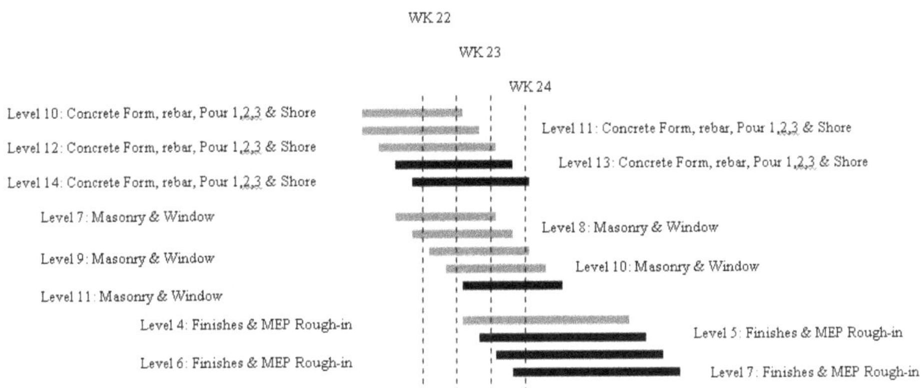

window subcontractors are working on Floors 7-11. The formwork, rebar, and concrete subcontractors are working on Floors 10-14. Limited space and rapid schedules allow for limited pre-placement of materials. The superintendent is required to place orders for materials and coordinate subcontractor space needs on a daily basis throughout the building. Because of variations in productivity rate as well as some variations in quantity of work across floors, changes in plans must be made every day.

The coordination difficulties faced by the superintendent are considerable. Limiting discussion to the activities depicted in Figure 2, in Week 23 there are seven subcontractors, each with at least one materials supplier, posing at least 14 firms that need coordination. With daily changes, there is a need to collect information from each of these firms to confirm/re-coordinate delivery and labor. This may need to be an iterative process if the subcontractors and suppliers cannot meet the initial schedule proposed by the superintendent. Further, it is difficult to decentralize the decision making or break it into smaller components. Typically, management focus is given to critical path activities, but when the schedule is fast paced, a delay to one activity can easily place it on the critical path. Each activity must be managed aggressively. Similarly, decentralizing data collection by having subcontractors manage their suppliers does little to reduce the information gathering needs of the superintendent. Each subcontractor must coordinate with their suppliers after initial discussions with the superintendent, requiring a return call to the superintendent. Hence, in the decentralized scenario, the total amount of communication is greater than in a centralized scenario, potentially leading to an increased rate of communication failure. Of course, the superintendent could take a more authoritarian approach and direct the subcontractors to perform as directed; such an approach can work, provided the subcontractors and suppliers can comply. If not, problems can be magnified, as they may not be detected until they occur in the field.

Of course, collection of data from the affected subcontractors and suppliers is only part of the problem. There is a need for analysis, and here research and commercial software can provide some help. Thabet and Beliveau (1997) and Akinci et al. (2002) describe research models to ensure there are no space conflicts among the trades. However, this research is only a partial solution to the superintendent's coordination challenge. The superintendent must also calculate materials needed and make dynamic adjustments to space needs based on materials and resource availability (information distributed across suppliers). Hence, the superintendent must be able to pull quantities from CAD files based on dynamic sizes of work packages. This is not an easy task, as even with object-oriented CAD models, breaking out quantities for specific work packages remains a manual task (especially if the model was generated with a different level of detail or purpose). These quantities must be augmented with current productivity data to provide a meaningful check of the schedule. Further, as schedules and perhaps work area breakdown changes, this information must be fed into the analysis tool as part of an iterative search for a feasible solution. Recommended schedule changes to avoid conflicts would have to be rechecked against materials delivery and resource constraints. With the difficulties of performing analysis and in collecting data from the participating firms, an iterative search to find the best space management solution is seldom performed. Rather, practitioners look for satisfying solutions to such coordination problems (O'Brien, Fischer, & Jucker, 1995).

Configurable Environments: Needs and Vision

The problems of space management are complex, posing opportunities for sophisticated analysis. At the same time, the information needs are great, requiring ready access to information sources. The presence of information tunnels (Tannenbaum, 2002) due to existing specialized applications and multiple firms makes data collection to support analysis troublesome. Information tunnels per se would not be a difficulty if projects had identical problems; walls between tunnels could be torn down or professionals could develop heuristic integration mechanisms that functioned with a high degree of reliability, speed, and accuracy. But each project faces unique challenges and team composition, making integration across tunnels difficult for both machines and humans. Further, despite common problems on projects, the uniqueness of each project design requires that appropriate analysis tools are utilized.

The scope of the problem faced by the superintendent on Tomorrow Square is not unique. Resolution of many design and construction problems could be aided by analysis tools (and many of those tools have been prototyped in research labs). But providing practitioners with practical tools requires an environment that provides comprehensive solutions set for a given problem. Abstracting from the discussion above, tools that practitioners can use to successfully analyze project-specific problems require a combined approach that enables users to: (1) utilize a variety of visualization and analysis tools, (2) access a range of data across many sources to support analysis, and (3) customize analysis to support the needs of the project and the use/user. We call such an approach a configurable environment (CE).

It is useful to consider how a CE would operate in contrast to existing tools. The CE provides the superintendent an integrated view of the site layout, schedule, resource utilization, and procurement and delivery schedule. The superintendent may access analysis tools to investigate the ramification of different schedules on site space use. Current 4D CAD tools (with some setup difficulty) provide much of the analysis and visualization capabilities envisioned for the site space management CE. Consider though the important differences: Through the CE, the superintendent has access to up-to-date procurement information from a variety of suppliers. In addition, the CE provides the information to the superintendent at a level of detail suitable for decision making (e.g., the shipment will arrive on date x as opposed to a detailed bill-of-materials that may be more suited to estimating or production planning). Rather than utilizing multiple pieces of software (each with its own interface), the CE provides an integrated, single toolkit for an important management function of the superintendent. Further, the CE provides all the information in a visualization environment customized to the superintendent's needs and project specifics.

Supporting Research and Technologies

The CE approach owes a significant intellectual debt to the stream of construction IT research. Current visions for flexibility in modeling construction artifacts (e.g., Clayton, Kunz, & Fischer, 1996; Zamanian & Pittman, 1999) in particular inform the CE approach. However, specifics of the CE approach have been developed from broader research in information technology, including problem-solving environments (PSEs), user modeling, and information integration. Each of these is discussed below in the context of the CE approach. The section on information integration also specifically discusses limitations of data standards to support information exchange required by CEs.

Problem-Solving Environments

PSEs have been defined as "an integrated system that provides all the facilities needed to solve a target class of problems conveniently and efficiently" (Rice & Boisvert, 1996). Inspired by the success of graphical user interfaces, PSEs were developed in the 1990s from a desire to make the large number of scientific computing tools more accessible and more useful. Several PSEs have been developed to help engineers and scientists solve problems. For example, MATLAB® (*www.mathworks.com*) is a successful product well known to many engineers and scientists. Other tools are more focused in application, designed to solve a narrow, domain-specific problem.

Most PSEs provide a useful but static toolkit to help solve a well-understood problem. Rice and Boisvert (1996) note that future PSEs will need to become more flexible to support a broader range of problems, and note an alternate definition of a PSE is the sum of a user interface, libraries, knowledge base, and integration. In this alternate definition, the knowledge base and integration aspects provide the user a more flexible toolkit than that provided by a static library with a visualization component. The range of environments classified as PSEs is large, although a review suggests that even flexible PSEs are designed for highly specific, generally technical applications. Concomitantly, most PSEs are designed for sophisticated users such as engineers and scientists who want to bring a range of tools to a problem. In this sense, the integration and knowledge base components refer to a PSE's ability to seamlessly connect its various library components with visualization tools and aid users in the selection of the appropriate library tools and interpretation of analysis results.

An example of a powerful and extensible PSE is the Environmental Decision Support System (EDSS, *http://www.baronams.com/EDSS/*). With a focus on air quality issues, it provides a library of analysis and simulation components to provide what-if analysis and visualization. The EDSS also provides a superset of its components called the Strategy Development Tool to help technical users evaluate the impact of air pollution control strategies. The EDSS further offers an API from which further tools can be integrated and adaptors to specific data sets can be built. The EDSS is an environment that is self-contained and useful as a toolkit for sophisticated scientists and engineers.

The PSE paradigm provides powerful tools to users. Commercial environments such as MATLAB® demonstrate the sophistication PSEs have reached in terms of user friendliness and ability to meet users' needs. Similarly, PSEs developed by researchers such as EDSS demonstrate the depth of analysis tools available to users in an integrated environment. However, PSEs have specific limitations. First, PSEs are not easily customized to include new tools or capabilities. Non-trivial customization requires the time of expert programmers. Second, PSEs are largely self-contained environments and do not easily integrate with heterogeneous data sources. While there are APIs, binding a legacy source requires expertise with both the data format of the PSE and the source. Third, as environments designed to aid in the solution of specific classes of problems, their analysis and visualization components are not easily applied to related problems that do not fall neatly within the designer's intent.

These three limitations—limited customization, limited integration with heterogeneous sources, and limited application to related problems—make application of PSEs to design and construction problems problematic. The truism that construction projects are unique has, perhaps, been misused by some practitioners to avoid adoption of software tools that aid generalized classes of design and construction problems. But the uniqueness of construction problems does pose problems for PSEs, as general analysis and visualization toolkits may not be fully applicable to (or fully support) project-specific problems.

Consider our sample project and the problem of managing space on construction sites. Much current research is application oriented and, with some reworking, could provide the core of an analysis library and knowledge base for a site space management PSE. But an application of these tools would have to be carefully mapped to the needs of specific sites. For example, the work of Riley and Sanvido (1997) is specific to multi-story buildings. Applying their research to other facility types could lead to problematic results. More broadly, each of the existing research projects makes several explicit and implicit assumptions about the problem being addressed. Without making these assumptions explicit in the PSE, users are likely to find analysis tools that make recommendations inappropriate for their specific site needs. Of course, such a critique is true for application of any formula or tool, but the capabilities provided by an integrated environment such as a PSE invites abuse. A component of the configurable environments vision is to include reasoning tools to aid users in adapting analysis tools to specific projects.

User Modeling

Exacerbating the problems that stem from tools of limited applicability to specific problems is the lack of easy customization. While software applications provide users some ability to filter results, customize menu bars, and so forth, it is generally impossible or impractical for users to make substantive changes in the capabilities provided to them. Proposed solutions to this problem are technologies based on user modeling. User models are software components that contain formal descriptions of users as well as reasoning tools to help describe preferences, enabling adaptive software that can be configured to user needs (Kobsa, 2001). User models are typically generated for a class of user (e.g., project manager, superintendent, etc.), although further customization is possible for preferences of individual users.

Much user modeling research has been directed to develop intelligent user interfaces that control the presentation of information and interaction with applications (e.g. Maybury & Wahlster, 1998). Beyond research implementations, several commercial systems have been developed that utilize user models of varying sophistication. As noted by Strachan, Anderson, and Sneesby (2000), these commercial implementations typically utilize simple or minimalist constructs, as more complex user models do not yet have the necessary robustness for industry application. That said, there are increasingly sophisticated commercial applications such as recommender systems (e.g., GroupLens; *www.netperceptions.com*; Terveen, Hill, & Amento, 1999) that make use of user models to learn about individual users and configure what resources they see. More broadly, the user modeling paradigm has permeated much of the software development community, where rather than an active agent, the user model is a written description used to guide development of software (Cooper & Reimann, 2003; Kulak & Guiney, 2000; Winograd, 1996). In this usage, a user model or user profile is broadly described as the mental picture a user has of how a particular software application should work (Payne, 2003).

Both the academic research literature and professional software development literature agree that specification of the user model—whether paper based or implemented in code—is a challenging task. Specification has dual challenges: understanding the needs of humans and translation of these needs to formal abstractions. The body of professional knowledge about development of user models (e.g., Cooper & Reimann, 2003) recognizes the importance of domain understanding; there is no simple knowledge elicitation formula. Similarly, the academic literature discusses the need for a user model and user modeling representation language that has the necessary expressiveness to represent the range of user needs, terminology, and concepts (Kobsa, 2001). Such a user model is non-trivial, and research to define requirements for user model abstractions is a necessary perquisite to further development. However, once expressive user models are developed, deploying them in applications to direct configuration and customization is straightforward. Given the need for domain knowledge to generate user models, the authors' believe that user modeling research is a high impact area for design and construction researchers and software vendors.

Data Integration: Limitations of Standards-Based Approaches

The problems of developing richly specified user models suggests that software development for the design and construction industry is an iterative affair, with applications being slowly improved over time within their established domain (e.g., scheduling, estimating, accounting, CAD, etc.). This reinforces the development of "information tunnels" (Tannenbaum, 2002) with integrated sets of data and applications designed to serve narrow specialties. However, these specialized applications do not directly support many project analysis needs. For example, managing site space allocation requires knowledge of scheduling, estimating, and CAD/spatial data at a minimum. The need to draw from multiple data sources or information tunnels to answer relevant business questions on projects is a general problem and has been argued by many (e.g., a recent paper by Froese, Yu, Liston, & Fischer, 2000, diagrams the information integration problems faced by professionals).

The presence of information tunnels has led to calls for data standards to support interoperability (e.g., Industry Foundation Classes, IAI, 1996). However, data standards are a partial solution to the problems of integrating data. Several authors have recently argued that the wealth of legacy applications and associated variation in level of detail (O'Brien et al., 2002) and range of functions (Amor & Faraj, 2001; Turk, 2001; Zamanian & Pittman, 1999) means that development and adoption of relevant data standards to support all project participants is unlikely. Hence information tunnels are likely to persist even with significant adoption of data standards. Consequently, the deployment of tools that require data from multiple sources will remain problematic.

Data standards are also limiting as the complexity of information to be shared increases. While in theory data standards should resolve issues with complexity, in practice the more complex the data exchanged, the greater the burden placed on the firm to integrate its existing systems with the proposed standard. Firms may prefer to retain the use of proprietary legacy sources for many if not all operations. Hence, asking those firms to link their existing systems with a data protocol that can be queried requires significant effort by the firm. Further, even if firms deploy applications built on a common data standard, this does not mean they will uniformly deploy those applications. Firms have different operating procedures, and different levels and types of data collected for internal management. (See Castro-Raventós, 2002, for a comparison of different information system implementations among medium-size firms.) Thus, even if a common data standard is employed, sharing complex data can require considerable effort to transform from a firm's internal representation to the requirements of the common standard.

Approaches to the problems of data integration are as broad as the problem itself, and the authors do not attempt to provide an exhaustive review. Rather, aside from security issues, we classify the main problem as one of discovery (i.e., identifying and connecting to data sources) and interpretation (i.e., making sense of what data is there and applying that data to the problem at hand). There is a wide literature in discovery and interpretation that the authors do not review here other than to mention a few key approaches: data reverse engineering (Aiken, 1996) to discover data, and mediation (Wiederhold, 1998) to interpret and analyze data. Many of these technologies are utilized by O'Brien and Hammer (2001) and O'Brien et al. (2002) in the SEEK: Scalable Extraction of Enterprise Knowledge project at the University of Florida. The reader is referred to these sources for practical approaches to the information integration problem in construction.

Configurable Environments Architecture

Configurable environments provide the analysis and visualization capabilities of PSEs while adding the ability to integrate information from multiple, heterogeneous sources (mitigating the problems of information tunnels). CEs also provide customization abilities not native to statically configured applications. The CE vision makes aggressive use of user models to direct configuration and customization. The data integration and customization aspects of CEs are challenging to implement for a given application. Knowledge to direct customization and integration for a range of users, who may need different levels of detail and have different information integration needs (and privileges), remains very much a research problem. However, the advent of automated approaches to information integration, together with advances in user modeling, make it possible to concretely envision an architecture and implementation approach for CEs.

A high-level CE architecture is shown in Figure 3. Reading from right to left, existing legacy sources for information (e.g., application data, databases) are accessed through

Figure 3. Configurable environments high-level architecture

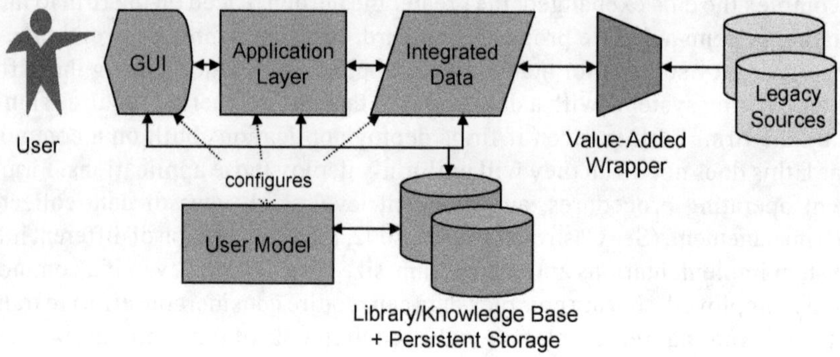

a value-added wrapper. The wrapper translates the data format of the legacy source into the internal format CE. The wrapper may have a value-added component of simple mediation tasks (Wiederhold, 1998) and may or may not be native to the CE. Several translators are being developed for applications. Toolkits such as SEEK, in development by the authors (O'Brien et al., 2002), provide extensive capabilities to semi-automatically configure wrappers and provide an advanced query interface to the legacy source. Alternate approaches to organizing data for flexible access by humans may also provide descriptors and wrappers that the CE can access (King, 2003; Lucas & Senn, 2002).

The integrated data component of the CE provides an organized view of the data specific to the needs of the user (e.g., superintendent, project manager, estimator, etc.). Depending on availability of data from the legacy source, the integrated data component may or may not cache data drawn from legacy sources. Rather, the primary purpose of the integrated data component is to provide organization of legacy data to support queries and analysis. (The separation between the wrapper and data component is subtle, and there is some overlap between their roles. Each is described separately, as the wrapper component may not be native to the CE while the integrated data component is a necessary part of the CE.) Shen, Issa, O'Brien, and Flood (2003) discuss a specific implementation of an integrated data component tied directly to a graphical user interface (GUI) to support queries. Shen et al.'s model enables effective calculation of quantities of materials needed under a dynamic schedule with frequent deliveries such as in the Tomorrow Square project; as such it demonstrates elements of the CE approach.

The application layer provides analysis and query capabilities on top of any mediation tasks contained in the wrapper and data components. The application layer is supported by the library of tools and knowledge base as in a PSE. For a space management CE, existing research tools provide a rich starting point for formation of a library. The application layer also supports information queries by users (and library components when executed) to take advantage of the CE's links to external data sources. The user accesses the CE through the GUI. The GUI component supports a variety of visualization tools and is customized to the needs of the user. Broadly, the GUI will likely look much like existing PSEs.

Underlying the GUI, application layer, and integrated data components is the user model component. The user model contains specific information about the class of user (e.g., superintendent) and may be further customized to the specific needs of an individual. The salient distinction of CEs compared to PSEs and other applications is that the user model directs configuration of the components of the CE. Thus a site management CE for a superintendent may provide a different level of detail and different analysis capabilities than does a site management CE for a trade foreman. For example, a superintendent may wish to view broad areas needed to look for space conflicts among trades, whereas the foreman may wish to view details of specific spaces to lay out production plans. As noted in the discussion above, a general requirement for the user modeling component is the necessary expressiveness to represent the needs and terminology of multiple classes of users.

Future Trends

The configurable environments architecture represents a medium term vision for construction practice. Current research in site space management and scheduling could be used as an analysis library. Augmenting this library with a simple user model and with rudimentary data integration capabilities (and by limiting data sources to well-understood formats) allows construction of an operable, albeit limited CE. Development of a commercially viable CE requires significant further research in user modeling and automatic approaches to information integration. At the same time, the potential range of problems in design and construction that can be addressed by CEs is broad; this range will help drive development of useful CEs.

Pathways to Configurable Environments

Likely commercial applications in the near term to bridge development to fully realized, commercially viable CEs will come from two areas: from next-generation Web-based information sharing tools such as project Web sites (O'Brien, 2000; Thorpe & Mead, 2001) and related Web-based applications. Current Web applications allow sharing of files and some posting of data into a database from which simple but useful reports (e.g., outstanding Requests-For-Information) can be generated. Next-generation Web tools will allow some limited search of attached files, as well as more sophisticated analysis of data posted to the application. These provide rudimentary tools for analysis and visualization, but will likely lack significant customization abilities. Dependent on manual posting of data, Web applications also lack data integration capabilities.

The second near-term bridge will be development of integrated applications that combine CAD data with cost and schedule data or facilities operations data. Many vendors now sell (or are developing) integrated application suites for project and/or facilities management. These integrated tools provide a form of problem-solving environment that have some data analysis and intelligent modeling tools, and can be augmented with user libraries. Built using data standards such as the IFC, these tools provide limited abilities to integrate data (especially data from reference sources that are well understood and hence there are few interpretation issues). While powerful, these integrated tools have limitations similar to existing problem-solving environments, including problems with mapping existing tools to specific projects, customization, and data integration with relevant project data distributed over many sources.

Developments in Web applications and integrated CAD/analysis tools will help generate practical knowledge that can augment university research in user modeling, tool building, and data integration. Analysis of current and near-term tools will help researchers and software vendors to directly observe the user experience, informing research in user models, visualizations, and user interfaces. Similarly, observing users allows a better understanding of specific integration problems and needs, guiding research in flexible schemas and reasoning tools that support information integration. Hence the authors believe that development of existing and near-term applications will provide the necessary knowledge base for implementing useful and robust CEs in the medium term.

Scope of Configurable Environments in Practice

The range of potential applications for CEs can be seen in the scope of the FIATECH technology roadmap (FIATECH, 2003). Figure 4 depicts the nine elements of the FIATECH roadmap. The central elements correspond to the project life cycle, from conception (scenario-based project planning) through design (automated design) through project execution (integrated, automated procurement and supply network) and facility operations (intelligent, self-maintaining/repairing operational facility). The real-time coordination and control element is proposed to support operations across the project life cycle. (Supporting other elements are enabling elements in materials and methods, technology-enabled workforce, and information integration.) While much of the FIATECH roadmap concerns the reuse of information across the project life cycle, each of the roadmap elements provides fertile ground for CE applications. A space and materials management CE would support the intelligent job site element for a superintendent, while a procurement and scheduling CE would support the supply network element for a project manager. There are likely to be several CEs for each element depending on the specific problem and class of user. For example, a procurement manager might use a different CE than a project manager, although each CE would fit within the supply network element. Of course, CEs would not necessarily be confined to individual roadmap elements. A procurement and scheduling CE could support both operations (supply network and intelligent job site elements) as well as be used to aid design decisions (automated design element). Potentially, a CE could be developed for any component of the FIATECH roadmap (or any aspect of a project) that requires analysis linked to data—a definition that fits most activities on a project. Of course, there must necessarily be a finite number of CEs; it is hoped that configuration control can help customize general classes to CEs for specific problems, projects, and users.

Figure 4. FIATECH technology roadmap elements (Several CEs could potentially support each of the roadmap elements.)

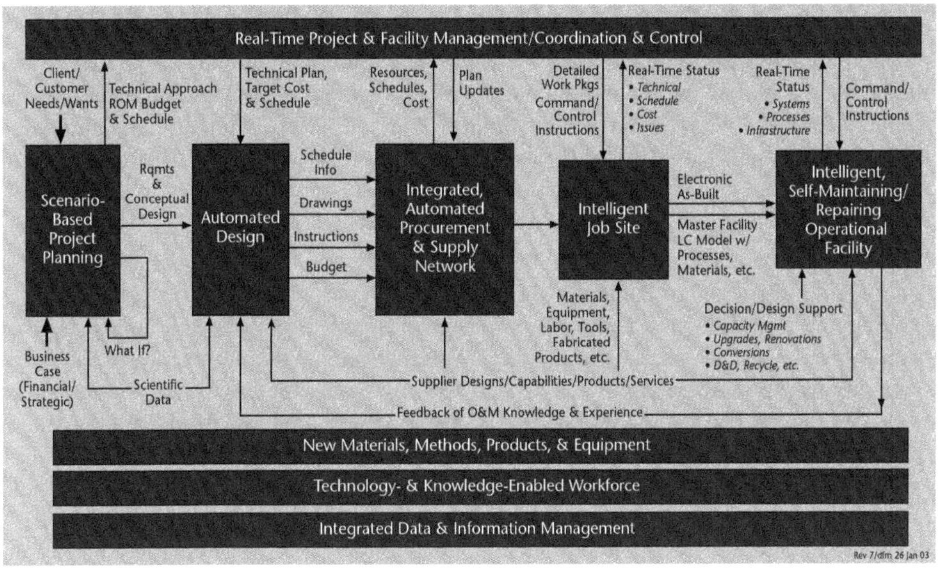

Implications of Configurable Environments for Knowledge Management

Configurable environments as a paradigm for work promise significant improvements to current practice. As noted by Liston, Fischer, and Winograd (2001) in the context of collaborative work on projects, information technologies for visualization and analysis promise to transform the nature of project work. Current collaborative/meeting work tends to be predominantly descriptive and explanative, whereas work enabled by IT tools promises to be predominantly evaluative and predictive. The benefit, then, from IT stems from an ability to focus on solutions to problems and explore alternatives, avoiding the current problems of practice that seek satisfying solutions to problems (O'Brien, Fischer, & Jucker, 1995). Liston et al.'s (2001) argument centers on group work, but it is reasonable to infer similar benefits will be found for individual work within a CE. Of course, a transformation of work tasks from descriptive to evaluative and prescriptive will not happen rapidly. Tools such as CEs will transform aspects of project practice for specific tasks. Only when individuals find themselves using CEs and related tools for the majority of their knowledge-intensive tasks will they find themselves truly enabled to spend much greater amounts of time solving problems.

There are several implications for the nature of work and knowledge management in design and construction that stem from freeing time for problem solving. On the level of individual participation in projects, there will likely be a culture shift from reactive problem solving to planning and analysis, with a corresponding change in the type of individual attracted to design and construction tasks. Particularly for construction, individuals who are action oriented but limited in their reflective and analysis capabilities may find themselves less able to contribute to projects. Of course, CEs and related technologies will not eliminate the need for good interpersonal and leadership skills. Rather, professionals will be augmented by better access to information and analysis. Hence, CEs also promise new roles for engineers and designers, who, through automated analysis and visualization capabilities, may be able to increasingly consider the construction cost and time implications of their design decisions (see, for example, Fischer, Haymaker, & Liston's (2003) discussion of the benefits of the 3D and 4D CAD).

On a broader organizational level, the analysis and time-saving capabilities of CEs promise organizational efficiencies and changes in the nature of competition. With efficiencies in descriptive and explanatory work tasks, organizations could see CEs primarily as a way to increase productivity without necessarily increasing the quality of work performed. It is hoped, however, that organizations will see the benefits of increased evaluative capabilities, and they will begin to compete on the basis of added value rather than cost. Practically, this means that CEs may help design and construction firms move away from fee-based competition (with corresponding pressures for ever lower fees and profit margins) to competition over value of completed projects. In turn, this may drive design and construction organizations to increasingly develop organizational capabilities that leverage knowledge collection and sharing, as opposed to many current practices that husband individual experience.

Conclusion

Projects have long been seen as information-rich environments with associated calls for information integration to support decision making. Yet even with advanced applications based on shared data standards, project professionals still face considerable difficulty accessing needed information in a form useful for decision making. In response, this chapter has presented configurable environments as an approach to supporting users in handling complex information processing and management tasks on projects. The CE approach provides data integration capabilities as an alternative to the information tunnels currently facing users. CEs also are highly customizable via the application of user models to direct configuration. In a simple sense, the CE approach is a natural evolution of the PSE paradigm to support the needs of projects. Similarly, CEs are a specific approach developed from broad visions for computing in design and construction.

However, CEs also represent a paradigm shift away from specific application development to broader research about environments to support human decision making. Despite calls for flexible systems, there remains considerable research in creating applications to optimize and automate solutions to specific problems. There is a lack of corresponding research to broadly understand problem context and human roles. Making the CE approach a reality will take considerable research about users, and hence a basic implication of the approach is that there is a need to reemphasize the role and value of documenting, understanding, and formalizing human information processing and decision-making needs in the design and construction domain.

The range of applications for CEs as depicted in the FIATECH roadmap is large, creating a need for CEs that will drive their development. Documents such as the roadmap provide detailed need assessments and targeted problems that can be addressed with CEs. Increasing use and development of current applications such as project Web sites and integrated applications will create awareness in the industry of the potential for more advanced tools like CEs. Such awareness will help to create a market for CEs, driving them from the laboratory to practice. As such, despite considerable research challenges, the authors are optimistic that CEs can be deployed commercially in the medium term.

Practical Tips and Lessons Learned

- The configurable environments approach is a medium-term vision for design and construction practice.
- Many of the constituent elements of CEs—including visualization, data standards, and analysis tools—exist in at least rudimentary form, allowing firms to begin incorporating the technology into practice.
- Firms should begin to consider the implications of CEs and related tools for job descriptions.
- As information technologies evolve, they will shift the balance of job tasks from information gathering to analysis. Firms should consider grooming younger employees for such changes.

Acknowledgments

The authors thank the National Science Foundation, which supports the SEEK: Scalable Extraction of Enterprise Knowledge project under grant numbers CMS-0075407 and CMS-0122193. The SEEK project is a direct antecedent of the configurable environments vision and approach.

References

Aiken, P. (1996). *Data reverse engineering: Slaying the legacy dragon.* New York: McGraw-Hill.

Akinci, B., Fischer, M., Levitt, R., & Carlson, B. (2002). Formalization and automation of time-space conflict analysis. *ASCE Journal of Computing in Civil Engineering, 6*(2), 124-135.

Amor, R., & Faraj, I. (2001). Misconceptions about integrated project databases. *ITcon, 6*, 57-66.

Amor, R., Betts, M., Coetzee, G., & Sexton, M. (2002). Information technology for construction: Recent work and future directions. *ITcon, 7*, 245-258.

Castro-Raventós, R. (2002). *Comparative case studies of subcontractor information control systems.* Unpublished MS Thesis, University of Florida, USA.

Clayton, M.J., Kunz, J.C., & Fischer, M.A. (1996). Rapid conceptual design evaluation using a virtual product model. *Engineering Applications of Artificial Intelligence, 9*(4), 439-451.

Cooper, A., & Reimann, R. (2003). *About Face 2.0: The essentials of interaction design.* Indianapolis, IN: John Wiley & Sons.

El-Mashaleh, M. (2003). *Firm performance and information technology utilization in the construction industry: An empirical investigation.* Unpublished Doctoral Dissertation, University of Florida, Gainesville, USA.

FIATECH. (2003). *Strategic overview: Capital projects technology roadmap initiative (version 1).* FIATECH, March, 70 pages.

Fischer, M., Haymaker, J., & Liston, K. (2003). Benefits of 3D and 4D models for facility managers and AEC service providers. In R. Issa, I. Flood, & W. O'Brien (Eds.), *Developments and applications: The visual construction enterprise* (pp. 101-124). Lisse, The Netherlands: A.A. Balkema.

Froese, T., Yu, K., Liston, K., & Fischer, M. (2000). System architectures for AEC interoperability. *Proceedings of Construction Information Technology 2000—the CIB-W78, IABSE, EG-SEA-AI International Conference on Construction Information Technology,* Reykjavik, Iceland.

IAI. (1996). *End user guide to industry foundation classes, enabling interoperability in the AEC/FM industry.* International Alliance for Interoperability (IAI).

Kam, C., & Fischer, M. (2002). *Product model and 4D CAD final report.* Technical Report 143, Center for Integrated Facility Engineering, Stanford University, USA.

King, J. (2003). Future watch: Taming data complexity. *ComputerWorld,* (June 30).

Kobsa, A. (2001). Generic user modeling systems. *User Modeling and User-Adapted Interaction, 11*, 49-63.

Kulak, D., & Guiney, E. (2000). *Use cases: Requirements in context.* New York: ACM Press.

Liston, K., Fischer, M., & Winograd, T. (2001). Focused sharing of information for multi-disciplinary decision making by project teams. *ITcon, 6*, 69-81.

Lucas, P., & Senn, J. (2002). *Toward the universal database: U-forms and the VIA repository.* MTR-02001. MAYA, March, 10 pages.

Maybury, M., & Wahlster, W. (Eds.). (1998). *Readings in intelligent user interfaces.* San Francisco: Morgan Kaufmann.

O'Brien, W.J. (2000). Implementation issues in project Web sites: A practitioner's viewpoint. *ASCE Journal of Management in Engineering, 16*(3), 34-39.

O'Brien, W.J., & Hammer, J. (2001). Robust mediation of construction supply chain information. *ASCE Specialty Conference on Fully Integrated and Automated Project Processes (FIAPP) in Civil Engineering,* Blacksburg, Virginia, , January 23-25, 2002.

O'Brien, W.J., Fischer, M.A., & Jucker, J.V. (1995). An economic view of project coordination. *Construction Management and Economics, 13*(5), 393-400.

O'Brien, W.J., Issa, R.R., Hammer, J., Schmalz, M., Geunes, J., & Bai, S. (2002). SEEK: Accomplishing enterprise information integration across heterogeneous sources. *ITcon—Electronic Journal of Information Technology in Construction—Special Edition on Knowledge Management, 7*, 101-124.

Payne, S.J. (2003). User's mental models: The very ideas. In J.M. Carroll (Ed.), *HCI models, theories, and frameworks: Towards a multidisciplinary science* (pp. 135-156). San Francisco: Morgan Kaufmann.

Rice, J.R., & Boisvert, R.F. (1996). From scientific libraries to problem-solving environments. *IEEE Computational Science & Engineering, 3*, 44-53.

Riley, D. R. (2000). The role of four-dimensional (4D) models in trade sequencing and production planning. *Proceedings of Construction Congress VI: Building Together for a Better Tomorrow in an Increasingly Complex World,* Orlando, Florida, February 20-22.

Riley, D.R., & Sanvido, V.E. (1997). Space planning method for multistory building construction. *ASCE Journal of Construction Engineering and Management, 123*(2), 171-180.

Shen, Z., Issa, R., O'Brien, W., & Flood, I. (2003). A proposed design component and schedule data integration model for construction. *Proceedings of the 10th European Group for Intelligent Computing in Engineering Workshop,* Delft, The Netherlands, July 3-4.

Strachan, L., Anderson, J., & Sneesby, M. (2000). Minimalist user modeling in a complex commercial software system. *User Modeling and User-Adapted Interaction, 10*, 109-145.

Tannenbaum, A. (2002). *Metadata solutions: Using metamodels, repositories, XML, and enterprise portals to generate information on demand.* Boston: Addison-Wesley.

Terveen, L.G., Hill, W.C., & Amento, B. (1999). Constructing, organizing, and visualizing collections of topically related Web resources. *ACM Transactions on Computer-Human Interaction, 6*(1), 67-94.

Thabet, W.Y., & Beliveau, Y.J. (1997). SCaRC: Space-constrained resource-constrained scheduling system. *ASCE Journal of Computing in Civil Engineering, 11*(1), 48-59.

Thorpe, T., & Mead, S. (2001). Project-specific Web sites: Friend or foe? *ASCE Journal of Construction Engineering and Management, 127*(5), 406-413.

Tommelein, I.D., & Zouein, P.P. (1993). Interactive dynamic layout planning. *ASCE Journal of Construction Engineering and Management, 119*(2), 266-287.

Tommelein, I.D., Levitt, R.E., Hayes-Roth, B., & Confrey, T. (1991). SightPlan experiments. Alternate strategies for site layout design. *Journal of Computing in Civil Engineering, 5*(1), 42-63.

Turk, Z. (2001). Phenomenological foundations of conceptual product modeling in architecture, engineering, and construction. *Artificial Intelligence in Engineering, 15*(2), 83-92.

Wiederhold, G. (1998). Weaving data into information. *Database Programming and Design, 11*(9).

Winograd, T. (Ed.). (1996). *Bringing design to software.* Addison-Wesley.

Zamanian, M.K., & Pittman, J.H. (1999). A software industry perspective on AEC information models for distributed collaboration. *Automation in Construction, 8*, 237-248.

About the Authors

Abdul Samad ("Sami") Kazi, Senior Research Scientist at VTT, the Technical Research Centre of Finland, is a civil engineer by profession, a construction manager by area of specialisation, and a knowledge management evangelist, construction IT specialist, groupware and collaborative tool explorer, and virtual organisation researcher by area of interest and expertise. Mr. Kazi has industrial experience and maintains close links with the construction industry throughout his research. He has been involved in several EU projects (each project having a budget ranging from 500,000 to 5,000,000 Euros) on construction IT, virtual organisations, and knowledge management. He has furthermore served the European Commission in various capacities, as an evaluator for knowledge management and e-government proposals in the IST (Information Society Technologies) programme; reviewer of IST projects on knowledge management, construction, and mobile devices; and expert evaluator for portfolio of knowledge management projects (more than 20) in the IST programme. He holds a Bachelor's of Science in Civil Engineering, and a Master's of Engineering in Construction Engineering and Management from the Middle East Technical University, Turkey, and the Asian Institute of Technology, Thailand, respectively. He is listed in the *Marquis Who's Who in Science and Engineering, 7th Edition (2003-2004),* in the *International Biographical Centre's 2000 Outstanding Scientists of the 21st Century, 2004 Edition,* and was chosen for distinguished standing and conferred with an honorary appointment to the research board of advisors of the American Biographical Institute. He is a proud member of the Entovation E100 Global Knowledge Leadership network.

****

Alaa Abdou is a lecturer in the Department of Architectural Engineering, College of Engineering, at the United Arab Emirates University (UAEU). He holds a Master's of Architecture and Certificate in Construction Management from the Illinois Institute of Technology (IIT) in Chicago, Illinois, USA. He has more than 15 years of academic and practical experience in the construction industry. Prior to joining the UAE University, he served as a project architect and project manager during the design and construction of many prestigious projects. He is currently a candidate for a PhD at Liverpool University, UK, in the field of project/design management. His research interests include design management, Web-based project management, and information technology applications in project/design management. He has published several papers in international journals and conference proceedings.

Tunca Ataoglu is a PhD student in the Civil Engineering Department of the Middle East Technical University, Turkey. He received his MSc (2002) while he was serving as a Graduate Research Assistant in the Construction Management Division (2000-2002). During his graduate studies, his main research areas were organisational learning in construction companies and strategic partnering. Currently, he is working as a project management team engineer in the Saudi ARAMCO project in Saudi Arabia, Jubail Industrial City. His current research areas are construction of project memory systems, development of decision support tools using artificial neural networks, and constructability analysis of heavy industrial projects.

Glenn Ballard is an associate adjunct professor, Civil & Environmental Engineering, at the University of California, Berkeley (USA). He is also research director for the Lean Construction Institute. Dr. Ballard is a founding member of the International Group for Lean Construction and conducts research in the management of project-based production systems. His contributions to the field include the Last Planner system of production control and the Lean Project Delivery System.

M. Talat Birgonul is an associate professor in the Civil Engineering Department of Middle East Technical University, Turkey. His main fields of interest are construction scheduling, engineering economy, construction management, and claim management. Dr. Birgonul has published various articles, conference papers, and book chapters on macro-economic aspects of the construction industry, international construction, organisational issues, as well as management of quality, time, and cost of construction projects. In addition to his research activities, he has been giving consultancy services to Turkish contractors on the legal issues of construction works, settlement of disputes, as well as economic and technical feasibility studies.

Alan Burke is a senior lecturer in Business Information Systems at Leeds Metropolitan University, UK. He teaches courses at both undergraduate and postgraduate levels and is particularly interested in the ways in which systems ideas and soft approaches to problem solving can be used in organisational and systems development contexts. He has been involved in research activities in the construction industry and has published a number of papers in the areas of knowledge management and organisational learning.

Patricia M. Carrillo is a senior lecturer, Department of Civil and Building Engineering at Loughborough University, UK. She has been working on knowledge management in construction since 1998. To date she is principal investigator on three UK government-funded knowledge management research projects, namely Knowledge Management for Improved Business Performance (2000-2003), Capture and Reuse of Project Knowledge in Construction (2003-2005), and A Knowledge Transfer Approach for Continuous Improvement on PFI Projects (2003-2004). She is also co-investigator on two further research projects, namely Cross-Sectoral Learning in the Virtual Enterprise (199-2001) and Creating, Sustaining and Disseminating Knowledge for Sustainable Construction (2001-2004). In addition, she is also the holder of a Royal Academy of Engineering Global Award. This funds the position of visiting professor at the University of Calgary, Canada, and the University of Colorado at Boulder, USA, to undertake research in knowledge management in Engineering Organisations in North America.

Zhen Chen started his academic career at the Tsingtao Institute of Architecture and Engineering in 1990. He then researched and lectured at Tongji University, Hong Kong Polytechnic University, and Loughborough University, before joining Massey University, New Zealand. During this period, he also worked with engineering design and construction firms, as well as investment banks, and also provided consultancy services to both private and public organisations in China and the UK. He has participated and conducted several funded research projects relating to the innovative application and transfer of construction information technologies for built environment, and he has published more than 10 journal papers in major journals and numerous conferences papers in proceedings. His research interests include innovative application of construction information technologies and operational research for built environment. For more information, visit: http://hk.geocities.com/at55379/index.htm.

Barbara Colledge is deputy dean in Innovation North Faculty of Information and Technology at Leeds Metropolitan University, UK. She is a chartered surveyor, accredited mediator, and has substantial academic experience in construction and related built environment fields and commercial experience in the construction industry. Her current role includes leadership in research and knowledge transfer in information management, computing, and technology-related disciplines and membership of the Steering Group for the New Technology Institute, West Yorkshire. In addition to her research in knowledge management, she has published more widely in the areas of construction law, partnering, and conflict resolution.

Andrew R.J. Dainty is a senior lecturer in Construction Human Resources Management, Department of Civil Engineering, Loughborough University, UK. His research interests are focused on human resource management issues in project-based environments, and particularly those related to major construction projects and organisations. He currently holds grants from the ESRC and EPSRC, together with funding from the Construction Industry Training Board, the European Union, the Learning and Skills Council, and the Regional Development Agency. Current projects include work on knowledge sharing, strategic human resource development, project team formation and deployment, performance management, and work-life balance. All of Professor Dainty's externally funded work has been widely disseminated in high-quality international refereed journals and conferences, as well as industry-based publications. He is also co-author of *HRM in Construction Projects*, published by Spon Press.

Irem Dikmen is an assistant professor in the Civil Engineering Department of the Middle East Technical University, Turkey. Her main research interests are construction planning, project risk management, strategic management, and decision support systems. She has published several articles and conference papers on strategic and organisational issues related with the construction industry, as well as artificial intelligence tools and decision support systems for construction project management. In addition to her research activities, she has been giving continuing education seminars and consultancy services to construction professionals about business development, international construction, construction risk management, and claim management.

Chen Swee Eng has had a long career as a construction manager and academic in both Singapore and Australia. He is currently professor of Building at the University of Newcastle, Australia. His research involves the understanding of project performance from a complexity perspective, the application of systems theory to the management and economics of the building procurement process, the problem-based learning approaches to professional education, and students' perceptions of their learning environment. He is Director of the Centre of Infrastructure and Property (CIP) at the University of Newcastle and has previous research experience in ARC- and CRC-funded projects.

Andrew Finegan is senior lecturer in e-Business in the School of Information Technology at Charles Darwin University, located in Darwin, Australia. He is currently developing curriculum in information systems and e-Business, and teaches and consults in these fields. His research interests encompass the areas of systems thinking, project management, and the management of innovation. He obtained a Master's of Engineering in Systems Engineering (1990) and a PhD (2001), both from RMIT University. Dr. Finegan has published a variety of papers that promote the use of soft systems thinking in a diverse range of applications, including knowledge elicitation, the adoption of technology, managing wicked problems, online learning, and knowledge management.

Patrick S.W. Fong is an associate professor in the Department of Building and Real Estate at the Hong Kong Polytechnic University. He is an adjunct professor at the Faculty of Design, Architecture and Building, University of Technology, Sydney, Australia. He has a BSc (Hons) in Quantity Surveying and Construction Economics and a master's degree by research in the area of value management from the University of Salford, UK. His PhD in Industrial and Business Studies was obtained from the Warwick Business School, UK. He has authored or co-authored many refereed research papers and received ANBAR citations for excellence in research, practice, and originality. He maintains close links with industry, and is vice president of the Hong Kong Knowledge Management Society, which is the main professional body disseminating information and knowledge of KM to all industries in Hong Kong. He is also a member of the international body, CIB W102 - Information and Knowledge Management in Building, and of the Knowledge Management Special Interest Group of the Association for Project Management in Hong Kong. He has a wide range of industry experience, which he gained in the UK and Hong Kong in contracting, consulting, and owner companies.

Gil Horsky is a management consultant in the Israeli consulting firm, Edna Pasher PhD & Associates, Israel. He holds a BA in Political Science from Tel Aviv University. He has vast experience in consulting projects with leading Israeli companies from a vast range of sectors: retail, construction, energy, and high-tech. He is also responsible for all of the firm's international activities, including its participation in EC-funded projects.

Raja R.A. ("Ray") Issa focuses his efforts on information technology integration in construction management. Currently, he is a University of Florida (USA) Research Foundation and Rinker professor and director of Graduate and Distance Education Programs in the Rinker School of Building Construction. Dr. Issa has many years of experience at various levels in the construction industry. He holds BS, MS, and PhD degrees in Civil Engineering from Mississippi State University, and a JD in Law from the University of Memphis, where he also completed graduate studies in Computer Science. He has chaired several committees and task groups in the areas of computing in education, intelligent computing, virtual reality, and information technology.

Marcus C. Jefferies is a lecturer in the School of Architecture and Built Environment at the University of Newcastle, Australia. He has recently completed a Master's by Research that involved the identification and analysis of risk and critical success factors of the Build-Own-Operate-Transfer (BOOT) approach to the private sector provision of public sector infrastructure. In 2003 he won the Australian Institute of Building Wilson Memorial Prize for the best refereed paper. He is a member of the Technical Committee of the International Council for Research and Innovation in Building and Construction (CIB) Working Commission W92 (Procurement Systems). He is currently involved in three CRC Construction Innovation research projects: Facilitating a Change in Project Culture, Value Alignment Process for Project Delivery, and Ethical Behaviour in the Construction Procurement Process.

Mark Jewell, based out of Brisbane, Australia, is a director of Addept Solutions and is responsible for Addept's business development within the Asia Pacific region. He started his career in the UK construction industry and has a BSc (Hons) in Construction Management. He served many years as a project manager and led teams on high-profile construction projects, including the redevelopment of The Royal Opera House of London. His career took a more strategic business direction when he became Head of Information Systems and Information Technology of one of the UK's largest construction services firms. Mr. Jewell placed knowledge management on the corporate agenda, and successfully implemented many people-driven and technology-supported knowledge management initiatives, including the establishment of communities of practice. He led the establishment of www.SigmaForums.com and is active in promoting knowledge management practices in his home state of Queensland Australia and beyond.

Stephen C.W. Kong obtained his BSc in Building Technology and Management from the Hong Kong Polytechnic University (1999). He then researched e-commerce application in construction under the supervision of Heng Li and obtained an MPhil from the same university in 2002. He is currently a researcher at the above university and is also a PhD candidate of Loughborough University under the supervision of Professor Chimay Anumba. He has provided consultancy services to construction IT firms in both Hong Kong and mainland China, and has substantial experience in developing Web-based applications for the construction industry. His research interests include virtual reality, e-commerce, Web-based project management systems, and knowledge management.

Steve de Kretser is a consulting chemical process engineer, as well as a process engineering manager in the Power & Industrial Group at the Auckland office of Sinclair Knight Merz Ltd., New Zealand. His interest in knowledge management developed during his studies for a Master's in Engineering Management from the University of Auckland.

John Lewis is a lecturer in the School of Architecture and Building Engineering, University of Liverpool, UK. He is currently engaged in research and undergraduate and postgraduate teaching on project management. He has spent some 35 years working in the construction sector, the time being approximately equally split between practice and education/research. He has published in excess of 50 conference and journal papers on the management of construction projects and organisations. Much of this work has focussed on computer-based modelling for decision support.

Heng Li started his academic career at Tongji University in 1987. He then researched and lectured at the University of Sydney, James Cook University, and Monash University, before joining the Hong Kong Polytechnic University. During this period, he also worked with engineering design and construction firms, and provided consultancy services to both private and government organisations in Australia, Hong Kong, and China. He conducted many funded research projects related to the innovative application and transfer of construction information technologies, and he has published one book, more than 100 journal papers in major journals of his field, and numerous conferences papers in proceedings. For more information, visit: http://bsun02.bre.polyu.edu.hk/~bshengli/.

Yu-Cheng (Vincent) Lin is a PhD candidate of the Construction Engineering and Management graduate program of the National Taiwan University in Taiwan. He received his MS in the Construction Management Program of Civil Engineering from the Polytechnic University, New York. Currently, he is researching the application and the use of knowledge management and e-learning in construction projects. His current research interests include construction knowledge management, construction supply chain management, lean construction, IT technology, and construction e-commerce-related topics. He has published several articles and papers on the role of academic support service in construction e-commerce.

Tayyab Maqsood is currently pursuing his doctoral degree at RMIT University (Australia) sponsored by CRC for Construction Innovation; he is conducting research on the topic of knowledge management, learning and innovation in the Australian construction industry. He is a civil engineer with a Master's of Engineering degree in Construction Engineering and Management from the Asian Institute of Technology, Bangkok, Thailand. He has worked in Pakistan, Thailand, the UK, and Hong Kong in various capacities as lecturer, project engineer, and research associate over the last nine years. His research interests include project management, risk management, partnering, knowledge management, information and communication technologies (ICTs), and supply chain management.

Fung Fai Ng is an associate professor in the Department of Real Estate and Construction, The University of Hong Kong. He teaches information management and knowledge management in construction. He believes that learning and knowledge sharing are essential elements in both knowledge management and education; therefore he employs a collaborative learning approach in his courses. His research interests include ontological engineering, knowledge mapping, and collaborative learning. He aims at integrating techniques and methodology of knowledge management in education, so as to enrich students' competence in knowledge management.

William J. O'Brien is an assistant professor in the Construction Engineering and Project Management program in the Department of Civil Engineering, University of Texas at Austin (USA). Dr. O'Brien is an expert on construction supply chain management and electronic collaboration. He is especially interested in the use of the information technologies to support multi-firm coordination, and works with leading firms to implement web-tools to support practice. Previously, he was CSR/Rinker assistant professor in the M.E. Rinker, Sr. School of Building Construction and assistant professor in the Department of Civil & Coastal Engineering at the University of Florida. Prior to returning to academia, Dr. O'Brien led product development and planning efforts at Collaborative Structures, a Boston based Internet start-up focused on serving the construction industry. Dr. O'Brien received a BS in Civil Engineering from Columbia University, an MS in Civil Engineering, an MS in Engineering-Economic Systems, and a PhD from Stanford University.

Patrick Onions is director and principle consultant at The Knowledge Studio, a knowledge management and business consultancy in Johannesburg, South Africa. His professional experience has included consulting to large public and private sector knowledge management projects, programme and project management in software and telecommunications, and software development across a broad range of industries. Holding postgraduate degrees in Science and Business, he has published papers on knowledge, knowledge management, and public sector optimisation. Mr. Onions is currently engaged in a PhD in Knowledge Engineering in Developing Countries.

Graham Orange is a principal lecturer in Innovation North - Faculty of Information and Technology of Leeds Metropolitan University (UK) and director of the research group, Centre for Information and Knowledge Management (CIKM). He has made a significant impact on the teaching of information systems development and systems development methodologies at both undergraduate and postgraduate levels. He has made valuable contributions to research, particularly in the areas of information strategy planning, business process redesign, knowledge management, and organisational learning, and has collaborated on a number of construction industry projects leading to the publication of his research.

Edna Pasher earned her PhD at New York University in Communication Arts and Sciences, and serves as a faculty member at the Tel-Aviv University. Dr. Pasher is a world pioneer in the field of knowledge management and is present on the Entovation 100 list. She is the founder and CEO of the Edna Pasher Ph.D. & Associates (Israel), one of the leading management consulting firms in Israel, existing since 1978. Her firm created the renowned Intellectual Capital Report of the state of Israel titled, "The Hidden Values in the Desert," which was one of the first national IC reports in the world. She is also the Chief Editor of *Status,* the leading Israeli management journal, and is also very active in the research and academic aspects of knowledge management and innovation, through her firm's active participation in EC-funded action research projects.

Vachara Peansupap has industrial experience as a construction engineer at Thai Obayashi Corp., Ltd. He earned a bachelor's degree in Construction Engineering from King Mongkut Institution of Technology Ladkrabang in Thailand, and an master's degree in Civil Engineering from Chulalongkorn University in Thailand. He is currently a PhD student at RMIT University (Asutralia) undertaking CRC research on the diffusion of information and communication technology within Australian construction organisations. Mr. Vachara is currently on secondment from being a lecturer at Chulalongkorn University in Thailand while completing his PhD. His research interests also include construction innovation, IT in construction, knowledge management, construction methods and techniques, simulation and virtual reality, and project planning and cost control. He has lectured in construction engineering, construction management, and information technology.

Ari Pennanen is a PhD candidate at the Department of Architecture, University of Tampere, Finland. He is also an experienced project manager and workplace planner at Haahtela Oy, a project management services company operating in Helsinki. Along with practicing in construction projects, he leads the Haahtela Research & Development group, which develops cost and project management software products for the Finnish Construction Industry.

Jidong Qin is a part-time PhD student in the Department of Civil Engineering, Loughborough University, UK. He has more than 10 years of experience working in the industry as both a civil and structural design engineer, and more latterly as a project manager in mainland China and Hong Kong. He is currently the deputy general manager of the Technology Department of a leading construction company in Hong Kong, where he focuses on new technology and management development and implementation, as well as providing staff training programmes.

Moh'd Radaideh is currently serving as coordinator of the Information Systems program at the College of IT, UAE University, United Arab Emirates. He previously worked for Verity (Sunnyvale, California) as senior consulting engineer. His projects involved content indexing and Web search enablement. Prior to Verity, he worked for IBM Canada as staff software developer. Dr. Radaideh holds a PhD in Software Engineering from McMaster University (Canada), a master's in Project Management from George Washington University, a master's in Computer Engineering from Jordan University of Science and Technology, and a BSc in Electrical Engineering from Yarmouk University (Jordan). His main research interests include knowledge/content management and e-commerce. Dr. Radaideh has published several papers in international journals, conference proceedings, and book chapters.

Zhigang Shen is currently studying for a PhD in Building Construction at the University of Florida (USA) while serving as an assistant project manager for Eccleston Signature Homes, Inc. His research focuses on the integration of design and schedule data. He worked as an architect for the Shanghai office of John Portman & Associates. Mr. Shen holds an MS in Computer Engineering from the University of Florida, an MArch (Professional) from Tongji University, China, and a BEng in Architecture from Zhengzhou Institute of Technology, China.

Hedley Smyth has extensive academic and industrial experience. His academic career started at Bristol University, then Oxford Brookes University and Reading University. He is currently part of the School of Construction and Project Management team within the Bartlett School of Graduate Studies at University College London. Mr. Hedley has worked in construction and for international architectural practices, specialising in marketing and business development, which has also informed the consultancy and training provided for international contractors and consultants. Research on relationship marketing developed into a focus upon behaviour, particularly trust as a competency. The externally provided KM is the other main area of current research. His most recent book is titled, *Marketing and Selling Construction Services,* and he is currently co-editing, *The Management of Projects: A Relationship Approach.*

H. Ping (Jack) Tserng is an associate professor in the Department of Civil Engineering, National Taiwan University in Taiwan. He holds a PhD in Construction Engineering and Management from the University of Wisconsin-Madison. His principal areas of research interest are information technology and management, enterprise resource planning, supply chain management, knowledge management, and construction automation. His papers on these areas have appeared in *ASCE Journal of Construction Engineering and Management, International Journal of Automation in Construction, Journal of Computer-Aided Civil*, and *Infrastructure Engineering*.

Derek H.T. Walker is professor of project management and program director of the Doctor of Project Management (DPM) program at the Faculty of Business, RMIT University, Australia. He worked in various project management roles in the UK, Canada, and Australia for 16 years before commencing his academic career in 1986. He obtained a Master's of Science from the University of Aston (Birmingham) in 1978, and a PhD in 1995 from RMIT University (Melbourne). He has written more than 100 peer reviewed papers and book chapters. His research interests centre on innovation diffusion of information and communication technologies, knowledge management, project management, and project procurement systems. For more information: http://dhtw.tce.rmit.edu.au/.

Michael Whelton is a PhD candidate in the Engineering & Project Management Program, Department of Civil & Environmental Engineering, University of California, Berkeley, USA. His research is on the management of project definition of construction projects. Specifically his research focuses on the management of group collaboration and learning in the front end of project development. He is trained as a civil engineer and holds a master's degree in Project Planning and Information Technology Support.

Alan Wild was a senior lecturer in Management at the University of Central England Business School in Birmingham, UK. He has worked with a wide range of managers from the public and private sectors. He now works as a freelance writer, and tutors on Henley, The Management College's International MBA by Distance Learning; Oxford Institute of International Finance MBA at Oxford Brookes University, and for other providers of continuing education. In connection with his work on the Executive MSc Suite in Construction Project Management, Property Development and Facilities Management at the University of Central England, he has written or co-authored journal papers, book chapters, and conference papers on various aspects of construction management, but with a focus on the problems of handling information and uncertainty.

Suzanne Wilkinson is a senior lecturer in Engineering Management in the Department of Civil and Environmental Engineering at The University of Auckland, New Zealand. She is a qualified civil engineer. Her current research interests include construction human resource management, project management, culture, and construction decision-making processes. She currently teaches undergraduate and postgraduate students in construction management, project management, and construction law.

Ka-chi Wong graduated from the Hong Kong Polytechnic University with a first class honours degree in Building Surveying. He is currently working in Raymond Chan Surveyors Ltd as an assistant building surveyor. His responsibilities include conducting condition surveys for buildings, diagnosis of building defects, contract administration and project management of major repair, and renovation works with on-site supervision; contract administration and project management of fitting-out works, including preparation of design drawings; and providing consultancy services to licensing applications.

Haiyan Xie is a PhD student in the Rinker School of Building Construction at the University of Florida (USA). She is also a project manager at Taylor Contractors of Florida, Inc. Her research interests are in the areas of analysis, design, and implementation of construction information systems, user modelling, and project management improvement. She holds an MS in Computer Science from the University of Florida, and a BE and an ME in Construction Engineering and Management from the Xi'an University of Architecture and Technology, China.

Qian Xu is currently a research student in Heng Li's Research Centre for Construction Management and Construction IT for an MPhil in Construction IT. His research focuses on the application of GIS-based decision support systems in construction.

Ralf Zenke holds a Bachelor's of Construction Management (Hons 1), an Advanced Diploma of Project Management, and a Diploma of Building. He is a certified master project director and chairman of the Tasmanian Building and Construction Industry Training Board, as well as the deputy chairman of the Australian Forum of Construction Industry Training Funds. He is employed by the Tasmanian Department of Health and Human Services as a senior consultant in the area of strategic facilities management and project management of capital works. His experience dissects all phases of capital works projects from inception through planning, definition, design, construction, commissioning, and operation. His special interest lies in the area of organisational learning and knowledge management, and its implications for construction.

Index

Symbols

4D CAD tools 349

A

absorptive capacity 113, 115, 244
accelerated subcontracting and procuring (ASAP) 291
acquisition 230
action research 130
action research project 130
active adoption 241
active server pages (ASP) 293
activity link activity 312
activity link knowledge 312
activity-based knowledge management (ABKM) 299, 303, 307
adaptive functions 209
adaptive management 251, 264
adaptive management action 263
adaptive management framework 253
adaptive management techniques 253
adaptive process 253
advanced electronic information 307
advanced knowledge 228
advertising 41
age of knowledge 54
analysis library 351
analytic network process (ANP) 329
application 4
application in construction 276
application-mature organisations 233
Architectural Services Department (ASD) 72
artificial intelligence (AI) 276, 286
artificial neural network (ANN) 287
asymmetry of information 36
Australian case studies 97
Australian construction contractors 90
Australian Standards Association 227
authoritarian approach 348

B

B-Hive project 132
benchmarking 226
bespoke 35
best practice programmes 122
branding 41
British Concrete Association (BCA) 44
building doctors 68
Building Industry Communications Research Project 204
building maintenance 67, 85
building maintenance knowledge 67
building maintenance knowledge management system 76

building maintenance projects 70
building organisational competencies 226
building surveyors 67
business community 20
business involvement 24
business performance 87
business process reengineering (BPR) 24

C

C-Sand project 328
CAD systems 140
capital 56
capital investment 252
captured knowledge 12
capturing feedback 9
Careers Education and Placement Centre 152
causal ambiguity 115
Cebus Rimon 63
central knowledge repository 142
central management tool 53
change management 226
CIB W78 345
client purpose-based knowledge 253
client satisfaction 87
client-specialist groups 251
coaching 27
codification 23
codification approach 5
cohesive fragmentation 213
COLA environment 138
COLA process 131
collaborative framework 253
collaborative group interaction 252
collaborative learning 155, 161
collaborative process 253
collaborative virtual environment 140
collaborative work 151
collective knowledge work space 252
collective responsibility 23
collegial support network 104
combination 21
combination of sociology and operational research 214
command processor 283
commercial environments 350
communication 151
communication channels 230
communities of practice (COPs) 39, 91, 112
company knowledge 300
company knowledge managers 1
competitive advantage 24, 54, 168, 226
competitiveness 87
complex adaptive process 252
complex systems 262
compound DSS 281
computer-based system 285
computer-supported-cooperative work (CSCW) 286
concept mapping 151
conceptual entity 136
conceptual interrelatedness 153
concrete information limited (CIL) 38
concrete society (CS) 44
configurable environments 343, 346, 355
configurable environments (CE) 343
configurable environments architecture 353
configurable environments for knowledge management 357
configurable environments in practice 356
configuration 4, 210
conflict tolerance 187
consistency and coherence 218
construction 213, 301
construction and demolition (C&D) 329
construction clients 130
construction consultancy firms 130
construction in society 215
construction industry 34, 70, 130, 150, 276
construction industry trade exchange (CITE) 107
construction knowledge 70
construction management 290
construction management decision support system 292

construction operation 300
construction organisations 20, 70, 91
construction pollution index (CPI) 332
construction project environment 19, 28
construction project organisations 18
construction project partners 130
construction projects 18, 299, 305
construction renewal 218
construction space management 343
constructive conversation 161
constructivism 153
constructivist learning environment 151, 155
constuction knowledge management 304
consulting construction companies 1, 2
consulting engineering 3
context-specific knowledge 305
continuance commitment 234
contract administration 12
conventional model-based systems 277
conversion 48
COP management tool 113, 118
COP performance 112
core competencies 59
core knowledge 228
core values 59
corporate knowledge assets 142
corporate team member 193
cost estimation 290
craft system 205
creation of the knowledge 255
creative chaos 232
creative thinking 228
creative thinking skills 228
cross organisational learning approach (COLA) 131, 132
culture 171
customer capital 57
customer knowledge 57
customer needs 252
customer success 57
customer's project purpose 252

D

Danya Cebus 53
Danya Cebus Report 60
data collection 71
data directory 282
data exchange 290
data integration 343, 352
data integration capabilities 343
data management subsystem 281
data protection group 123
data reverse engineering 353
data visualisation systems 277
data warehouse 292
data-collection exercise 141
data/information needs 234
database 282
database management system 282, 291
database-oriented DSS 281
day-to-day planning 290
decentralize the decision making 347
decision making 87
decision programming language (DPL) 329
decision support system technologies 284
decision support systems (DSSs) 276, 278, 280, 288, 294
decision support systems 276
decision-making approach 232
decision-making process 277
demand chain 36
departmental meetings 26
design 207
development arrangements 233
dialog mapping 151
direct participation 67
directive functions 209
discursive penetration 213
disjointed incrementalism 215
dissemination 4
distinctive competence 208
diverse stakeholders 251
document management 140, 290
dynamic capability 261

dynamic conservatism 216
dynamic profiling 120

E

E+ 328
E+ system (software environment) 330
e-commerce 291
e-commerce facilities 290
E-library 27
e-mail 140
e-mail system 290
effective communication structures 19
effective knowledge creation 156
effective management of knowledge 228
effective management principles 252
effective training 29
effective training and development systems 29
electronic data management 123
electronic information exchange standards 291
emergent purposes 262
empirical study 173
engineering consultancy 2
engineering consulting business 5
enterprise information systems (EISs) 285
enterprise resource-planning (ERP) 285, 294
environmental decision support system (EDSS) 350
environmental impacts assessment (EIA) 324
environmental management (EM) 323
environmental management system (EMS) 325
environmental-conscious construction 323
evaluative capabilities 343
executive and enterprise information systems 285
executive information systems (EISs) 277, 285
executive support systems (ESSs) 285
expert consultancy 213
expert systems (ESs) 277, 287
explicit knowledge 5, 20, 21, 142, 232, 299, 305
explicit-explicit interaction 5
explicit-tacit transfer 5
exploitation 59
eXtensible Markup Language (XML) 291
external environment 228
external provision 36
external provision of KM 39
external provision of learning 35
external sources 170
externalisation 5, 21, 232

F

face recognition 115
face-to-face interactions 26
face-to-face interviews 175
face-to-face training 27
fast track management programme 123
FIATECH 345
FIATECH technology roadmap 356
filtering 4
financial balance sheet 54
financial capital 55
financial management 170
financial reports 54
flexible modular construction 63
fluid mix 4
fuzzy logic 284

G

gatekeepers 34, 35, 38, 42
general theory 218
generalised maturity model 226
global construction 227
global network 92
graphical user interface (GUI) 354
group collaboration 252
group commitment 259
group creation process 252
group decision support system (GDSS) 286
group emphasis 187
group knowledge 252
group knowledge-sharing processes 28

group support systems (GSSs) 277, 286
groupware ICT applications 91
groupware technologies 285
guided keywords 10

H

harvesting 4
high-tech enterprise 60
higher education 150, 151
holistic management of knowledge 226
holistic strategic approach 225
Hong Kong 18, 24
Housing Department (HD) 72
HRM 19
HRM mechanisms 22, 24
HRM strategies 18
human capital 55
human resource management (HRM) policy 19

I

ICT diffusion 94
ICT infrastructure 233
ICT innovation 92
ICT innovation diffusion 94
ICT knowledge 104
implementation 116
improvisation 205
inadequate consideration 255
inadequate involvement 255
inadequate management 255
incentive reward program (IRP) 329, 333
industry foundation classes 352
informal knowledge workshops 26
informal networks 123
information 35
information and communication technologies (ICT) 19, 90, 225
information delivery 119
information processing 343
information sharing 142
information society 152
information standards 345
information system 138
information systems and technology 140
information technology 6, 276
information technology development 343
information tunnels 344, 348
infrastructure 232
initiation 116
innovation 59
innovation diffusion 226
innovative knowledge 228
institutional network 103
instructional goals 153
insufficient time 255
intangible assets 54
integrated prescription 218
integration 116
integrative knowledge management prototype 323
integrative knowledge management system 323
integrative methodology of KMS 324
intellectual assets 54, 69
intellectual capital (IC) 53, 55
intellectual capital (IC) report 53
intellectual capital report 53
intelligent agent (IA) 284, 287
intelligent decision support systems 286
inter-project learning 24
internal distance 41
internal generation 34
internalisation 21, 232
Internet-based decision support systems 288
interoperability 345
interpretative knowledge 141
interrelationships 334
intranet 27
intranet discussion groups 26
intranet-based knowledge repository 18
intranet-based project management 291
intrinsic motivation 228
Israeli construction company 53

K

K-Adv concept 228
K-Adv deployment 234, 242
K-Adv model 229
K-Adv vision 242
key management 253
KM packages 40
KM strategy 19, 30
KM will 328
knowledge 2, 19, 54, 67, 115, 130, 134, 225
knowledge acquisition 151, 304
knowledge acquisition phase 304
knowledge advantage (K-Adv) 225, 227
knowledge and experience management 79
knowledge and innovation network 122
knowledge assets 230
knowledge attribute 309
knowledge attribute phase 310
knowledge bank 308
knowledge base 6, 351
Knowledge Café 61
knowledge centre 122
knowledge content 305
knowledge creation 232
knowledge creation 21, 151, 252
knowledge depositories 234
knowledge determination phase 310
knowledge diagram 308
knowledge emergence 251
knowledge exchange 26, 114
knowledge exchange seminars 26
knowledge extraction 304
knowledge extraction phase 304, 310, 317
knowledge forum 122
knowledge hoarding 3
knowledge holders 9
knowledge leadership 227
knowledge link knowledge 312
knowledge linking phase 312
knowledge management 3, 19, 37, 40, 68, 87, 95, 115, 130, 138, 150, 170, 186, 203, 213, 217, 226, 252, 299, 303, 323, 345, 357
knowledge management challenges 24
knowledge management competencies 151
knowledge management market 37
knowledge management mechanisms 170
knowledge management portal 299
knowledge management portal system 299
knowledge management problem 3, 8
knowledge management services 35, 36
knowledge management strategy 5, 121
knowledge management system (KMS) 1, 155, 284, 325
knowledge management theory 4
knowledge map 299, 308
knowledge networks 232
knowledge perspective 226
knowledge production 39
knowledge sharing 3, 18, 22, 28, 156, 178, 232, 305
knowledge sharing culture 23
knowledge sharing phase 305, 319
knowledge society 150, 151
knowledge source 44
knowledge sources 170
knowledge stickiness 117
knowledge storage 178, 304
knowledge storage phase 305, 318
knowledge transfer 117, 142
knowledge transfer partnerships 44
knowledge update 304
knowledge update phase 305, 319
knowledge validation phase 312
knowledge vision 227, 229, 230
knowledge vision options 230
knowledge warehouse (KW) 325
knowledge work 2
knowledge worker 36, 234
knowledge-at-large 6
knowledge-based economy 87
knowledge-based management sub-system 283

knowledge-based systems 277
knowledge-deployment 177
knowledge-enterprising culture 21
knowledge-hoarding culture 77
knowledge-intensive organisational environment 303
knowledge-relationship-gatekeeper management 50
knowledge-sharing barriers 19
knowledge-sharing environment 21
knowledge-sharing junctions 61
knowledge-sharing process 233
knowledge-transfer opportunities 28
knowledge-transfer techniques 26

L

large construction company 130
learning capability 167
learning culture 189, 196
learning disabilities 169
learning environment 156
learning from others index 177
learning histories 185, 200
learning history 185
learning index (LI) 173
learning mechanisms 166
learning organisation 137, 167
learning process 189, 200
learning sources 170
learning sources 166
learning styles 190
legacy systems 234
legal requirements 70
leisure services provider 130
life-cycle assessment (LCA) 329
lifelong learning 150
local area network (LAN) 27

M

management 278
management of project changes 139
management steering 253
management strategies 24
management system 2, 231, 253
market reconfiguration 45
market value 53, 54, 60
marketing 34, 40
marketing mix 41
MATLAB® 350
maturity level assessment 244
maturity level development 119
maximum return 142
means-ends orientation 187
mediation 353
member identity 187
mentoring 27, 233
meticulous quality in construction 62
Microsoft Access database 141
minimum charge service 47
model base management system (MBMS) 283
model directory 283
model execution 282, 283
model management subsystem 282
multi-organisational 216
multimedia 140
mutual behavioural change 136
mutual learning 136
myth of certainty 209

N

nature of knowledge 74
negotiated order 213
neural network 287
New Zealand 1
normative (ought to) commitment 234

O

object-oriented CAD 348
OL competency 171
OL strategy 171
on-line analysis processing (OLAP) 292
online business process models 290
online database 10
open communications 28
open dialogue 96
open process transparency 259
open-system focus 187
organic learning style 186, 198
organisation 300
organisation development 216

organisation knowledge base 7
organisational activity 35
organisational conservatism 215
organisational culture 21, 186, 189
organisational decision support systems 277
organisational hub 40
organisational knowledge 138, 169, 170
organisational knowledge acquisition 177
organisational knowledge-sharing practices 26
organisational learning (OL) 40, 130, 138, 167, 175, 188
organisational learning ability 166
organisational learning competency 166
organisational learning framework 170
organisational learning process 169
organisational performance 197
organisational setting 170
organisational structure 29, 171
organisational sustainability 226
organisations 278
organisation's learning culture 185, 197
organizational learning 135
over-the-counter (OTC) retailing market 41

P

packages 38
peer assessment 162
peer learning 153
people focus 187
people infrastructure 232
performance 40, 173
performance driver 166
performance measurement 173
performance measurement systems 176
performance rating 174, 175
personal assistance 233
personal communication 104
personal experience 76
personalisation 23
personalization approach 5
policy framework 19
poor communication 25, 255
post-project evaluation 175
post-review event 141
pre-project workshop 193
pre-review event 141
pre-review event information 141
prescriptive teaching 136
problem solving 35, 37, 39
problem structuring methods 139
problem-based learning 153
problem-solving environments 350
problem-solving group 253
problem-solving steering 268
process capital 56
process innovation 92
product 37
production information 39
professional building surveyors 72
professional competency 301
professional development 150
professional development 150
professionalism in customer service 62
programmed internal training 233
project award scheme 27
Project Constraints 258
project database 290
project definition 251, 252
project definition activity 252
project definition group 253, 265
project delivery 87
project Eagle 123
project end user 42
project environment 40
project generated knowledge 1
project information technologies 343
project knowledge retention 87
project learning 170
project life cycle 7, 87
project management 7, 12, 290, 325
project management knowledge base 7
project manager 192
project manager/engineer and collegial support net 104

project needs 251
project organisation 34
project planning 290
project purpose 252
project stakeholders 251
project structure 192
project working environment 37
project-driven environment 6
project-generated feedback 10
project-generated knowledge 1, 2
project-generated knowledge management systems 1
project-specific drawings 9
project-specific knowledge 37
project-specific knowledge base 7
proliferating boundaries 217
proposed process improvement 192
psycho-social system 209

Q

qualitative competitive advantage 93
quality of living 63
query facility 282
questionnaire 166, 175
questionnaire survey 67, 71

R

ramp-up 116
real-world environments 153
real-world problems 153
reciprocity 22
recursive-ness 205
reflection 130
reflective improvement 193
reflective practice 137, 212
reflexivity 205
relationship marketing 41
remote login 290
renewal and development capital 57
repository 39
resource controllers 208
resource-based view 169
resources 34
response rate 72
reuse of knowledge 67, 73
reusing Knowledge 68
review event 141
reward mechanisms 176
rich pictures 230
risk tolerance 187
routine meetings 26
rule-oriented DSS 281

S

safety 62
sampling 72
search engine 290
self-actualisation 228
self-learning 152
senior management team member 192
senior managers 1
shared 3D model 290
shared vision 22
sharing of knowledge 130
Sigma Connect 117
site space management PSE 351
site visit programme 27
site visits 27
site-based problems 39
Skandia market value scheme 58
Skandia model 55
smart construction 53
smart construction techniques 61
social interactions 235
social network 26
socialisation 5, 21
socialisation of tacit knowledge 232
society in construction 215
socio-technical perspective 67
socio-technical solution 37
socio-technical systems 261
sociology of values 210
soft systems methodology (SSM) 226, 230, 235
software management 112
solution strategy 241
solver-oriented DSS 281
sophisticated mold-based construction 63
space management 346
spatio-temporal fragmentation 218
speciality knowledge 25

speeding access 142
spreadsheet-oriented DSS 281
spreadsheets 140
stability 203
stakeholder environment 241
stakeholder identification 235
stakeholder interests 251
stakeholder K-value 230, 241
stakeholder knowledge 230
stakeholder needs 252
stakeholder values 254
standards-based approaches 352
state-of-the-art technology 63
steering 265
sticky knowledge 113
stock fluctuation 53
strategic options development and analysis (SODA) 139
strategic planning 241
structural capital 5
structured/unstructured fields 119
student-centred learning 155
studio space 154
subjective reporting approach 174
summary reports 27
superintendents 344
supply chain 285
supportive culture 19, 21
survey findings 73
surveying studio 153
system development life cycle (SDLC) 236
systematic examination 53

T

tacit coordination 213
tacit knowledge 5, 20, 21, 113, 299
tacit knowledge explicit 5
tacit knowledge sharing 20
tacit order 213
tacit-explicit interaction 5
tacit-tacit interaction 5
Tasmanian State Public Health Sector 186
Tavistock (TIHR) studies of mining 204
Tavistock Institute of Human Relations (TIHR) 204
Tavistock research and social thought 206
taxonomy 119
teaching company scheme 44
teamwork 23
technical support network 104
technology fusion 215
Tel Aviv Stock Exchange 60
text-oriented DSS 280
theory of construction 206
total customer service 57
total information transfer system 290
total quality management (TQM) 168
traditional working class community of practice 204
transaction costs 36, 40
transfer of knowledge 137
transformation-flow-value theory 254
transmission model of learning 153
Turkish construction companies 175
Turkish construction industry 174
Turkish Contractors Association (TCA) 174

U

UK coal mining 203
unit integration 187
unit manager 192
usability 119
user authentication 119
user interface subsystem 283
user modeling 351
user models 343

V

valuable knowledge 300
value chain 235
value delivery 254
value identification 254, 262
value proposition 230, 241, 254
video cam 290
virtual knowledge marketplaces 6
virtual teams 231
virtually linked sub-organisations 91
vision realisation 229
visualization 343

W

Web-based commercial systems 290
Web-based cost control system 291
Web-based knowledge management system 85
Web-based prototype system 67, 71
Web-based tools 290
Webfill approach 333
white boarding chat 290
wisdom 54
word processors 140
work content 29
working model 255
workplace planning 264
workplace planning management system 264
workplace planning project 268
workplace planning system 252
workshop-based approach 130
workspace planning 253

www.ingramcontent.com/pod-product-compliance
Lightning Source LLC
LaVergne TN
LVHW062110200926
848484LV00007B/1637